HOW
TO
INVENT
EVERYTHING

HOW TO INVENT EVERYTHING

A Survival Guide for the Stranded Time Traveler

RYAN NORTH

RIVERHEAD BOOKS • NEW YORK • 2018

RIVERHEAD BOOKS
An imprint of Penguin Random House LLC
375 Hudson Street
New York, New York 10014

Copyright © 2018 by Ryan North

Illustrations © 2018 by Lucy Bellwood. Public domain images of *The Last Supper*, *The School of Athens*, and the mill painting were found on Wikimedia Commons. And yes, our trigonometric values were verified with NASA. We're not messing around here.

ISBN 9780735220140

Printed in the United States of America
10 9 8 7 6 5 4 3 2 1

BOOK DESIGN BY LUCIA BERNARD

The information and instructions in this book involve materials and activities that may be dangerous. The publisher and author are not responsible for any injuries or damage that may result from the use of such information and instructions. That's right: you're about to read a book *so badass,* it has to include a legal disclaimer at the front. And by the time you finish this book and know its secrets, you'll be able to apply that exact same badass disclaimer to yourself. "Hi," you'll say, holding out your hand whenever you meet someone new, "I'm [Your Name Here]. And just so you know: *the person you are about to meet knows materials and activities that may be dangerous.*"

It's going to be amazing.

A NOTE FOR READERS

I didn't write this guide. I found it. It was wholly encased in bedrock, and I know that because I was the one who broke that heavy granulite stone open. I was working construction for a few weeks because I heard it paid well.

It did not.

I can tell you that I personally don't have the technology to place a book inside solid stone. Nobody does. I've tried to have the book carbon dated, but that's impossible: whatever exotic polymer the guide was printed on, it's one that doesn't include carbon. The stone I found it in can be dated, of course: it's Precambrian, which means it predates humans, dinosaurs, and most life on Earth. Precambrian rocks are some of the oldest rocks on the planet.

So, no help there.

There is obviously the possibility that the text you're about to read is part of a well-constructed and incredibly expensive prank, accomplished using technology unknown to the rest of the world, including the technology to insert objects inside solid rock while maintaining a profile tolerance of under 10^{-4}mm. It seems unlikely. But the alternative—that time travel is possible, that somewhere it is being practiced, and that our entire universe is but a copy spun off from their original at some unknown point in the past—also seems impossible.

I've researched all the claims made in this guide. Everything that can be verified has been, and the text appears to be an honest, sincere, and accurate effort to explain how to rebuild civilization, from scratch, in any time period in Earth's history. All historical events mentioned in the text line up with our own, though with the guide's focus on technology and civilization instead of nations and people, there are fewer dates and individuals to compare against than you might expect. "Their" world appears to be much like ours, only better: they have a higher level of technology, a greater understanding of history, and, of course, consumer-market rental time machines. There's a chance that we also might one day invent time travel, in which case the claims made here could finally be verified and we could discover when, and how, this impossible book ended up embedded in the solid rock of what would eventually become the Canadian Shield.

On the other hand, we might not.

The guide that follows is presented in its original and unaltered format, except for the endnotes, which I added in two cases: when I thought clarification or references to supplementary texts would be helpful, or when a claim was being made in the original text that reached beyond our current science, engineering, or historical knowledge. Footnotes are presented as in the original text, and no other changes have been made to content or presentation. The original illustrations in the guide, credited to one "Lucy Bellwood," have also been included. There is an artist working under that name in our world; she claims to have no knowledge of this book or its origins, and I have no reason to doubt her.

Finally, I should touch on what is perhaps the most unlikely note of all in this. The technical writer responsible for this guide shares his name just once, and then only in a footnote. It is the same as mine. Part of me knows I can't read much into this: there are a lot of Ryan Norths out there, and I've emailed most of them. Our writer could be an alternate-timeline version of any of us. Or he could be someone new, someone with no parallel in our world. Maybe a time-travel accident left this book embedded in stone somewhere in our distant past, stranding the traveler there, or in some other time, changing our world in small but significant ways that we may never tease apart. Maybe that's why we don't have time travel.

Or, again, maybe this is all just part of an incredibly expensive prank.

I know what I believe. I know how incredibly, cosmically unlikely it is that I would ever find this guide *and* share my name with its author *and* know a Lucy Bellwood too. And if you think perhaps I'm involved in some deception with this text, I will repeat what I said at the start of this section: I didn't write this guide.

At least . . . not in this timeline.

I'm thrilled to share, for the first time, a complete and unabridged copy of what was originally titled *The Time Traveler's Handbook: How to Repair Your FC3000™ Time Machine, and Then How to Reinvent Civilization from Scratch When That Doesn't Work.*

Those who cannot remember the past are condemned to repeat it.

—George Santayana, philosopher, essayist, and poet
1905 CE

↔

Those who cannot remember the past are cordially invited to revisit it.

—Jessica Bennett, CEO of Chronotix Solutions,
proud manufacturers of the FC3000™
2043 CE

INTRODUCTION

Congratulations on your rental of the FC3000™! The FC3000™ is a state-of-the-art personal time machine that allows *you* to experience the entire range of human endeavor, from the earliest chimpanzee-human divergence (in 12,100,000 BCE, the backward limit of this rental unless you have purchased the *Protoprimate Encounter Pak*) to the latest mass-market portable music players (present day).

Note that travel to any times *later* than 1.5 seconds after the instant you left your personal present ("the future") is not permitted with this rental, and a sensitive chronometer has been installed to detect and disable any attempts to visit these periods.

Please carefully study the features of the FC3000™, depicted on the following page. Federal regulations require us to inform you that due to the nature of genetic and acquired immunity, there are a large number of diseases to which present-day humans are immune but which have not yet been encountered by past humanity. For your safety, and the safety of those around you, multiple biofilters installed throughout the FC3000™ work to ensure that your appearance in the past will not obliterate all human life with the introduction of dozens of deadly plagues and pestilences in a single instant.

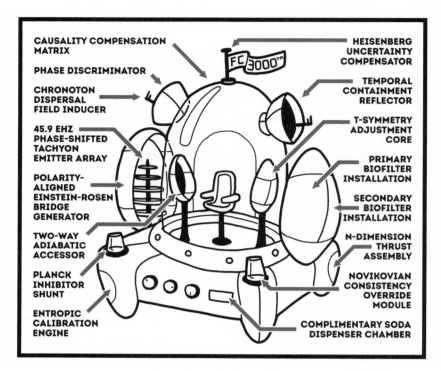

Figure 1: The FC3000™.

All other features of the FC3000™ depicted above are self-explanatory.

FREQUENTLY ASKED QUESTIONS BY NEW TIME TRAVELERS

Q: Will traveling to the past destroy the present, due to the "butterfly effect," which they made several movies about (2004, 2025, 2034, etc.)?

A: No. Those films were based on a speculative understanding of time travel that, thankfully, is not accurate. In reality, any temporal machinery—including the state-of-the-art FC3000™ rental-market time machine—creates a new "timeline," or sequence of events, with each trip back in time. Observe the following illustration:

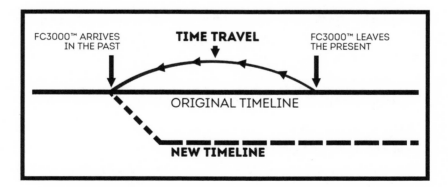

Figure 2: Traveling through time with the FC3000™.

Each trip to the past creates a new event sequence for the world, beginning with the intrusion of the time machine into history. In effect, with each trip back in time, you are creating a "what-if" universe, all proceeding from the premise of "what if a time traveler came back and visited this particular time in a state-of-the-art FC3000™ rental-market time machine?" When you return home, your FC3000™ will travel through space, time, *and* timelines, always returning you to your original, unaltered history.

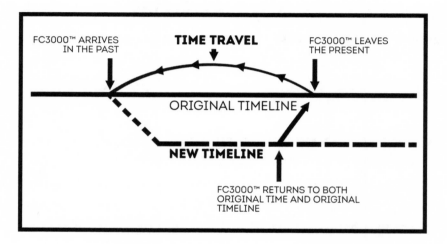

Figure 3: Returning back home with the FC3000™.

Put simply, even the most egregious time traveler cannot affect *the* present, but merely an *alternate* one created by their time traveling. Feel free to step on as many butterflies as you see fit.

Q: Can I interact with my past self?

A: Yes. It is not recommended. You will likely notice you do not look as good from behind as you thought. Please note that despite the FC3000™ offering travel to any point in human history, the first instinct many clients have is to arrange an encounter with their past selves. We respectfully suggest that the FC3000™ was built to explore time, to better understand the origins of humanity and the potential of ourselves and our world, and that choosing to visit yourself suggests you sincerely believe you are the most interesting person on the planet *in any time period*. By definition this can only be accurate in one case and is therefore likely not accurate in yours. We invite you to reconsider.

Q: Can I give my past self lottery numbers?
A: Any lottery numbers you give will benefit another you, and not you personally.

Q: Can I give my past self lottery numbers, and then kill my past self and take their place, so that the lottery winnings go to me instead?
A: Yes. However, you may have to answer to the authorities in that time period.

Q: Will being rich in the past make me happy?
A: It might.

Q: If not my past self, then whom should I visit?
A: The expanse of human history spreads out before you, awaiting your curious and empathetic gaze. That said, as part of our legally mandated commitment to customer satisfaction at Chronotix Solutions, we have created several *Chrononaut's Choice* pamphlets, which you will find stored beneath your seat in the FC3000™. Each includes not just background information and space-time coordinates for one of our many artisanally selected points in history, but also the descriptions of the specific historical personages and precise sentences you must say to them in order to be swept up in an epic adventure. Popular pamphlets include *How to Get Michelangelo, Rembrandt, and Vincent van Gogh to Paint Your Portrait for Free; Choose Your Side in the Battle of Marathon!; Join the Roanoke Colony and See What Happens;* and *1001 Wacky Places to Shoot Adolf Hitler.* Follow our guide, or feel free to go off-script whenever you want.

Q: If every time I travel back in time it creates a new alternate timeline, so nothing I can do affects my own timeline, then isn't time travel pointless?
A: If going back in time did affect the original universe we all come from,

then it would be wildly irresponsible to rent out time machines willy-nilly to members of the general public. Alterations are not pointless, however: remember that these alternate timelines you create in your travels are identical to ours in every way, except with the addition of you, the time traveler. The people in these new timelines are, by any measure, just as real as the people you know in your own timeline.

Q: Wait. If that's true, aren't there hugely staggering ethical implications to the idea that we can create whole alternate realities—entire universes just as valid as our own and filled with just as many people (more, actually, since now there's an extra time traveler there!)—simply for the purpose of entertainment?

A: We have several ethicists on staff who have assured us, in no uncertain terms, that this is totally fine. In addition, please keep in mind that these alternate realities are not just for the purpose of entertainment. They have also been used for mining and resource extraction.

Q: What if something goes wrong with my FC3000™ time machine?

A: The FC3000™ is the most reliable time machine on the rental market today. However, as with any activity involving unstable Einstein-Rosen bridges constructed across disparate spatial/temporal reference frames, there are always risks. In the event of a catastrophic failure of your FC3000™, please refer to the convenient Repair Guide, which follows this page and makes up the bulk of this volume.

REPAIR GUIDE

There are no user-serviceable parts inside the FC3000™.

The FC3000™ cannot be repaired.

OH.

Yes. This is a problem. If you are reading this Repair Guide, then you will not be returning to the future, and we apologize for any alleged failures in the FC3000™, real or implied, that facilitated this scenario.

If you would like to make peace with the idea that you will never return to your friends and family, please do so now. It helps to focus on the things you didn't like about them, such as their irritating habits or weird smells. Do not focus on the things you will miss, like cheap, convenient, clean, and safe drinking water, or the latest mass-market portable music players.

And now that you have accepted the fact that you are stuck in the past, we would like to offer a suggestion. Since you can no longer go back to the future . . .

. . . *we invite you to bring the future back to you.*

Allow us to explain that intriguing ellipsis-filled sentence.

The rest of this guide contains all the science, engineering, mathematics, art, music, writing, culture, facts, and figures that are required for *one human*— without any specialized training—to build a civilization from the ground up. You may be under the impression that modern civilization took several million humans and protohumans several thousand millennia to construct. It did, but that was only because we didn't know what we were doing the first time around and had to invent everything as we went.

You, in contrast, hold all the answers in your hands already.

This guide will allow you to create a world like the one you left, but better. It will be one in which humanity matured quickly and efficiently, instead of spending 200,000 years stumbling around in the dark without language (Section 2), not knowing that tying a rock to a string would unlock navigating the entire world (Section 10.12.2), and thinking disease was caused by weird smells (Section 15).

We make no assumptions about what period you're trapped in or what you already know. Everything you need is built from scratch, making this text nothing less than a complete cheat sheet for civilization.

We at Chronotix Solutions are excited to have accidentally provided you with this opportunity, and wish you all the best.

HOW TO USE THIS GUIDE

This guide is divided into seventeen equally interesting sections. While you are encouraged to read it from start to finish before referring to individual sections as required, you may still jump ahead to the areas that interest you the most. If you are curious about a particular technology, please refer to the technology tree in Appendix A to see what prerequisites are required, then prioritize those inventions in order to unlock that technology as soon as possible.

A word of warning: while it's fair to say that it would be pretty ridiculous to strand you in the past without any knowledge of how to produce the technologies, inventions, and chemicals you might need, it's also pretty ridiculous how many of those technologies, inventions, and in particular chemicals are extremely dangerous to produce, store, inhale, touch, or even just be around. So in a legally mandated compromise, we are hereby compelled to tell you that while what you will find in this book is all you need to rebuild civilization from scratch, you should not attempt to produce anything dangerous, especially the chemicals, unless you *really* need them, and you also hereby legally agree, affirm, and attest that you will definitely not blow yourself up trying to make them happen.

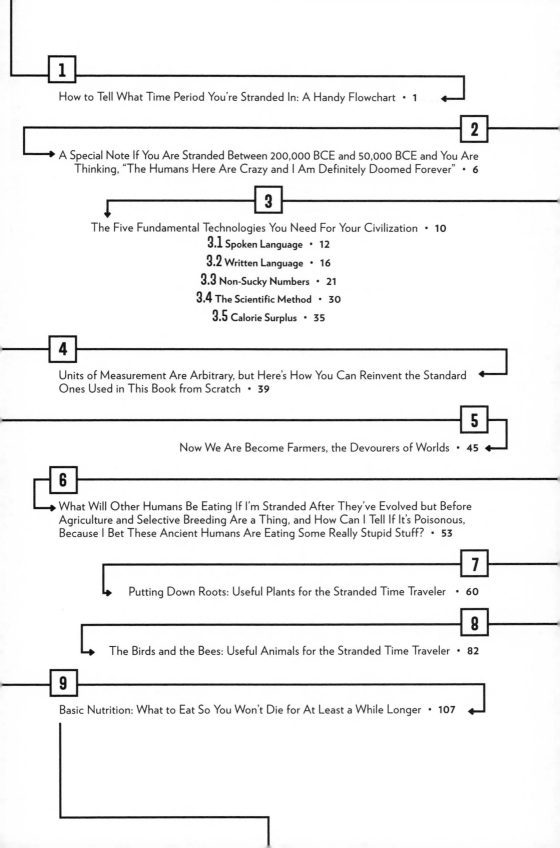

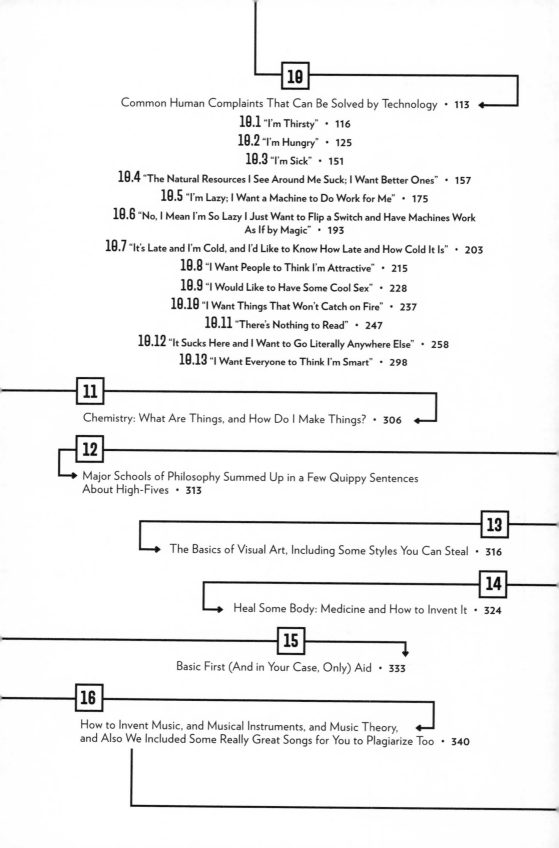

1

HOW TO TELL WHAT TIME PERIOD YOU'RE STRANDED IN: A HANDY FLOWCHART

There is a small chance that, in the process of your FC3000™ suffering a catastrophic failure for which no legal liability can be assigned, you may arrive in a time frame different from the one you were expecting. We recommend proceeding through this chart first, to better orient yourself in your new period in history.

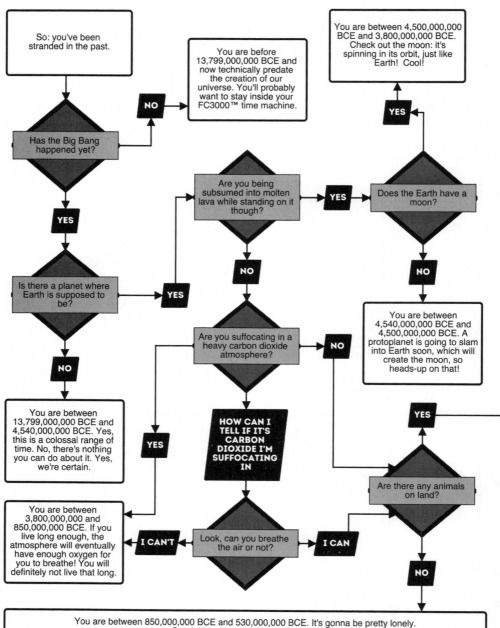

So: you've been stranded in the past.

Has the Big Bang happened yet?

NO → You are before 13,799,000,000 BCE and now technically predate the creation of our universe. You'll probably want to stay inside your FC3000™ time machine.

YES ↓

Is there a planet where Earth is supposed to be?

YES → Are you being subsumed into molten lava while standing on it though?

YES → Does the Earth have a moon?

YES → You are between 4,500,000,000 BCE and 3,800,000,000 BCE. Check out the moon: it's spinning in its orbit, just like Earth! Cool!

NO → You are between 4,540,000,000 BCE and 4,500,000,000 BCE. A protoplanet is going to slam into Earth soon, which will create the moon, so heads-up on that!

NO (from molten lava) → Are you suffocating in a heavy carbon dioxide atmosphere?

NO → (arrow to "Are there any animals on land?")

YES → HOW CAN I TELL IF IT'S CARBON DIOXIDE I'M SUFFOCATING IN

↓ Look, can you breathe the air or not?

I CAN'T → You are between 3,800,000,000 and 850,000,000 BCE. If you live long enough, the atmosphere will eventually have enough oxygen for you to breathe! You will definitely not live that long.

I CAN → Are there any animals on land?

YES → (arrow leading off page)

NO → You are between 850,000,000 BCE and 530,000,000 BCE. It's gonna be pretty lonely. Good thing this book is so long, huh?

NO (from Big Bang planet) → You are between 13,799,000,000 BCE and 4,540,000,000 BCE. Yes, this is a colossal range of time. No, there's nothing you can do about it. Yes, we're certain.

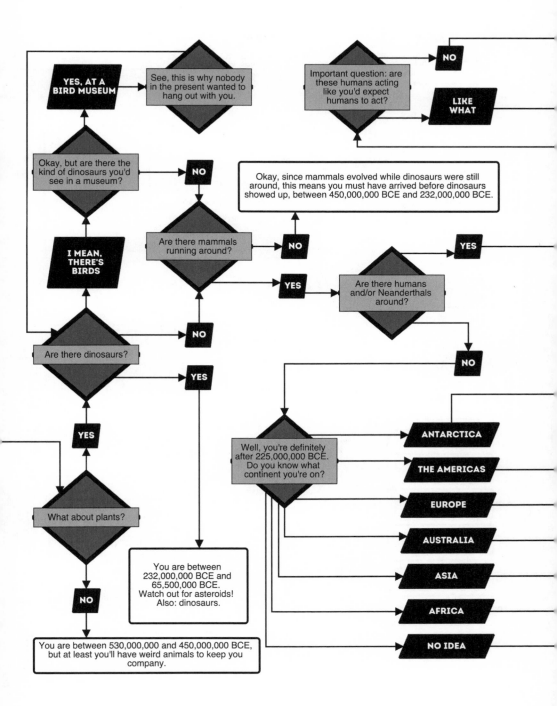

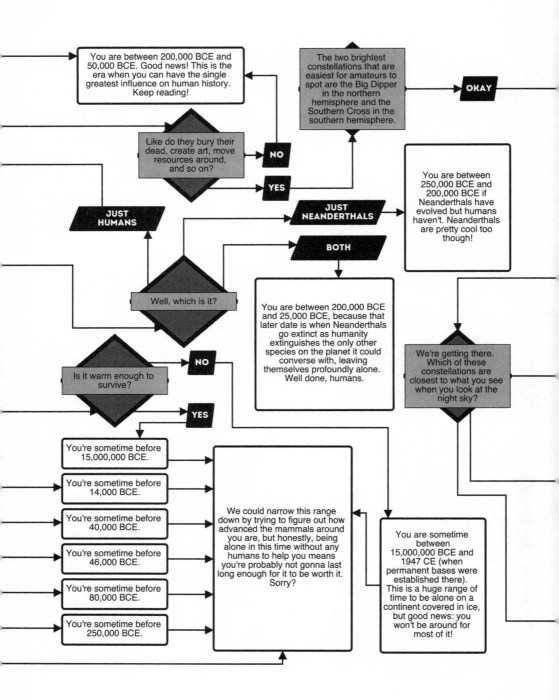

You are between 200,000 BCE and 50,000 BCE. Good news! This is the era when you can have the single greatest influence on human history. Keep reading!

The two brightest constellations that are easiest for amateurs to spot are the Big Dipper in the northern hemisphere and the Southern Cross in the southern hemisphere.

OKAY

Like do they bury their dead, create art, move resources around, and so on?

NO

YES

JUST HUMANS

JUST NEANDERTHALS

BOTH

You are between 250,000 BCE and 200,000 BCE if Neanderthals have evolved but humans haven't. Neanderthals are pretty cool too though!

Well, which is it?

You are between 200,000 BCE and 25,000 BCE, because that later date is when Neanderthals go extinct as humanity extinguishes the only other species on the planet it could converse with, leaving themselves profoundly alone. Well done, humans.

We're getting there. Which of these constellations are closest to what you see when you look at the night sky?

Is it warm enough to survive?

NO

YES

You're sometime before 15,000,000 BCE.

You're sometime before 14,000 BCE.

You're sometime before 40,000 BCE.

You're sometime before 46,000 BCE.

You're sometime before 80,000 BCE.

You're sometime before 250,000 BCE.

We could narrow this range down by trying to figure out how advanced the mammals around you are, but honestly, being alone in this time without any humans to help you means you're probably not gonna last long enough for it to be worth it. Sorry?

You are sometime between 15,000,000 BCE and 1947 CE (when permanent bases were established there). This is a huge range of time to be alone on a continent covered in ice, but good news: you won't be around for most of it!

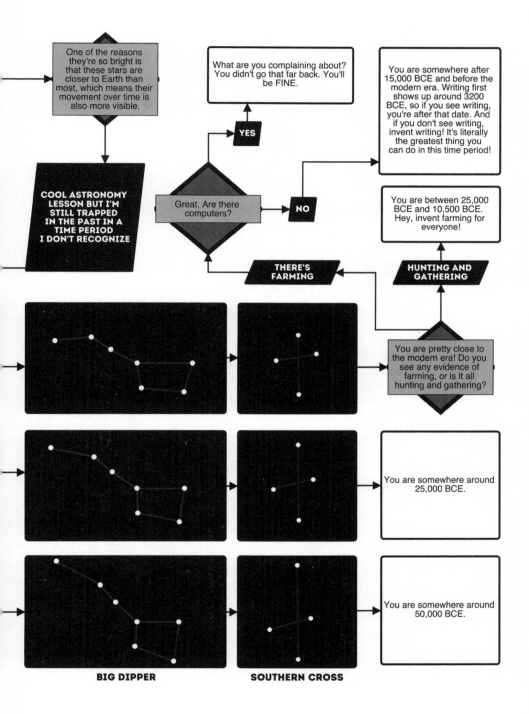

One of the reasons they're so bright is that these stars are closer to Earth than most, which means their movement over time is also more visible.

COOL ASTRONOMY LESSON BUT I'M STILL TRAPPED IN THE PAST IN A TIME PERIOD I DON'T RECOGNIZE

What are you complaining about? You didn't go that far back. You'll be FINE.

YES

Great. Are there computers?

NO

You are somewhere after 15,000 BCE and before the modern era. Writing first shows up around 3200 BCE, so if you see writing, you're after that date. And if you don't see writing, invent writing! It's literally the greatest thing you can do in this time period!

You are between 25,000 BCE and 10,500 BCE. Hey, invent farming for everyone!

THERE'S FARMING

HUNTING AND GATHERING

You are pretty close to the modern era! Do you see any evidence of farming, or is it all hunting and gathering?

You are somewhere around 25,000 BCE.

You are somewhere around 50,000 BCE.

BIG DIPPER

SOUTHERN CROSS

2

→ A SPECIAL NOTE IF YOU ARE STRANDED BETWEEN 200,000 BCE AND 50,000 BCE AND YOU ARE THINKING, "THE HUMANS HERE ARE CRAZY AND I AM DEFINITELY DOOMED FOREVER"

Great news! You can actually be the most influential person in history!

As your careful study of the flowchart on the previous pages likely revealed, humans first evolved around the year 200,000 BCE.[1] We call them "anatomically modern humans," and they mark the moment when humans with skeletons exactly the same as ours first appeared. As an experiment, we could put your skeleton beside that of an anatomically modern human from 200,000 years ago and it would be impossible to tell them apart.

We will not be performing this experiment, but we could.

But what's fascinating is despite the fact that modern human bodies were now available, nothing really changed. For more than 150,000 years, these humans behaved pretty much the same as any other protohuman species. And then, around the year 50,000 BCE, something happened: these anatomically modern humans suddenly started acting like us. They began to fish, create art, bury their dead, and decorate their bodies. They began to think abstractly.

Most important, they began to talk.

The technology of language—and it is a technology, it's something we've had to invent, and it took us over 100,000 years to do it—is the greatest gift we humans have ever given ourselves. You can still think without language—close your eyes and imagine a really cool hat and you've just done it—but it limits the kinds of thoughts you can have. Cool hats are easy to imagine, but the meaning of the sentence "Three weeks from tomorrow, have your oldest stepsister meet me on the southeast corner two blocks east from the first house we egged last Halloween" is extremely difficult to nail down without having concrete words for the concepts of time, place, numbers, relationships, and spooky holidays.* And if you're struggling to express complex thoughts even in your own head, it's pretty evident that you won't be having those complex thoughts as often, or at all.

It was language that gave us the ability to imagine better, grander, more world-changing ideas than we otherwise could, and most important, it gave us the ability to store an idea not just in our own heads but inside the minds of others. With language, information can spread at the speed of sound, or, if you're using sign language instead of speaking, at the speed of light. Shared ideas lead to communities, which are the basis of culture and civilization, and which brings us to our first Civilization Pro Tip:

CIVILIZATION PRO TIP: Language is the technology from which all others spread, and you've already got it for free.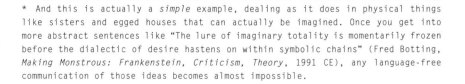

---

* And this is actually a *simple* example, dealing as it does in physical things like sisters and egged houses that can actually be imagined. Once you get into more abstract sentences like "The lure of imaginary totality is momentarily frozen before the dialectic of desire hastens on within symbolic chains" (Fred Botting, *Making Monstrous: Frankenstein, Criticism, Theory,* 1991 CE), any language-free communication of those ideas becomes almost impossible.

This huge expanse of time—the 150,000 years between 200,000 BCE, when humans first appeared, to 50,000 BCE, when they finally started talking—is where you can have the single greatest effect on history.[2] If you can help humans of this era become behaviorally modern as soon as they became anatomically modern—if you can teach them to talk—then you can give every civilization on the planet a 150,000-year head start.

It's probably worth the effort.

We once thought the change from anatomical to behavioral modernity was due to some physical change in our brains. Perhaps a random genetic mutation in one human—who suddenly found themselves able to communicate in ways no animal had done before—provided us with the huge advantage of a new capacity for abstract thought? However, the historical record doesn't support the idea of this great leap forward. The things we most associate with behavioral modernity—art, music, clever tools, burying the dead, making ourselves look cooler with jewelry and body paint—all appear before the breakthrough around 50,000 BCE, but in fits and starts, appearing locally and then disappearing. Much like the magic that rhetorical wizards have long revealed was actually inside us *all along*, so too have humans had the capacity for language. We just needed to unlock it.[3]

The unique challenge facing you in this era is how to teach a language to people when the very *idea* of spoken language may be new to them. It's important to remember that most humans you encounter may not have language, but they'll still *communicate* with one another, through grunts and body language. All you need to do is move them from grunts to words, and don't worry: a complicated language like English with things like "subjunctive clauses" and "imperfect futures" (used here in the grammatical sense, not the time-travel sense) is not necessary, and you can get by with a simplified version of the language you already know, called "pidgin." You will also have better results if you focus on teaching children. The older humans are, the harder it is for them to learn languages, and fluent acquisition of a first language becomes much more challenging—if not impossible—after puberty.

CIVILIZATION PRO TIP: Babies begin to focus on the noises used in language around them after about six months of age, so if you're inventing a language from scratch, you'll likely have more success incorporating whatever sounds the baby is already hearing from its parents.

Remember: evolution happens very slowly, and even 200,000 years ago the people you'll encounter are humans, just like you—indistinguishable at the biological level. They just need to be taught.

You can teach them.

And you will be remembered as a god.

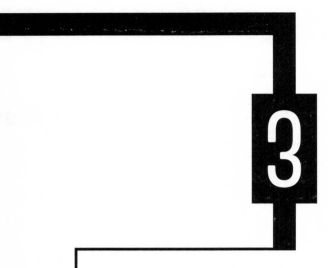

THE FIVE FUNDAMENTAL TECHNOLOGIES YOU NEED FOR YOUR CIVILIZATION

No, the list is not "a really good computer" five times.

Your civilization is going to be founded on five technologies. Each of these technologies is information-based: once you have the idea of them, the rest pretty much follows. Because these technologies are conceptual rather than physical, they are extremely resilient: they are ideas, and ideas cannot be destroyed as long as members of your civilization survive (or at least some of their books do, see Section 10.11.2: Printing Presses).

While the five technologies listed on the following pages are all but invented once you understand the ideas behind them, they each nevertheless took an embarrassingly long time for us, as humans, to figure out.

Please carefully examine the following extremely embarrassing table.

Technology	First invented	When we could've invented it	Years spent not having this technology when we easily could have	This same time period, now expressed as how many colossal 500-year Roman Empires could've both risen *and* fallen in the huge expanse of time humanity spent sitting around not inventing this technology
Spoken language	50,000 BCE	200,000 BCE	150,000 years	300
Written language	3200 BCE	200,000 BCE	196,800 years	393
Non-sucky numbers	650 CE	200,000 BCE	200,650 years	401
The scientific method	1637 CE	200,000 BCE	201,637 years	403
Calorie surplus	10,500 BCE	200,000 BCE	189,500 years	379

Table 1: A table any human should be embarrassed to even be in the same room with.

As these are the absolute technological foundations of civilization, we will now go over the specifics of each.

3.1

SPOKEN LANGUAGE

Listen to those voices in your head.

Before spoken language,* humans communicated through grunts and body language. This allowed us to do the following things:

- draw attention to ourselves
- make noises or gestures expressing emotions like "fear" or "anger"
- cry

Unfortunately, these expressions are easily misunderstood. As an example, babies—famously pre-linguistic—are notoriously difficult to understand. A baby's cry could indicate "I'm sad" or "I'm hungry" or "I'm tired" or "I'm frustrated" or several other emotions, but there's no way to tell what the child actually wants besides giving the baby different things to see if that satisfies it (a short-term solution) or, if you prefer a long-term solution, by gradually teaching the child a language over the course of the next several years until you can finally ask, "Hey, what was all that crying about when you were sixteen weeks old?"

In contrast, spoken language allows us to do the following things:

---

* We refer to spoken language here for simplicity's sake, but this includes signed languages, which are just as expressive as spoken ones. Interestingly, no signed language has developed without a spoken language developing first in our world's history, but of course, you can change that fact in your world *at your merest whim*.

- draw attention to ourselves
- make noises or gestures precisely expressing nuanced emotions, like "fear of one day being trapped in the distant past," or "distinct anger at having now become trapped in the distant past"
- cry (with words)
- have ideas survive the death of their host
- conceive of more complex ideas than we could otherwise express
- transmit complicated sentiment with a reasonable confidence of minimal loss, corruption, or misunderstanding of intent

We tend to think of language as something natural, some property of the universe that we're exploiting. But it's not: we made it up, and it's arbitrary.* However, while the sounds you choose, the order you put words in, and the ways words can interact and change one another are all up to you, there are some recurring patterns that you may want to keep in mind.

These "linguistic universals," as they're called, are found in every natural language on Earth, and while they're not mandatory—people can and have constructed artificial languages that don't use them—they may make it easier for people to use your new language. Please commit the following table to memory:

---

* One of the hallmarks of human language is how completely arbitrary it is. The sounds and letters in the word "cat" actually have nothing at all to do with cats. Since words are arbitrary, the word for "cat" in unrelated languages can be markedly different: *kucing* in Indonesian, *pisică* in Romanian, *kedi* in Turkish, *macska* in Hungarian, *pusa* in Filipino, and *saka* in Malagasy. In contrast, the few non-arbitrary words we have are usually based on mimicking sounds, which is what makes a cat's *meow* identical in English and Filipino, and makes it the very similar *meong* in Indonesian, *miau* in Romanian, *miaú* in Hungarian, *miyav* in Turkish, and *meo* in Malagasy. Revealingly, baby-style words for parents (like *mama*, *papa*, and *dada*) are (a) very close to the babbling noise that all babies make, even deaf ones, (b) composed of speech sounds especially easy for babies to produce, and (c) markedly similar across unrelated languages. One thing that unites most parents across both space and time is how willing they are to read *just a little* into baby's first "words."

Universal property	Description of this property	Example phrases using this property	A grim vision into a dystopian world where this property does not exist
Pronouns exist in all natural languages.	Pronouns are words that let us refer to something without repeating the name of that thing.	*I rented the FC3000™ time machine. It is as reliable as it is well designed, and I am happy to recommend it to everyone without reservation.*	*I rented the FC3000™ time machine. The FC3000™ time machine is as reliable as the FC3000™ time machine is well designed, and I am happy to recommend the FC3000™ time machine to everyone without reservation.*
No "thbbbth" sounds.	Spoken languages are built from the noises our bodies can make, but no natural language uses the "blow a raspberry" tongue-out-of-the-mouth thbbbth sound.	*To be, or not to be: that is the question.*	*To thbbbth, or not to thbbbth: that is the questhbbbbbbbttbbbbth.*
If the language has a word for "feet," it also has a word for "hands," and if it has a word for "toes," it also has a word for "fingers."	Hands are generally more useful to most humans than feet, so if we've reached a point where we're naming body parts and gotten around to naming our feet, we've definitely already named our hands too.	*I have ten toes and ten fingers. Yes, Chad, I know technically I only have eight fingers. Chad, yes, I know thumbs aren't fingers. Everyone knows, I was just . . . Chad. Chad. Chad, listen to me. See, Chad, this is why we don't hang out anymore.*	*I have ten toes and ten, uh . . . extra-bendy upper toes? Yes, Chad, I know two of my extra-bendy upper toes are opposable and therefore should be classified differently. Chad, listen to me. Chad. Chad. I'm doing the best with the words I've got, Chad.*
All languages have vowels.	Vowels are sounds produced with an open mouth and often form the core of a syllable. For example, "cat" uses a as a vowel and c and t as consonants. It's hard to speak without vowels.	*Chad, can we please talk about something else? Anything, Chad. Please.*	*Thhhbbbttth*

Universal property	Description of this property	Example phrases using this property	A grim vision into a dystopian world where this property does not exist
All languages have verbs.	Verbs are action words, which allow us to talk about things happening to other things. Since things tend to happen a lot on Earth, they are useful words to keep around.	*The quick brown fox jumps over the reliable FC3000™ time machine and is happy to recommend it without reservation.*	*The quick brown fox. The reliable FC3000™ time machine. Happy without reservation.*
All languages have nouns.	Nouns are people, places, or things. They are objects or ideas in the world. Since there are a lot of those on Earth, they're useful to keep around too.	*The quick brown fox jumps over the reliable FC3000™ time machine and is happy to recommend it without reservation.*	*The quick brown. Jumps. Reliable. Is happy to recommend.*

Table 2: One advantage of being trapped in the past is you will have finally escaped Chad.

Which language you choose to build your civilization on is a matter of personal preference, and there are no wrong answers here. But while you have your choice of languages to build your civilization on, this also means you have the opportunity to *fix* these languages. Don't like English's pronoun system or French's insistence on giving every object in the universe its own *entirely imaginary* gender? Well, now's your chance to fix them forever.

Spoken languages solve a lot of problems with very few downsides, and they're a technology you're already carrying around in your head. However, they still share one tremendous vulnerability: they rely on human beings to transmit information. If a group of humans dies together, so too do their ideas. You can do better.

You are about to.

WRITTEN LANGUAGE

The technology that made the spelling mistake possible.

While the spoken word is great, it still suffers from significant limitations. It frees ideas from their original host, but it allows ideas to be transmitted only as far as the speaker can travel, or can shout, or can travel while shouting. Most critically, it depends on an unbroken chain of humanity for ideas to survive. Break this chain even once, and all information in it is lost forever.

Writing solves this problem. It allows ideas to become resilient, stronger than our fragile human bodies, which tend to get old and die all the time. It allows ideas to become fixed, immune to changing memories and historical revision. It allows ideas to be broadcast, reaching a much larger audience than could ever listen to your spoken words. Writing even allows ideas to survive not only when their original host has died, not only when everyone who has ever heard them has died, but even when everyone who has ever spoken their *language* has died too: the deciphering of Egyptian hieroglyphs being the greatest example of this. Most incredibly, writing allows information to be shipped around the world with no more difficulty or expense than you'd encounter shipping grain: less, actually, since books don't go bad nearly as quickly. Despite its huge advantages, humans have spent most of their time on Earth—over 98 percent of it—stumbling around without this technology.

Like spoken language, which written language you choose to base your civilization on is not particularly important, but we do recommend (assuming you are multilingual or feeling ambitious) choosing a language that is not English. That prevents you from accidentally teaching others how to read this text, which may be something worth considering, especially since your current temporal circumstances have conspired to make this book the most insanely valuable and dangerous item on the planet.

Though the idea behind writing is simple—store invisible noises by transforming them into visible shapes—the invention of writing was actually an incredibly difficult thing for humans to do. It's so difficult, in fact, that across all of human history, it has happened a grand total of two times:

- in Egypt and Sumer around 3200 BCE.
- in Mesoamerica between 900 and 600 BCE.

Writing shows up in other locations, such as China in 1200 BCE, but this is a result of the Egyptians culturally contaminating the Chinese.[4] Similarly, Egyptian and Sumerian script developed at very close to the same time, and while visually quite distinct, they share many of the same influences. One of these cultures invented writing while the other just lifted the idea, probably after seeing what a super useful invention it was.

There are two other times when writing *may* have been invented: in India around 2600 BCE, and on Easter Island after 1200 CE but before 1864 CE. (We say "may" because this is one of several historical mysteries still unresolved. Confirmation *could* easily be obtained with an incident-free visit to the times and places in question, but for some reason most time travelers have historically been more interested in "experiencing the colossal breadth of human experience" rather than "settling obscure linguistic debates by running controlled temporal observation with an eye to publishing peer-reviewed research.")

The older Indian script (called "Indus") is pictographic and has never been deciphered. Most messages written in Indus script are short (just five characters) which does not suggest an actual language, but rather simpler pictograms or ideograms. What are pictograms and ideograms? We're very glad you asked:

- Pictograms are when an item is represented by a picture of that thing: an image of fire, for example, means "fire." Along similar lines, the little icon of an envelope on the latest mass-market portable music player you purchased represents "email." When used in protowriting, pictograms can function as a memory aid to help remember an event or story, or simply as decoration.

- Ideograms are when a *collection* of ideas are represented by a single picture: an image of a water drop could represent rain, but also tears or sadness. An image of sunglasses could represent extremely cool sunglasses, but also sunlight, fashion, or popularity. An image of a peach shaped so it looks like buttocks could represent either peaches, buttocks, or any number of activities humans have discovered they can perform with either.

It's important to note that neither pictograms nor ideograms are language, because there is no *one-to-one correspondence* between them and their meaning. Pictograms and ideograms are *interpreted* rather than read. As an example, consider the following images:

Figure 4: An extremely compelling narrative.

There are several different ways to interpret those images. If you know the story they're trying to tell, these pictures can remind you of it, but if you don't, you will have to make lots of assumptions. Perhaps it is the story of a very cool woman eating a peach. Perhaps it is the tale a regular woman eating a very cool peach. We will never know.

In contrast, the sentence "Cynthia waved, her hair catching in the warm ocean breeze, and in her sunglasses I saw reflected a horrible, monstrous giant peach: it was my body, forever transformed by those hateful scientists I'd once cut off in traffic" has a meaning that's much more clearly defined. While there is ambiguity in any language,* the non-ideographic version has a much more particular and specific meaning than the alternative.

The Easter Island script, called "Rongorongo," has also never been deciphered. It's a pictorial language, comprised of stylized images of animals, plants, humans, and other shapes. It was written by the Rapa Nui people who inhabited their island, and it looks like this:

Figure 5: Possibly language, possibly cool pictures, possibly . . . both?

If the Rapa Nui independently invented writing here, it would be only the third confirmed time in human history this was done: a colossal achievement.

---

* Constructed languages like Lojban are a possible exception here. Lojban is an artificial language designed to allow only syntactically unambiguous sentences. In English you can say "I want to party like Joey," but this sentence has two readings: you could want to party like Joey parties, or you, like Joey, could want to party. By making such a construction ungrammatical—basically turning it into *an illegal sentence*—Lojban forces speakers to be very clear about who is doing what to which objects, when, why, and how.

However, it's also possible this writing was invented only after European contact with the island: Spain annexed the island in 1770 CE and induced the Rapa Nui to sign a treaty. That could've introduced the *concept* of writing to the island, which then quickly evolved into Rongorongo.

There is a dark note here: early visitors to Easter Island were told reading and writing was a skill reserved for a privileged few among the ruling elites. And if Rongorongo script is writing—if the Rapa Nui did come up with the idea of turning invisible ideas into visible shapes, an idea so groundbreaking that it had only ever occurred *twice before* in human history—then they also forgot it. Within the space of a century—a century, it should be noted, defined on Easter Island by European diseases, catastrophic slaver raids, a smallpox epidemic, deforestation, and cultural collapse—the island population had been reduced from thousands to just two hundred individuals, and none of those survivors had ever been taught to read their island's script. Words and sentences had decayed into meaningless shapes and squiggles, part of a cultural tradition that nobody left alive could understand.

This, by the way, should terrify you. Writing is not something humanity gets for free, and like all technologies, it can be lost.

We recommend building writing into your civilization as soon as possible.

3.3

NON-SUCKY NUMBERS

Because everyone wants their civilization . . . to really count.

The story of numbers in human history is the story of countless* missed opportunities and unnecessary delays. While written numbers first appear in 40,000 BCE, predating written language by tens of thousands of years, they were simply tallies: one mark for one count. They looked like this:

Figure 6: Some tally marks.

They're great for small numbers, but once they get to any size, they're a huge pain to deal with. Quick, what number is this?

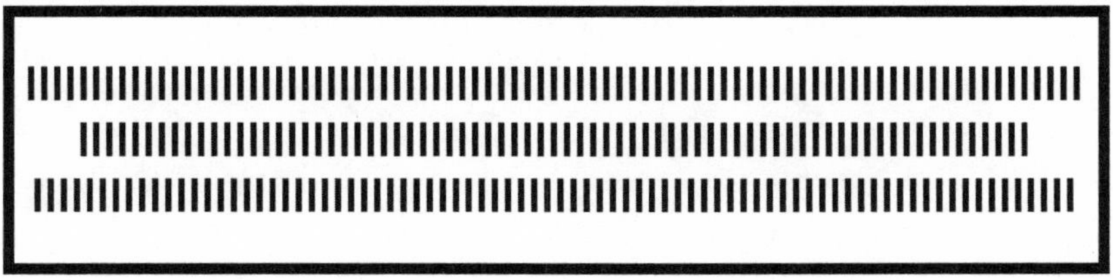

Figure 7: Arguably, too many tally marks.

---

* Pun unintended, *but welcome.*

The answer is "It doesn't matter, because nobody has time to sit around and count that up, come on, *we're literally trying to reinvent civilization in the past here."* This is what makes tallies "sucky numbers." There were other sucky number systems used throughout history, but instead of wasting time with them we're going to skip right to the endgame: your civilization is going to (a) use Hindu/Arabic numerals (b) in a positional value system (c) based on the number 10.

Here's what each of those means and why they're great!

a) **Hindu/Arabic numerals:** these are the numbers you're familiar with: 0, 1, 2, 3, 4, 5, 6, 7, 8, and 9. You can make up different shapes to represent these numbers if you want; they're completely arbitrary. Also, as it is now you and not the Hindu and Arabic cultures that are inventing these numerals, you can call them "[Your Name Here] numerals."

b) **In a positional value system:** this is where a number's value is communicated by the position of each digit in the number. For example, 4,023 means "4 thousands, no hundreds, 2 tens, and 3 ones." This probably sounds pretty familiar, and that's because positional numbers are the ones you've known since you were a kid! Everyone uses them because they rule, and are an extremely efficient and flexible way to represent numbers.*

---

* For an example of a much less efficient number system, just look at Roman numerals, which large sections of humanity wasted time messing around with for thousands of years, and which a smaller section of humanity continues to mess around with to this very day. This is a number system that isn't positional, but instead has you adding up numbers again—just like the tally system we started out with. But here, instead of having just one mark (|), you've got a bunch of them, each representing a different amount: I for 1, V for 5, X for 10, L for 50, and so on. You represent the number you want by adding (and sometimes subtracting) those base numbers together: 2 is II (or 1 + 1), 3 is III (or 1 + 1 + 1), and 4 is IV (or 5 - 1; you always subtract when a smaller letter is in front of a larger one). So a number like "LXXXIX" is equal to 50 + 10 + 10 + 10 + (10 - 1), or 89. The length of a number in Roman numerals doesn't correspond to its value, Roman numerals require you to do mathematics in your head just to figure out what number you're dealing with, and we should all stop talking about them right away. Here are their only uses: making the numbers on clocks look nice, or putting them after your name to distinguish yourself if your parents couldn't even think of a single name for you that wasn't the exact same name one or both of them already had. Avoid.

c) **Based on the number 10:** our positional system is based on 10, which means each digit in a number is 10 times different than the one beside it. When you move from left to right, each column is 10 times smaller, and when you move from right to left, each column is 10 times bigger. Here's 4,023 again:

1000s (i.e., 100 × 10)	100s (i.e., 10 × 10)	10s (i.e., 1 × 10)	1s
4	0	2	3

Table 3: There are 4,023 good reasons to study this chart. No, we're kidding, there aren't that many. But you will probably still want to briefly glance at it so you can know what a number is.

The fun part is, you can actually build a positional number system around any number! Base 10 is the most common throughout history and human cultures—probably because 10 is the approximately average number of fingers per human—but it's far from the only base humanity has experimented with. Babylonians used 60 as their base (which remains with us today when we talk about each hour having 60 minutes and each circle having 360 degrees; see Section 4) and we built computers to use binary, which uses 2 as its base. In base 2, each column is only two times different from the one beside it, instead of ten:

8s (i.e., 4 × 2)	4s (i.e., 2 × 2)	2s (i.e., 1 × 2)	1s
1	0	1	1

Table 4: Numbers in base 2, or binary. You've got 1,011 good reasons to study this chart. Yes, we're aware that's markedly fewer reasons than the last table.

So 1,011 in base 2 is equal to 8 + 2 + 1, or 11. As you're probably guessing, the same sequence of digits can represent different numbers in different bases. If you weren't told "1,011" was in base 2, you'd probably read it in base 10, where it represents "a thousand and eleven." If you read it in base 5 it'd represent 131, in base 7 it means 351, and in base 31 you're looking at a number that represents 29,823. Experiments in other timelines have suggested that building a number system around a weird number like 31 is a bad idea, but guess what: you're trapped in the past, so *none of us can stop you.*

Now that we've got the basics of how we write numbers down out of the way, here is a sad fact: inventing the rest of a number system, with all the features we take for granted, took humans over forty millennia. It actually took us most of that just to invent fractions, an idea so basic that we now teach it to literal babies. Because of that, the following table of features to add to your number system is actually *the most time-saving table in history.*

Feature	Example	What you can do with this	Why you want it	When it was first invented (approx.)
Written numbers	|||||	• Not have to remember numbers in your head all the time.	• Because you have finite space in your brain. • It's hard to do long division in your head.	40,000 BCE
Abstract numbers	5	• Conceive of numbers as abstract ideas (i.e., "one" and "five"), instead of being limited to counting things that actually exist (i.e, "one sheep" and "five goats"). • Numbers that exist outside the things they're counting allows you to access new planes of abstract numerical thought, instead of just reminiscing about specific sheep or goats all the time.	• Numbers as a purely abstract object are necessary to unlock future innovations in numbers, like irrational and imaginary numbers. It is not a coincidence that both these numbers are named in such a way as to make their very existence sound *completely crazy*, but they're also very practical.	3100 BCE

Feature	Example	What you can do with this	Why you want it	When it was first invented (approx.)
Fractions	½	• Represent things that aren't whole numbers, like 1, 2, or 3. • Be able to talk about parts of things.	• Because sometimes you have 4 apples but then Chad ate 3 and a half of your apples, and you'd like to say, "Hey, Chad, you owe me three and a half apples" and not have him get out of it by replying, "Based on our understanding of what numbers can do, what you have just said is nonsensical."	1000 BCE
Rational numbers	0.5	• Represent things that aren't whole numbers *and* not have to mess around with fractions. • Any fraction can be written as a rational number, and vice versa.	• Because adding 201 hundredths and 3 halves together is kind of a pain, but adding 2.01 + 1.50 is easy. We just did it. It's 3.51. No big deal.	1000 BCE
Irrational numbers	$\sqrt{2}$, pi	• Represent numbers that are a lot like rational numbers, except when you write them out they actually go on forever, never repeating or terminating.	• Because there are an infinite number of irrational numbers, so it'd be nice to have a number system that can handle that without exploding. • Also pi (the ratio of a circle's circumference to its diameter) is one of the fundamental constants of the universe that gets used all the time when building things, so irrational numbers are practical too! Bonus.	800 BCE
Prime numbers	2, 3, 5, 7, 982451653: any positive number that's greater than 1 and that is divisible only by 1 and itself	• Nothing save for appreciating the beauty of pure mathematical study, until of course you invent public-key cryptography, which absolutely relies on prime numbers and is extremely useful.	• Because there are an infinite number of prime numbers, though there is no way to know which numbers are prime until you test them. This makes prime numbers one of the only infinite, inexhaustible natural resources in the universe! Don't you want access to an infinite and inexhaustible natural resource? • Yes, you do.	300 BCE

Feature	Example	What you can do with this	Why you want it	When it was first invented (approx.)
Negative numbers	−5	• Conceive of the entire other half of the number system so your numbers don't just stop at 1. • Deal with a concept and its opposite concept in a single number, like heat and cold, income and expenses, expansion and contraction, acceleration and braking, and so on.	• To capture change (in both directions) in a single number. • Negative signs give numbers an emotional connotation for the first time (negative numbers are usually seen as "bad"), which can be useful when you want people to react emotionally to a number. • Also, it's nice to be able to say what 1 minus 2 is without your head exploding.	200 BCE, but listen, European mathematicians as late as 1759 CE were still arguing that negative numbers were "nonsensical" and "absurd," which should tell you all you need to know about European mathematicians as late as 1759 CE.
Zero	0	• Talk about nothing. • Have a number system that is place-based, so you can write numbers like "206" and not have them confused with "26."	• Because otherwise you have a number system that doesn't know what zero is, and that is extremely embarrassing. • Zero works both as a place-holder (as in 206, showing 2 hundreds, 0 tens, and 6 ones) and as a number that can be used like any other in mathematical operations (with some caveats for division; see sidebar on page 28).	Zero existed as a placeholder back in the 1700s BCE, but it took until 628 CE for a concept of zero that you can add, subtract, and multiply with to be realized. You will save so much time right now by just telling everyone "5 plus 0 equals 5, and you should write that down."
Real numbers	3.1, 3.111, 3.1111, 3.11111, 3.11111111, 3.111111111, 3.1111111111, there's more but we'll stop because we could literally be here forever	• Merge both rational and irrational numbers into a single number system. • Describe any number with a (potentially infinite) decimal notation. • Explore the infinitely many numbers that exist between each and every whole number.	• To blow your mind by considering that while there's an infinite number of numbers between 3 and 4, these also include numbers like pi, which are themselves also infinite. • You now have a whole lot of numbers, so congrats.	1600s CE

Feature	Example	What you can do with this	Why you want it	When it was first invented (approx.)
Imaginary numbers	√–1, i, 3.98i	• Manipulate numbers that include the square root of –1 in them. The square root of a number is a smaller number that, when multiplied by itself, produces your original number. The square root of –1 is impossible in the real number system, since any number multiplied by itself has to be positive. So mathematicians said, "Okay, well, let's imagine it *is* possible and call that number *i*, and that's whatever the square root of –1 is."	• This may sound like a complete waste of time (and the name "imaginary numbers" was originally intended as a *brutally sick burn* for just this reason), but they actually have practical applications in everything from modeling electrical flow to the swing of a pendulum in the air.	10 CE, but generally considered "fictitious" or "useless" (like negative numbers were) until the 1700s CE.
Complex numbers	3 + 2i	• Imaginary numbers and real numbers mixed together.	• Useful in fluid dynamics, quantum mechanics, electrical engineering, and calculations in special and general relativity. • It's really useful to have these numbers in your back pocket in case anyone in your civilization invents these things we just mentioned.	1800s CE

Table 5: *Homo sapiens sapiens*, a species that considers itself so smart that it put "smart" in its own name, *twice*, and in *Latin*, took more than 40,000 years to figure this chart out.

See all these ideas? We covered them in a single chart that took you a few minutes to read, at most. You can introduce them in an *afternoon*, saving thousands and thousands of years that humans wasted wandering around not knowing what a zero is. *You're welcome.*

As for the other things you can do with this number system, well, that's up to you. There are a bunch of useful mathematical formulas that took humanity a while to figure out sprinkled throughout this guide, but here's the deepest, darkest secret of math: you can build up the fundamentals of mathematics however you choose.

It may surprise you to hear this, but mathematics is actually built on a foundation of things we can't prove but assume are true. We call them "axioms," and we think they're safe assumptions, but at the end of the day, they remain beliefs we cannot prove. Axioms include ideas like "2 + 1 gives the same result as 1 + 2" and "if

Sidebar: Why Can't You Divide by Zero?

Division by zero is famously* impossible. The reason is not that it will create a black hole but rather because it exposes a contradiction at the heart of our mathematical system. Consider taking a number (let's use 1) and dividing it by progressively smaller numbers that approach, but never quite reach, zero.

Zero marks the point where negative numbers end and positive numbers begin. If we approach it from the positive side, we'll see that 1 divided by 1 is 1, 1 divided by 0.1 is 10, and 1 divided by 0.001 is 1,000. The smaller the number you divide by gets, the larger the result gets. Therefore, one divided by zero should equal infinity.

But there's a problem if we approach zero from the opposite end, dividing by progressively smaller *negative* numbers, again getting closer and closer to zero. Here we see that 1 divided by –1 equals –1, 1 divided by –0.1 is –10, and 1 divided by –0.001 is –1,000. So here the smaller the number you divide by, the closer it gets to *negative* infinity. So by this logic, 1/0 should equal negative infinity.

But a number can't be equal to infinity and negative infinity at the same time. That is, in fact, just about as unequal as two numbers can possibly get. So we've got a contradiction. And it's this contradiction that makes us say, "You can't divide by zero, because the answer doesn't make sense, and nobody knows how to fix that yet."

---

* For definitions of "famous" liberal enough to include properties of mathematical systems.

a equals *b*, and if *b* equals *c*, then *a* equals *c*." These assumptions are useful because they match up with reality—and building math on a foundation that matches up with reality has proven reasonably practical—but there is nothing stopping you from building different mathematical systems. While we do recommend building math that is practical first, it can be quite entertaining to figure out how multiplication would work in a universe where *a* + *b* is not the same as *b* + *a*.*

Now that you have invented non-sucky numbers—and the basics of math to go along with them—you have unlocked several perks. Numbers obviously let you precisely quantify the world around you, which will be the basis of everything from cookbooks to accounting to science. Physical resources like sheep or trees and abstract resources like money, popularity, and *time itself* are all managed, understood, and communicated through numbers. Most universally, numbers act as a sortable set of labels: page 123 of a book is intuitively between pages 122 and 124, and if you know how many pages that book has, you have a pretty good idea of where that page will be. The context a sorted set of numbers provides will be very useful to members of your civilization down the road, whether they one day use them to label the hours in a day, the days in a year, the buildings on a street, or the floors in those buildings. They can also be used to label temperatures, radio frequencies, vitamins, and maybe, one day, if your civilization is *very* lucky, the strength of unstable Einstein-Rosen bridges constructed across disparate spatial/temporal reference frames.

---

* If you're interested, the answer to the question of "But seriously, how does multiplication work in such a system?" is "Quite well, thank you."

3.4

THE SCIENTIFIC METHOD

Much improved over the earlier scientific approximation.

People who build time machines tend to love science, as they generally tend to be scientists by training, or at least well-intentioned amateurs who have no idea of the powers they are about to unleash until a bunch of their future selves come back to warn them. But it's important to remember even science has its limitations and is not an oracle of truth. In fact, science is merely:

1. provisional,
2. contingent, and
3. our best effort so far.

Here is the bad news: the scientific method can produce knowledge that is wrong. Here is the good news: the scientific method is still our best technology for uncovering, verifying, and refining correct knowledge, because what the scientific method allows us to do is make wrong knowledge *gradually more correct*. Usually, this refinement results in progressively more accurate theories—classical physics leading to relativity, leading to quantum physics, which leads to metaquantum ultraphysics[5]—but it does sometimes result in entire theories being thrown out.

For example, in the 1700s CE we thought things burned because they were *phlogisticated*: that they were filled with "phlogiston," an invisible and intangible substance that you couldn't see or touch or distill but which was still required to make things burn. Things that were highly phlogisticated—like wood—burned quickly, while those that had less phlogiston in them burned less well, and ashes—having already been completely dephlogisticated—wouldn't burn at all. Phlogiston theory even explained why things got lighter when they burned: the phlogiston was floating away into the air. It also predicted that a match placed in a sealed glass jar would eventually stop burning: the air in the jar would absorb all the phlogiston it could, and then the fire would go out. Matches in sealed glass jars do go out, so this all looked great. Done! Thanks, science! *Now we know what fire is.*

Phlogiston theory started to fall apart when we did more experiments and found some results that didn't quite make sense. Sure, wood gets lighter when you burn it (the ashes left over clearly weigh less than the original wood did) but some metals (like magnesium) actually *gain* mass when burned. And now we had a problem: results that did not agree with our theory. More science was needed!

Some scientists tried to revise phlogiston to try to get it to match up with results: maybe phlogiston could sometimes have *negative* mass, so the less you had of it in something, the more it would weigh? But this was a big leap—especially since negative-mass matter was a completely new form of matter that was being invented simply to solve this phlogiston problem. Other scientists looked for a more conservative explanation, and the oxygen theory of combustion was what resulted: the idea that fire wasn't phlogiston leaving matter but rather a *chemical* reaction between matter and oxygen, one that produces both heat and light. This theory also predicted that a match in a sealed glass jar will eventually stop burning, but for different reasons: the oxygen in the jar is consumed and so the fire goes out, because oxygen fuels the chemical reaction that is fire. This is the more accurate theory of combustion that we still operate under today, but we could still be wrong.

Or, more likely, we could still be *more correct.*

Here's how you produce knowledge using the scientific method.

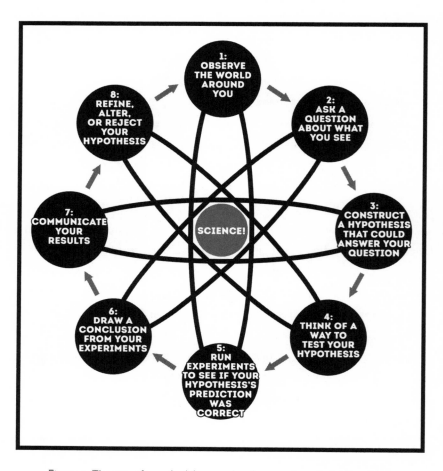

Figure 8: The scientific method, here rendered as a cool atomic-looking thing.

An example: maybe you notice (as per step 1) that your corn didn't grow well this year. For (2), you might ask, "Hey, what the heck, everyone, how come my corn didn't grow well this year?" You might suspect the drought affected the corn's growth (3), and so (4) decide to grow corn under controlled conditions, giving each plant different amounts of water but equal amounts of everything else you can think of (sunlight, fertilizer, etc.). After carefully doing that (5), you might conclude (6) that a precise amount of water grows the best corn plants, and (7) let your farmers know. And when your corn still doesn't grow as well as you want, you might explore (8) and wonder if there's more to growing great corn than just making sure your corn isn't thirsty.*

The more ways a hypothesis has been tested, the more likely it is to be correct, but nothing is certain. The best case you can hope for by using the scientific method is a theory that happens to fit the facts as you understand them so far: science gives you an explanation, but you can never say with absolute certainty that it's the correct one. That's why scientists talk about the *theory* of gravity (even though gravity itself clearly exists and can cause you to fall down the stairs), *theories* of climate change (even though it's obvious our environment is not the same one our parents enjoyed, or that you're enjoying right now), or the *theory* of time travel (even though it's a fact that you're clearly trapped in the past for reasons that cannot have any legal liability assigned).

Note that the scientific method requires you to keep an open mind and be willing—at any time—to discard a theory that no longer fits the facts. This is not an easy thing to do, and many scientists have failed at it. Einstein† himself hated how his own theory of relativity argued against his preferred idea of a fixed and stable universe, and for years tried in vain to find some solution that reconciled them both. But if you succeed at following the scientific method, you will be rewarded, because you will have produced knowledge that is *reproducible*: that anyone can check by doing the same experiment themselves.

---

* There is. See Section 5: Now We Are Become Farmers, the Devourers of Worlds, ideally before you are starving from a lack of delicious corn.

† Albert Einstein was a scientist who, among other achievements, realized that matter and energy were equivalent and could be described by the equation "energy is equivalent to mass times the speed of light squared," or "$E = mc^2$." Hey! Now as far as anyone knows, you're as smart as Einstein!

Scientists are often seen as turbonerds, but the philosophical foundations of science are actually those of pure punk-rock anarchy: never respect authority, never take anyone's word on anything, and test all the things you *think* you know to confirm or deny them for yourself.

Figure 9: A typical scientist.

CALORIE SURPLUS: THE END OF HUNTING AND GATHERING, AND THE BEGINNING OF CIVILIZATION

Hunting and gathering better ways to live your life.

Beginning with our ancestors in prehistory, and continuing even past the appearance of anatomically modern humans around 200,000 BCE, humans spent all their time fully immersed in the fabulous hunting-and-gathering lifestyle. This is, as you might suspect, a situation in which hunters hunt while gatherers gather. You live off the earth, survive by your wits, and follow food wherever it might take you, leaving areas when you've exhausted their local resources. There are many benefits to this lifestyle: you get to eat a wide variety of foods (a varied diet helps ensure good nutrition) and you get to visit lots of interesting places, eat whatever lives there, and get lots of exercise. But this means food doesn't come to you: you have to go to it. And this is expensive.

This expense manifests in several ways: It costs calories to search for food. It costs you years of your life, as you're much more likely to eat something new that happens to be toxic, or to get injured or killed by the very animals you're trying to gobble. Plus, you're constantly introduced to new bacteria and parasites as you chase after a food supply that's never assured. But its greatest expense comes from constantly moving: when you never know how long you'll be living somewhere, you're not going to build expensive and labor-intensive infrastructure. Anything you can't take with you can turn into wasted effort overnight. You won't store any resources long-term, because there *is* no long term.

And for almost 200,000 years—the vast majority of human experience—this was all anyone did. Hunt, gather, maybe build some temporary settlements, and move on whenever the going got tough or someone saw a tasty-looking herd of animals over the next hill. It was only around 10,500 BCE[6] that anyone thought to suggest that rather than taking the planet as it was when we come across it, we could instead *change* the planet to better suit our needs.

This idea represents both the invention of farming (the process of raising and caring for plants and animals in a convenient location, so that food supplies are more reliable), and with it, full domestication (the process by which these plants and animals, once kept in a convenient location, are transformed into more convenient versions of themselves).* There was no reason this idea couldn't have occurred to us sooner, except that we either didn't think of it or were too lazy to make it happen. Humanity wasted almost 200 millennia *not* having this idea. But it's already occurred to you, because you just read it. Look at you. You're already doing great!

Once you begin farming and domesticating animals, you have entered a new phase of humanity, one in which *a single human can reliably produce much more food than they need to survive.* Humans run on food energy—calories—and you have just produced a surplus. In fact, a farmed field can produce anywhere from 10 to 100 times more calories than what you'd get by hunting and gathering an equal area of unfarmed land! And when you add more farmers and more arable land, you just add to that pile of extra food. It's on calorie surpluses—and therefore farming—that civilizations are built.

How? Well, more food obviously lets you have more people. But it also lets those people stop worrying about where their next meal is coming from, freeing them up to start worrying about different, more productive things: why stars seem to move across the sky or how come things fall down instead of up. Farming also formalizes the idea of an economy in your civilization, since now farmers can regularly trade their food with others. With an economy comes specialization: instead of each human having to do everything necessary to survive (or splitting it among a family group), now someone who's particularly talented at farming can focus all their efforts on that. A hunter and gatherer

---

* As you'll see in Section 8, while humans hadn't domesticated any plants before farming, they had managed to domesticate a few animals, including the dog.

simply does not have the time to invent integral calculus, but a professor or philosopher—someone who can both conceive of and devote themselves to such problems—does.

Specialization gives the people in your civilization the opportunity to go further in any direction of study than any other human has gone before. It unlocks doctors who can devote their entire lives to curing disease, librarians who can devote their entire lives to ensuring the accumulated knowledge of humanity remains safe and accessible, and writers who, fresh out of school, take the first job they find and devote the most productive years of their lives to writing corporate repair manuals for rental-market time machines that their bosses almost certainly don't even read,* ironically for so little money that they can't possibly afford to go back and fix that one horrible, horrible mistake.† Specialization goes hand-in-hand with a civilization's development, because the greatest resource of your or any civilization is not land, power, or even technology. It's human brains—yours and those of the humans around you—that will be the creative, inventive, brilliant engines that drive your civilization forward. And it's specialization—supplied by a calorie surplus—that allows those human brains to reach their full potential.

Unfortunately, the advantages we've just outlined do come with several challenges. While we believe the benefits do outweigh the drawbacks, you should be aware of the following *Extremely Garbage Features of Farming*:

---

* That's right: you're not fooling anyone, *Chad*. I know you just run a word count on whatever I write and call it a day. Honestly, I'd just paste in some of your more embarrassing mass emails and be done with this project, except for one little thing: time travelers really *could* get stuck in the past, and I don't want to leave them hopeless and alone. So here's the deal, stranded time traveler. We're both trapped: you in the past, and me in a job I hate. You and I are going to get through this together, okay? We're going to do well. I'm toning down the dry corporate language that Chad thought was *so important*, and you're going to promise to build into your civilization the cultural tradition that years down the road, if anyone should happen to meet my boss, they should tell him he's a tool. They'll be able to recognize him instantly: his name will be Chad "The Chad" Packard, and he will have the world's most punchable face. Listen: I'm rooting for you.

† Oh, and one more thing: maybe also build in the tradition that if one day someone should meet a Ryan North, fresh out of school and about to take the first job he finds . . . give him a heads-up.

- When wild food is abundant, farming is *way* more work than hunting and gathering. What farming offers, though, is the promise of a more reliable food source and, through domestication, more *convenient* food sources too.
- Farming also requires technologies for food storage, because its whole point is to produce more food than you can eat at once. This again is more work, but thanks to Section 10.2.4: Preserved Foods, you at least have the advantage of knowing exactly what to do.
- Farming creates the first income inequality, because not everyone can be farmers or share equally in the land required for it. Farmers have the most food and (initially) the most to trade, and everyone who doesn't want to become a skeleton needs to continue eating food. You have just created rich and poor people, or at the very least the potential for them.
- Farming requires infrastructure (fences, etc.), which means you are no longer mobile. Your civilization has just become a giant and stationary target. While this text does not include *explicit* instructions for weaponry, we're certain that, should the need arise, you can probably adapt several of the technologies included here to that purpose.
- Animals carry diseases and can transmit them to humans. Worse, some of our deadliest diseases don't bother animals at all. Over 60 percent of all human diseases originate from close contact with animals, including such all-time champion diseases as anthrax, Ebola, plague, *Salmonella*, listeriosis, rabies, and ringworm. We understand if after reading that list you want to go back to hunting and gathering, but we promise civilization will *eventually* be worth it. Just don't be surprised if people start getting sick, and maybe read Section 14 (Heal Some Body) before it's critical.

In light of these downsides, we would like to take this opportunity to remind you that it is inarguable that farming leads to calorie surpluses, which leads to specialization, which leads to innovations like apple pies, time machines, and the latest mass-market portable music players. If you work hard, you will produce these. If you hunt and gather, you will not. Instead, you will eat bugs you find under a rock.

Best of luck with your decision.

4

→ UNITS OF MEASUREMENT ARE ARBITRARY, BUT HERE'S HOW YOU CAN REINVENT THE STANDARD ONES USED IN THIS BOOK FROM SCRATCH

Can you really reinvent measurement while trapped in the past?

We wouldn't . . . rule it out.

↗ All units of measurement are arbitrary, but the vast majority of humanity* agrees that you should at least make them practical by having your units scale predictably, combine intuitively, and be easy to reproduce while

---

* We say "the vast majority" because there remain precisely three holdout countries that insist on not using the otherwise predictable and global standard of weights and measures: Liberia, Myanmar, and the United States. The United States has had *actual spacecraft* (the Mars Climate Orbiter) collide with *actual planets* (Mars) because they insisted on using their archaic units while the rest of the world has agreed on a more practical standard, then forgot that they did that, and then messed up their orbital trajectories because some calculations used metric and others didn't. Even the *327.6 million American dollars* wasted on that 1999 CE impact did not give the United States enough motivation to join the rest of the world in standard units of measurement. They wouldn't budge an inch!

trapped in the primordial past. Accordingly, in this guide we use the metric system of measurement (which is based on units of ten) along with the centigrade temperature system, which ensures you can replicate those measurements no matter what time period you're stranded in. *All you need is this book and some water.*

The centigrade temperature system is defined by locking in 0 degrees centigrade (or 0°C) as the point water freezes, and 100°C as the point where it boils. You can therefore reproduce it easily no matter what time you're in: just mark those two points on your thermometer (see Section 10.7.2), divide what's left into 100 equal segments, and you've done it.* The "competing" Fahrenheit system, in contrast, locks in 0 degrees as the temperature of a weird slurry of ice, water, and salt that Mr. Fahrenheit threw together before calling it a day, and we will discuss it no further, except to say 32 degrees Fahrenheit is about where water freezes, 212 degrees is about where it boils, and *you can do so much better.*[7] If you'd like to operate without negative numbers, you can invent the Kelvin temperature scale, which is just degrees centigrade, but with 0 Kelvin set to −273.15°C: the coldest temperature possible in our universe. Water freezes at 273.2 degrees Kelvin and boils at 373.2 degrees.

So that takes care of temperature.

Weight we'll base around the kilogram, which until 2019 CE was still tied to actual prototype kilograms: physical masses of platinum safely stored worldwide in nested bell jars so that humans can point to it and say, "A kilogram is however much that hunk of platinum weighs."[8] There was the canonical copy—stored in France—and dozens of duplicates placed throughout the world for both convenience and safekeeping: after all, you don't want anyone stealing the only kilogram in a dramatic and elaborate heist, leaving the entire world unsure of precisely how much one kilogram weighs.

There are several downsides to this idea, even factoring in the undeniable allure of *heisting the kilogram.* These backup kilograms were occasionally returned to France to verify that they were all still the same weight,

---

* Well, you've *almost* done it. See Section 10.7.2 for some important details. Water also behaves differently at different pressures, so these numbers are calibrated for sea level (as opposed to at the top of a mountain or at the bottom of a mine).

and the thing is: they're weren't. Different kilograms stored throughout the world—even those made from the initial batch of forty cast in 1884 CE—were becoming different weights, drifting ever so slightly apart over time—and until time travel was invented, *we didn't even know why*.[9] It gets worse: since the measurements comparing the weight of prototype kilograms are all relative to one another, this left open the possibility that *all* kilograms were gaining or losing mass, and some were just better at it than others. As kilograms are a central measure in the metric system, contributing to the definitions of units of force (newtons), pressure (pascals), energy (joules), and power (watts, amperes, and volts), not to mention all the myriad other units spun off from these, you can easily see how even a minor change in the mass of the official kilogram would redefine a (metric) ton of other units across disparate fields of measurement.

CIVILIZATION PRO TIP: There are downsides to building the foundations of modern science and measurement around the mass of an old hunk of metal in a jar in France.

Luckily for you, you won't need such precise measurements for a very, very long time, and the amount that those prototype kilograms were *supposed* to weigh was defined simply as the weight of 10 cubic centimeters of water at 4°C. You've already got water and you've got temperature, so all you need to know now is how big a centimeter is and you can easily reproduce a kilogram.

Before we do that, some terminology. All metric units scale up and down by powers of ten, which are indicated by prefixes. Here are some common ones, from smaller to larger:

Prefix	Symbol	Scale
Nano-	n	1000000000× smaller
Micro-	μ	100000× smaller
Milli-	m	1000× smaller
Centi-	c	100× smaller

Prefix	Symbol	Scale
Deci-	d	10× smaller
[none]	[none]	actual size
Deca-	da	10× bigger
Hecto-	h	100× bigger
Kilo-	k	1000× bigger
Mega-	M	1000000× bigger
Giga-	G	1000000000× bigger

Table 6: A true megachart.

Centimeters are 1/100th the size of a meter, and you know that because of the "centi-" prefix. Similarly, the word "kilometer" tells you it's 1,000 times longer than a meter: 1,000 meters. We usually abbreviate meters as "m," making centimeters "cm" and kilometers "km." So how long is a meter anyway?

The meter started in 1793 CE and was defined as "one ten-millionth of the distance from the equator to the North Pole." It was redefined in 1799 CE to be tied to a physical prototype (like the kilogram), redefined again in 1960 CE to be the wavelength of the emissions of one particular isotope of the element krypton, and again in 1983 CE to be the precise distance light travels in a vacuum over 1/299792458th of a second. Given your current circumstances, you've probably already noticed how these definitions started out completely useless to you but then somehow *still* managed to get worse. Luckily, we noticed that too, and so we have printed a handy 10cm ruler in this section—as well as an arguably more convenient one on the dust jacket—so you don't need to mess around with any of that. From that you can construct meter sticks that will be pretty darn close to accurate.

So now you've got standard measurements for length, weight, and temperature. The only major unit left for you to define is time, which is based on the second. While the modern definition of the second is the frankly ludicrous "however long it takes 9,192,631,770 periods of the radiation corresponding to the transition between the two hyperfine levels of the ground state of the

Sidebar: Handy Measurement Templates

Here's a 10cm ruler.

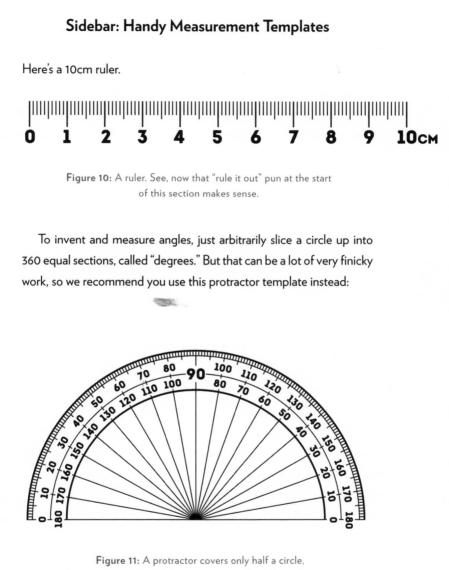

Figure 10: A ruler. See, now that "rule it out" pun at the start of this section makes sense.

To invent and measure angles, just arbitrarily slice a circle up into 360 equal sections, called "degrees." But that can be a lot of very finicky work, so we recommend you use this protractor template instead:

Figure 11: A protractor covers only half a circle, but you can use two of them to measure the full 360 degrees.

cesium 133 atom," you intuitively know how long that takes—*it's one second*—and all you need is a convenient reference. To produce a device that can indicate one second without requiring cesium 133, you'll need to construct a simple harmonic isolator, which in non-time-machine-repair jargon means "tie a rock to a string."

A rock tied to a string that can swing freely is called a pendulum, and it turns out that one second is the time it takes any pendulum on Earth—regardless of weight—to swing from one end to the other, as long as the pendulum is 99.4cm long. This cool property of pendulums—that they always take just about the same amount of time to complete a swing no matter how far back you pull the weight before letting it drop—makes this a particularly easy experiment to perform. This property was discovered by a guy named Galileo Galilei in 1602, but you're going to take the credit. It's your property now.

As we saw, other units can be built up from these measurements. We've got weight and length, but to measure volume you'll need liters. Conveniently, a liter is the area enclosed by a cube 10cm wide on each side—the same sized cube that, when filled with water, weighs exactly one kilogram. For sound you'll want to measure frequency, which is simply the number of vibrations per second. A hertz (or "Hz" for short) is one complete cycle per second, so a frequency of 20Hz just means 20 vibrations per second. For physics, the amount of force required to accelerate a 1kg mass by 1 meter per second squared is called a "newton," the amount of energy transferred to that object over 1m is called a "joule," and a watt is simply one joule per second. These units might seem abstract, but they'll come in handy with some technologies you'll be inventing later on.

And with that, the tiny centimeter printed in this section has helped unlock measurement of not just length but volume, mass, force, energy, and time itself. If you're using the pages of this guide for toilet paper (and you shouldn't, why would you be doing that, *use anything else*) maybe save this page for last.*

---

* We recommend measuring and then memorizing the width of your pinkie finger now: it'll be a useful approximation if you ever lose your reference ruler. And don't worry: if you do lose the centimeter but have already created a one-kilogram reference mass, you can recover the centimeter by making different-sized cubes and filling them with water until one matches its weight perfectly: you can use the balance scale you invent in Section 10.12.6 to test them. Metric is the time traveler's friend!

5

NOW WE ARE BECOME FARMERS, THE DEVOURERS OF WORLDS

How you too can be out standing in your field.

Wouldn't it be nice if you had machines that ran on water and light and turned gross dirt into delicious food and fun chemicals? Also, wouldn't it be nice if the machines were self-duplicating, self-improving, and—most exciting of all—not all of them wanted to kill you?

Good news: these machines exist! They're called "plants," and they will be some of your greatest resources in your new civilization. Think of them as *free technology*: machines you can use, even by accident, to transform the inedible dirt around you, the boring light above you, and the tedious water that falls from the sky into all sorts of useful materials, medicines, chemicals, and foods your civilization requires. If we did not already have plants, we would think they are magic. But they are everywhere and evolved before we did, so most of us think they're pretty boring.

It took humans almost 200,000 years to realize we could do more than just gather them up when we were hungry. We could instead *domesticate* them, growing them in an environment safer from predators, where we could be choosy, picking the plants that did more of what we wanted and less of what we didn't.

This is called "selective breeding," by the way, and you just invented it.

SELECTIVE BREEDING

Here's all you need to do:

1. Find a plant (or animal; this also works on animals) that has exceptional properties you like. Maybe it produces more kernels of delicious and nutritious corn than your other plants, or lasts longer in storage, or resists pests or drought, or maybe even . . . all of the above?
2. Plant seeds from that plant instead of seeds from your other, crummier plants. (If you're working with animals, let only the ones you like reproduce.)
3. Repeat.

By doing this over successive seasons, you will produce crops that emphasize whatever properties you choose, almost by accident. Here are three examples of what humans have wrought with the mighty power of selective breeding alone:[10]

Fruit or vegetable	First domesticated	Modern variety you used to enjoy like it wasn't even a big deal	Incredibly disappointing ancient ancestor
Corn	7000 BCE	• 190mm long • peels easily • sweet and juicy • 800 soft kernels	• 19mm (10 times smaller, 1,000 times less volume) • peels by being smashed into pieces • tastes like a dry, raw potato • 5–10 very hard kernels
Peach	5500 BCE	• 100mm long • 9:1 flesh-to-stone ratio • soft and edible skin • sweet and juicy	• 25mm long (4 times smaller, 64 times less volume) • 3:2 flesh-to-stone ratio • waxy skin • tastes earthy, sour, and slightly salty

Fruit or vegetable	First domesticated	Modern variety you used to enjoy like it wasn't even a big deal	Incredibly disappointing ancient ancestor
Watermelon	3000 BCE	• 500mm long • available in seedless varieties • easily opened, you can just punch your way in • almost fat- and starch-free • delicious taste, sweet smell	• 5mm long (100 times smaller, almost 1,000,000 times less volume) • 18 bitter, nutty seeds • requires hammer or smashing to open • high in starches and fats • bitter taste, unpleasant smell

Table 7: Food used to suck.

And these were all bred before we knew what genetics were, before we knew we could induce plants and animals to evolve for us, and before we knew that directed selective breeding could have effects within a single human lifetime. But you already know all this stuff. You're ahead of the game already!

There is, however, a downside to planting the same thing over and over. Rather than letting you find out by surprise and then starving to death like countless other humans have throughout time, we thought we'd tell you. Growing the same plant repeatedly will kill your soil (slowly) and then you (more quickly). Luckily, you can solve this problem with a technology called "crop rotation." What's "crop rotation," you ask, as allured as you are entertained? We're more than happy to answer.

CROP ROTATION

There are three simple but extremely critical things to keep in mind about plants:

1. Plants use the sun's energy to become big and delicious.
2. The chemical they use to extract that solar energy is called "chlorophyll."
3. Nitrogen is a critical ingredient in chlorophyll.

It's an oversimplification to call nitrogen "magic plant food," but it's not far off. It's the most commonly supplied plant nutrient around the world, and the whole reason plants like the Venus flytrap and pitcher plants evolved *actual carnivorous mouths* was simply to harvest the nitrogen from passing insects. The good news: if you're alive enough to read this, then Earth's atmosphere is full of nitrogen. The bad news: plants can't access it there. Instead, they take it from the soil. And since plants remove nitrogen from soil and don't replenish it, if you plant the same thing over and over, you will have some very, very bad problems.

Specifically, you will have these particular very, very bad problems:

1. The soil will exhaust its nitrogen, along with the other nutrients required by your crop, meaning your crop will grow progressively less well every year until it fails entirely.
2. Pests and diseases attracted by the crop you keep constantly planting will thrive, as there is no interruption to their life cycles or habitat.
3. By not having different crops in rotation at the same time, you are vulnerable to complete starvation if your crop fails.
4. Crops with shallow root systems cause dirt to not stick together as well, and your soil will erode away.
5. Crops with shallow root systems leave less biomass in the soil after harvesting, which leaves it less nutrient-rich for the next crop.
6. Your farm will be sad and boring, and you'll have the same thing for dinner every night.

To avoid these problems, you need to allow the soil to recover. Unless you've got an extreme case, recovery is easy: just till the ground but don't plant anything for a year (in farmer talk, this is letting the field "lie fallow") while your animals hang out there. Tilling kills weeds, and your animals' poop and pee are full of nitrogen, so it restores the soil.* So, great! Your soil will be fine, as long

---

* We know what you're thinking: "Wait, if animal poop has nitrogen in it, then human poop does too, right? Can I truly poop my way to a fun and productive civilization?" This is a bad idea for several reasons. First, it's unpleasant: human poop tends to smell worse to humans than other animal poops. But more critically, human poop—which when used as fertilizer is charmingly and euphemistically referred to as "night soil," and also less charmingly and less euphemistically referred to as "fecal sludge"—contains all sorts of pathogens humans are vulnerable to, including other people's *parasitic worms*, which can then easily infect you, the farmer. If you don't have the technology to remove parasites from poop yet (in Section 10.2.4: Preserved Foods you'll figure out a way to do it, but you'll literally

as you don't mind *not eating anything for a full year.* And if at this point you're thinking you could easily improve this system by farming only half your fields each year while the other half recover, congratulations: you just invented crop rotation!* Specifically, two-field crop rotation. It looks like this:

	Field 1	Field 2
Year 1	Plant whatever food you want.	Lie fallow, let animals graze here so their poop fertilizes the land.
Year 2	Lie fallow, let animals graze here so their poop fertilizes the land.	Plant whatever food you want.

Table 8: The two-field crop-rotation system, featuring both foods *and* poops.

This system leaves 50 percent of your fields unproductive, but it's simple, it's reliable, and it lets you eat every year. But if you want to get fancier and/or want to answer people's complaints about how nice it'd be to farm at greater than 50 percent efficiency, you can invent three-field crop rotation. It works like this:

	Field 1	Field 2	Field 3
Year 1	Lie fallow, animals poop here.	**FALL**: plant wheat and rye (human food).	**SPRING**: plant oats and barley (animal food), plus legumes.
Year 2	**SPRING**: plant oats and barley (animal food), plus legumes.	Lie fallow, animals poop here.	**FALL**: plant wheat and rye (human food).
Year 3	**FALL**: plant wheat and rye (human food).	**SPRING**: plant oats and barley (animal food), plus legumes.	Lie fallow, animals poop here.

Table 9: The three-field crop-rotation system. Now you're planting twice a year and working twice as hard! What a world!

---

be *pasteurizing your poop*), it's probably not worth the risk. But the good news is that you don't need to worry about pee: human pee is fine! Pee in a field all you want!

* Congratulations: you also just invented "working half as hard while pitching it as a net gain for civilization"!

You're now planting and harvesting twice as much food, which requires labor or better plows (historically the moldboard plow does the trick: See Section 10.2.3), but this does bump you to 66 percent productivity. But with fields being used twice before they're harvested, won't the soil get exhausted?

The answer lies in the legumes you were planting. Legumes are dry fruit contained within a shell or pod, and they include plants like chickpeas, regular peas, soybeans, regular beans, alfalfa, clover, lentils, and peanuts. We've listed them out just now because you'll definitely want to farm at least one of these bad boys! Why? Besides being reasonably tasty, legumes are one of the few plants that can host certain bacteria (called "rhizobia," though of course you can call them whatever you want) in their roots, and these bacteria do something extremely valuable, something that no plant on Earth can do on its own.

They add nitrogen back into soil.

Specifically, when rhizobia infect plants, they act as a symbiote, taking some carbon produced in the plant during photosynthesis, and in return converting nitrogen gas (N_2) into a form plants can use (NH_3, or ammonia), which is stored in the roots of that plant as nodules. And when you harvest these plants—leaving the roots in the ground—both nitrogen and rhizobia return to the soil to wait for the next planting.

The legumes—or rather, the bacteria that infect them—are the glue that holds this whole "three-field crop rotation" thing together. Civilization lasts only as long as people are kept fed, and while three-field crop rotation allows you to increase your food output—thereby increasing the maximum size of your civilization and therefore the number of human brains available to it—it also means that everything you do, from your smallest victories to your greatest accomplishments, will depend on a bunch of *invisible single-celled microbes that live in the dirt.* If they fall, so too falls your civilization.

CIVILIZATION PRO TIP: Don't forget to plant your legumes.

But can we get even more efficient? Can we be so bold as to invent a *four*-field crop-rotation system, bringing efficiency up to 75 percent, or—*dare we dream—*

100 percent? It took hundreds of years for humans to muster up the courage to even consider doing such a thing, but you bet your calorie surplus we can:

	Field 1	Field 2	Field 3	Field 4
Year 1	Wheat	Turnips	Barley	Clover
Year 2	Turnips	Barley	Clover	Wheat
Year 3	Barley	Clover	Wheat	Turnips
Year 4	Clover	Wheat	Turnips	Barley

Table 10: Finally, a way to farm that doesn't require anyone to take any time off ever. Progress!

Here is a system of crops that works to support both the land and the farmer. Wheat is for humans, barley and turnips are intended for both humans and livestock, turnips keep well over winter to feed animals, and clover restores the soil: any legume works, but clover is especially good at it.* In addition, animals can graze on the fields during the turnip and clover phases, which helps control weeds. Each field goes three years before the same crop is grown again, which starves pests that could survive if the plant they fed on was grown more frequently. If you don't have all these crops in your area, you can substitute other plants, as long as the nitrogen keeps being restored.† Be careful, though: when you don't let a field rest, you can overplow it, which causes problems (see Section 10.2.3: Plows).

You might be feeling like humans are pretty smart for figuring all this stuff out, but here's something embarrassing: the science for this "nitrogen" stuff all came later. Instead we just used trial and error over thousands and thousands of years, which meant even the most basic two-field crop rotation didn't show up until 6000 BCE, and four-field crop rotation arrived only in the 1700s

* Clover's so good for soil that it's sometimes referred to as "green manure": one of the few instances in history in which that phrase is intended as a compliment.

† It's hard to tell you've exhausted the soil until it's too late, so you may want to keep simpler two- and three-field systems going until you're confident that your selection of plants works in a four-field rotation.

CE. That's more than 20,000 years just to invent non-crappy farming! And it gets worse: the symbiosis between rhizobia and legumes, which is what makes advanced crop rotation possible, first evolved over 65 million years ago. That's so far back that *actual dinosaurs* could've invented our most complicated system of crop rotation, if only they were smart enough to, and had tried to, and had also not been horrifically killed by asteroids.*

Besides nitrogen, plants also need calcium and phosphorus. You can get phosphorus from bones and calcium from teeth, so recycling animal skeletons is a good idea. You can crush and boil them to produce bone meal—an easier-to-spread alternative to chucking bones into a field—and by reacting the bone meal with sulfuric acid (see Appendix C.12) you produce a phosphate that's easier for plants to use and therefore a more effective fertilizer.

Now that you know about selective breeding and crop rotation, you (or members of your civilization, if you're not "the farming type") are ready to efficiently farm. However, depending on where and when you are, different plants and animals will be available to you: these are detailed in the next two sections. Most of the biomass on Earth is useless for feeding humans: it's either indigestible, poisonous, dangerous, too tedious to gather or prepare, or not nutritious enough to make it worthwhile. But don't despair: a small minority of the plants and animals on this planet are really useful for humans, providing shelter, food, and treatment for diseases!

Hey, fingers crossed you're near some of those ones!

---

* Not all species of dinosaur were horrifically killed by asteroids. Some made it to the present day by evolving into birds, and others made it to the present day by being taken there in time machines. Upon arriving in the present day, dinosaurs are usually placed in special "Jurassic parks," which, like the FC3000™, have only rarely suffered catastrophic failures for which no legal liability can be assigned.

6

WHAT WILL OTHER HUMANS BE EATING IF I'M STRANDED AFTER THEY'VE EVOLVED BUT BEFORE AGRICULTURE AND SELECTIVE BREEDING ARE A THING, AND HOW CAN I TELL IF IT'S POISONOUS, BECAUSE I BET THESE ANCIENT HUMANS ARE EATING SOME REALLY STUPID STUFF?

Great news: you can eat anything once!

We do not know a lot about fruits and vegetables from this era (from between 200,000 BCE and 10,500 BCE), as most temporal research on this period has focused on more interesting questions, like "What did Mitochondrial Eve—who even without time machines we know lived between 99,000 and 148,000 years ago, and is the most recent common female ancestor of every single human alive today—look like?" In case you're curious, she looked *pretty great*. The male equivalent, "Y-chromosomal Adam," is the most recent ancestor from which all humans in the present are descended through their fathers. He too was a *stone-cold fox*.

But getting back to fruits and vegetables, here's what we know:

- 780,000 BCE: figs, olives, and pears are being eaten. This predates even the evolution of anatomically modern humans.
- 40,000 BCE: dates, legumes, and barley are being enjoyed.
- 30,000 BCE: apples, oranges, and wild berries are being consumed.
- 10,500 BCE: farming is invented, and selective breeding of plants and animals.

Luckily for you, you'll find edible fruits and vegetables in any era humans exist in, because if there weren't any, then humans would've died out and we'd never have lasted long enough to take the very fabric of time and space and tear a hole in it big enough to shove an FC3000™ through. The bad news is, these edible plants, leafy greens, and vegetables are likely be different from the ones you're familiar with.

And they will almost certainly be worse.

As we have seen, selective breeding made plants better (from a human perspective: higher yields, bigger fruits, hardier breeds, etc.), which necessarily means the further you go back in time, the worse fruits and vegetables you're going to find. Smaller yields, more disgusting flavors, worse harvests, inconvenient packaging: all these frustrations and more await you in your future, by which, of course, we mean the distant past. Remember the early ancestors of corn, peaches, and watermelons that we saw in the previous section? Well, here's what they looked like before and after selective breeding:

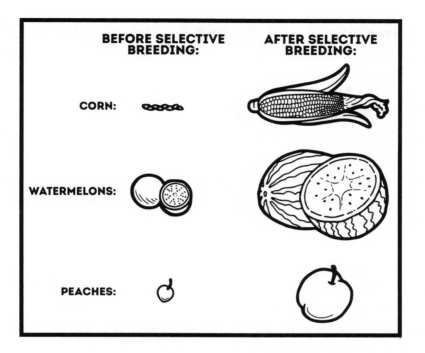

Figure 12: Prepare for disappointing salads.

If you ever want to eat watermelon and peach fruit salad (with . . . corn on the side?) again, you're going to have to selectively breed these plants yourself. You won't find a decent head of corn before 900 CE, which is too bad, because a single kernel of modern corn has more nutritional value than an entire ear of its early ancestor. You won't find orange carrots before 1600 CE, every avocado you've probably ever eaten derives from a seed found under mysterious circumstances in 1926,* and the red grapefruit you're familiar with didn't exist before

---

* This is the Hass avocado, which produces fruit year-round that are larger, tastier, and last longer before spoiling than other avocados. The eponymous Hass bought the seed from a Mr. Rideout, a man who took seeds from wherever he could, including restaurant scraps. Hass planted the avocado seed, intending to use the random plant only as a base upon which he could graft seedlings of other, more popular avocado breeds. After two grafting attempts failed, Hass was going to cut the now-useless tree down, but he was convinced to let it stand by a Mr. Caulkins, who believed the young tree was strong. It soon afterward began producing the fruit enjoyed worldwide today. The seed may have been a cross-pollination, or it may have just been a rare evolutionary

government-sponsored radiation experiments performed the 1950s.* However, you *will* find you have the distinct advantage of knowing that you can coax these plants you remember into existence through selective breeding, and that what you want to do is possible, achievable, and totally worth it.

But before you can do any of that, you'll need to forage for plants and animals—both to eat and to start farms with—and you'll probably encounter potential foods that both you and humans native to this era aren't familiar with. How can you tell if they're safe to eat? A bad answer is "Eat a bunch and see if you die," a better answer is "Eat just a little bit and see if you die," but the best answer is "Read this section and then remember what you read, because there's actually a way to eat strange foods relatively safely."

First off: there aren't any mammals whose flesh is naturally poisonous, so they're generally safe to eat (food allergies aside)—but if you're dining on platypus, maybe don't go to town on the venom sack.† There are no birds with naturally poisonous flesh either, but there are a handful of them—like quail and the African spur-winged goose—whose flesh, skin, or even feathers can on the right diet *become* poisonous. These birds can eat plants and animals toxic to humans and then incorporate those toxins into their bodies, so don't gobble

---

leap forward, but all Hass avocados today are grown from grafted seedlings derived from that original tree. Without the coincidence of a Mr. Hass, a Mr. Rideout, and a Mr. Caulkins in your timeline—or their equivalents—you will not have this avocado.

* The red grapefruit eaten today is a product of a 1950s program in the United States called Atoms for Peace, whose goal was to promote practical uses for nuclear power outside of a wartime context. One of the things Atoms for Peace came up with was the *gamma garden*, which is exactly as amazing as it sounds. Radioactive material was put in the middle of a garden, around which concentric rings of plants were planted. The plants closest died of radiation poisoning, the plants farthest away were largely unaffected, but the plants in the middle mutated. Some of those mutations were useful, and among them was the modern red grapefruit: a sweeter, atomically induced mutation of the existing red grapefruit, whose flesh often faded to a less-desirable pink. Most red grapefruit today comes from the descendants of those atomically mutated plants.

† It won't kill you, but it does cause excruciating pain that can last for months and cause stiffness that lasts for years. And it doesn't even taste that good! Don't waste your time!

all birds willy-nilly.* It gets even dodgier when you move on to eating snakes, reptiles, fish, spiders, and dinosaurs, all of which have naturally venomous species. In fact, *all* spiders are technically venomous, but only a few have venom strong enough to kill you.

Plants are an even greater toxic threat than animals. Since plants can't move like animals to escape predators, they've evolved several defensive strategies, many of them operating on the principle of "I'll make whoever's eating me so sick that they never bother me again, but on second thought, why take the risk, *better to just kill them the first time.*" Some plant toxins are irrelevant (apple seeds contain cyanide, but you'd have to eat tons of them to be affected), while others are brutally horrible. Among the worst is the Australian stinging bush, also known as the "suicide plant" after stories of both humans and animals killing themselves to escape the pain it produces. When you touch this plant, the neurotoxin-coated hollow hairs covering it pierce your skin, causing an unbearable pain described as similar to being burnt with acid and electrocuted at the same time, and the only treatment is to *soak the affected area of your body in hydrochloric acid* and then remove the plant hairs with tweezers—carefully, because if they break off inside the skin, that only increases the pain.

CIVILIZATION PRO TIP: Even the plants in Australia want to kill you.

Needless to say, you'll want to stay away from any plants 1m to 3m tall with hairy, heart-shaped leaves ranging from 12cm to 22cm long that you find in Australia. Luckily, you can put any food, whether plant or animal, to the universal edibility test described in the sidebar on page 58. Before submitting a new food to the test, make sure to have lots of water nearby (fresh and

---

* In rare circumstances this can also happen to mammals: see Section 9 for an example of polar bear livers becoming toxic.

salted), and some charcoal (Section 10.1.1). The fresh water is for drinking and to clean affected areas of your skin, lips, or tongue if something bad happens, the saltwater is to induce vomiting if something bad happens inside you, and a teaspoon of charcoal can be mixed with the water to produce a paste that will either induce vomiting when swallowed or, if you can keep it down, help absorb toxins. Bon appétit!

Sidebar: The Universal Edibility Test

Test each part of a potential food separately (seeds, stem, leaves, buds, fruit, etc.), and in the same state you intend to eat it in (raw or cooked). Cooked is always safer. Don't eat for eight hours before each test, and remember that testing just a single part of a single potential food will take the better part of an entire day. To save yourself from wasting time eating poisons, remember that bright colors are usually (but not always) used in nature to mean "I'm easily seen, which means I'm not worried about predators, which means if you eat me you're probably going to have a bad time."*

1. Smell the food: a strong, unpleasant smell is usually a bad sign. Don't eat it if it smells rotten—because it probably is—and if your food smells like almonds but is not, in fact, almonds, it might contain cyanide.

---

* If an animal has bright colors to warn of its toxicity, and that warning works, then there's an evolutionary incentive for other non-toxic animals to copy the coloring to stop predators from eating them too. This strategy generally works as long as the actually toxic animals outnumber the ones faking it.

2. The skin inside your elbow or wrist is sensitive: gently rub a piece of the food there and wait 15 minutes. If your skin burns, itches, goes numb, or reacts in any way, don't eat it.

3. If nothing bad happens, briefly touch the food to the corner of your mouth, then wait 15 minutes.

4. If nothing bad happens, briefly touch it to your lip and tongue, and wait another 15 minutes.

5. If nothing bad happens, place a piece on your tongue and wait with it in your mouth for, you guessed it, 15 minutes.

6. If nothing bad happens, chew it once, but don't swallow. Hold it in your mouth for 15 minutes.

7. If nothing bad happens, swallow the piece you chewed 15 minutes ago, and consume nothing but water for the next 8 hours.

8. If nothing bad happens, eat a small handful of the food and wait another 8 hours.

9. If you've gotten this far without any reaction, the food is probably safe to eat! You can add it to your diet slowly over the next week.

This whole process takes seventeen and a half hours, and you can't eat anything else during the test, but it is a viable way to experiment with new foods and also not die! And while it's not perfect—a rash from poison ivy, for example, can sometimes take days to manifest—it does limit your exposure.

7

PUTTING DOWN ROOTS: USEFUL PLANTS FOR THE STRANDED TIME TRAVELER

A taste for these plants is definitely worth . . . cultivating.

The following is a list of the plants that have been most useful to humanity, as well as the things you can do with them. Plants predate humanity and their natural evolution generally proceeds slowly, which means that in any time where you can find humans, you will also find plants that match or approximate the entries on this list. It is important to note, however, that the plants you encounter may be slightly different—or in some cases markedly different—from the kinds you are familiar with. For more information on why this is, and how you can recover the plants you're familiar with from their weird ancestors you're seeing, see Section 5.

Examine the following pages to find the most useful plants native to your region: plants are listed alphabetically, and each includes the area they first evolved in. If you don't know what region of the planet you're on, you can try to find some plants you recognize around you and then locate them in these pages. In the unlikely event you don't know what any one of these plants look like, at the very least you'll know what sorts of things you can expect out there, and while entire books have been written about each of these plants, a few sentences about each species is *technically* better than nothing. Depending on your time period, you may get lucky and find instances of these plants outside their native ranges.[11]

7.1: APPLES

ORIGIN

Central Asia

USES

• The apple tree was among the first trees cultivated, and its fruit has been improved for thousands of years, so if you're back before selective breeding was invented, expect some disappointing and tiny sour apples consisting mostly of seeds and a core. Enjoy!

• Apples picked in fall and stored in a cool place keep well over winter.

NOTES

• Apple cider can be made by leaving out apple juice and allowing natural yeast to ferment it. It might not be delicious, but it *will* be alcoholic!

7.2: BAMBOO

ORIGIN

Warm, moist tropical regions

USES

• For a long-lasting writing surface (especially useful before paper), scrape off the outer green skin, split them open on one side, and flatten them. Multiple shoots can be joined together!

• Great for flutes (also, blow guns), plus the shoots are edible.

• Can be used to construct arrows, baskets, scaffolding, furniture, walls, flooring, electric light filaments, water pipes, and as a reinforcing agent in concrete if steel is not invented yet, since bamboo has a tensile strength (the ability to withstand heavy loads without snapping or being pulled apart) that's almost as good as steel.

• Attracts adorable giant pandas.

NOTES

• Bamboo is an incredibly versatile plant, and if you are stranded in an area of the planet where it grows, you will have a lot of what you need for a civilization from this one plant alone!

• Bamboo is one of the fastest-growing plants in the world, and that's useful if you want plants, like, *yesterday*.

7.3: BARLEY

ORIGIN

Temperate regions worldwide

USES

• Humans eat it! Animals eat it too! This is an essential crop available over most of the world!

NOTES

• Barley beer is among the first alcoholic drinks ever made by humans!* To find out how to make your own beer, see Section 10.2.5.

• One of the oldest extant recipes is for a barley beer, and you know we're going to give it to you (see sidebar, page 80). Cheers!

---

* Mead and wine predate beer, because you can make both of those by accident. For mead, just dilute honey (Section 8.8) in water, and it'll start to ferment when it's colonized by yeast (if you don't want to rely on wild yeast, Section 10.2.5 has instructions on how to farm your own). And alcohol occurs naturally in rotting fruit, which is the same process—but under human control—used to make wine.

7.4: BLACK PEPPER

ORIGIN

Southern Asia and Southeast Asia

USES

• Harvest the red fruits of the pepper vine and let them dry in the sun, and you'll have black peppercorns that you can then grind into your food for that spicy, peppery flavor.

• A ubiquitous spice in the modern era, it remains the world's most traded spice!

NOTES

• In Medieval Europe, pepper was worth ten times more than any other spice! People either love this stuff or hate bland food. Maybe both? Probably it's both.

• Pepper was also thought to cure constipation, insomnia, sunburn, and toothaches, as well as other diseases. It doesn't, so don't waste your time.

7.5: CACAO PLANT

ORIGIN

Rain forests of Central and South America

USES

• Chocolate is delicious, and the cacao bean is where it comes from. Scoop the beans from the pods, let them ferment beneath banana leaves, then dry them in the sun, roast them, and remove the shell. Grind what's left and you've got pure chocolate.

• Naturally bitter, for centuries chocolate was roasted, ground, and then added to stews or wine, but chocolate really took off in Europe when paired with sugar and made into a delicious beverage!

NOTES

• The pulp from the bean pods can also be eaten (or, like any sweet fruit, fermented), and it was the pulp—rather than the beans—for which the plant was originally cultivated![12]

• Chocolate is one of the most sought-after flavors in the world, so get ready for your civilization to be *pretty popular.*

• The most universally agreed-upon-to-be-delicious chocolate form is milk chocolate, which you can produce by adding milk, sugar, and fats to your chocolate over heat, and then allowing it to cool. Milk chocolate keeps well and is high in calories, making it a useful food on long voyages (see Section 10.12.5: Boats).

7.6: CHILI PEPPER

ORIGIN

Central and South America

USES

• Good for making food spicier and, therefore, more delicious. Also: extremely useful in making chili.

• The active component is called "capsaicin" and can be used in very light concentrations for temporary pain relief: it works by overloading pain receptors.

NOTES

• Capsaicin is the second-most widely consumed condiment, after salt. People love this stuff!

7.7: CINCHONA

ORIGIN

Bolivia, Peru

USES

• The bark of this tree contains quinine, and guess what? Quinine is a treatment for malaria!

NOTES

• To fight malaria, strip the bark from the tree, dry it, powder it, and gobble it. Side effects include headaches, trouble seeing, ringing in the ears or deafness, and an irregular heartbeat, so don't eat quinine unless you need it!

7.8: COCONUT

ORIGIN

Indo-Pacific regions

USES

• An extremely versatile plant: its fronds can be used for fuel, baskets and mats can be woven from its leaves, its stalks can form brooms, the hair from the coconut shell can be woven into rope, and of course coconut flesh is extremely delicious!

NOTES

• Coconuts are airtight, which means the water they contain is actually germ- and bacteria-free. They're a great source of safe and clean drinking water, and one that doesn't require any technology to produce! You might even say they're "coco-nuts," but you won't, because you're too busy being stranded in the past to make puns right now.

7.9: COFFEE

ORIGIN

Africa

USES

• Dry the beans, grind them up, and run water through them to produce a black liquid that a lot of people like to drink for some reason.

• Unrelatedly, coffee is high in caffeine, which is the world's most-consumed psychoactive drug!

NOTES

• Caffeine prevents the symptoms of drowsiness and stimulates parts of the central nervous system.

• You can overdose and die from consuming too much caffeine, so maybe take it easy with the coffee there, champ.

7.10: CORN

ORIGIN

The Americas

USES

• A staple crop of American civilizations, corn is a convenient and efficient way to feed both humans and animals. Everyone loves corn! Or at least, everyone will eat it when there are no other options.

• An extremely versatile vegetable, corn can be boiled, baked, steamed, eaten raw, crushed into corn powder, heated to make popcorn, baked into bread, or brewed into beer.

NOTES

• Domesticated corn (available after around 7000 BCE) does not naturally reproduce: to produce more corn, you must conserve kernels until next spring, then bury them in the ground. It's been so thoroughly domesticated that it can no longer survive without human aid and interference. Thanks for the trust, corn!

7.11: COTTON

ORIGIN

The Americas, Africa, India

USES

• Even in the modern era, it's one of the most important non-food crops in the world.
• Cotton can be used to make soft, breathable clothing and textiles, along with sails and fishing nets for boats, paper, coffee filters, tents, and even fire hoses.
• Cotton fiber is very high in cellulose, which makes it great for papermaking: see Section 10.11.1.

NOTES

• To make cotton yarn—which can then be woven in fabric—first pick the fluffy balls off the tops of cotton flowers. Pull them across a rough board to separate the seeds from the lint. The lint can then be combed to straighten out its fibers and spun into yarn. See Section 10.8.4 for more.

7.12: EUCALYPTUS

ORIGIN

Australia

USES

• Bark resins can be used to produce mouthwash.
• The flowers attract honeybees that make delicious honey.
• The leaves produce oil that's actually surprisingly useful in medicine (allow your eyes to drift over to the notes section for details).
• Also, eucalyptus oil can be used to make food taste deliciously spicy and can be added to soaps as a perfume.

NOTES

Eucalyptus oil . . .

• applied topically is antiseptic and anti-inflammatory. Apply it to wounds to help prevent infection!

• swallowed helps relieve cold and flu symptoms such as sore throat.

• inhaled as a vapor is a decongestant and treatment for bronchitis.

• applied to skin also works as an insect repellant! Thanks, eucalyptus oil, you're super useful.

• is also extremely flammable, so much so that burning eucalyptus trees sometimes *explode*, so be careful with it.

• can be toxic if too much is consumed: a lethal dose is in the range of 0.05 mL to 0.5 mL per kg of body mass.

7.13: GRAPES

ORIGIN

Western Asia

USES

• The fruit can be eaten raw, sun dried to produce raisins (which keep for longer), or fermented to produce wine, which humans, historically, have enjoyed getting pretty buzzed on.

• While grapes will ferment on their own, the art of the wine maker is to stabilize the drink after fermentation stops, preserving it at the right moment to produce a more delicious beverage.

NOTES

• If you're around when steam travel is beginning between Europe and America, be aware that the unprecedented speed of these ships will allow pale yellow American "phylloxera" insects—all of whom used to perish on the previous, longer oceanic crossings—to survive the voyage. When these insects arrive in Europe, they will become an epidemic, destroying vineyards there for generations. The eventual solution will be to graft European grape plants onto the rootstock of phylloxera-resistant American grape species: the sooner you come up with this, the more you will alter the history of the world (at least the wine-drinking parts of it).

7.14: OAK TREES

ORIGIN
Northern hemisphere

USES
• Oak is a very dense, strong, but flexible hard wood, resistant to insects and fungus.
• Everything from boats to buildings can be made from oak!
• Oak bark also contains tannins: chemicals that let you turn gross animal skin into flexible, wearable leather. See Section 10.8.3 for instructions.

NOTES
• Oak trees can live for more than 1,500 years, and it takes about 150 years for an oak tree to grow to the point where it can be harvested for wood, so oak farming is a situation in which you'll want to plan ahead.

7.15: OPIUM POPPY

ORIGIN
Eastern Mediterranean

USES
• A pretty plant that also happens to produce a sap from its seed head that contains opium: the source of morphine (a painkiller), codeine (another painkiller, also used to treat coughs and diarrhea), and heroin (a highly addictive narcotic).
• If you're not interested in drugs, poppy seeds can also be used as a tasty spice!

NOTES

• Harvest opium by scouring the surface of a ripening poppy head in the evening, collecting the sap that oozes out in the morning. Let it dry in the sun and you've got raw opium.

• Morphine can be extracted from poppies by cutting up the dried plants and boiling them in three times their weight of hot water until a paste forms at the bottom. Add lime (see Appendix C.3), repeat the process, and then add ammonium chloride (Appendix C.6) to precipitate out morphine, which is then purified by hydrochloric acid (Appendix C.13).

7.16: PAPYRUS

ORIGIN

Egypt, tropical Africa

USES

• A great thing to write on before you've either invented paper or discovered that if you dry and stretch animal skins you can make parchment to write on instead.

NOTES

• Make papyrus paper by peeling away the outer layer of papyrus stalks, then cutting them into long strips. Let them soak in water for a few days. Align them side by side, edges slightly overlapping, then put another layer on top, perpendicular to the first. Press them flat for a few days, and ta-da!: you have produced a single sheet of papyrus paper!

7.17: POTATO

ORIGIN

The Andes in South America

USES

• Potatoes are one of the few plants that contain all the nutrition humans need! You can live entirely off potatoes (but shouldn't, because then you're extremely vulnerable to crop failure).

• All parts of the potato are poisonous until cooked, so don't eat raw potatoes. Their poison gives you a slight advantage: humans are the only animals that cook their food,* so animals that also find potatoes toxic won't steal them from your fields.

• Boil them, mash them, stick them in a stew, even cook them in oil to make delicious fries and potato chips (which are not healthy, but which are extremely delicious, and listen, sometimes a civilization just wants a bunch of chips for dinner, okay?). Potato chips don't normally get invented until the 1800s CE, but they're so easy to make that you might as well enjoy them now.

NOTES

• Potatoes can be grown almost anywhere except in the tropics, and they produce more calories per square kilometer than any cereal crop!

• Historically there was European resistance to the potato: Protestants thought they were an evil thing from the "New World" that multiplied underground and whose curvaceous shape was "suggestive." Yes: if you can trace your ancestry back to the early days of the United States, there's a fair chance your ancestors were turned on by potatoes. This resistance was overcome in many ways, most notably in France, where potatoes were planted on the grounds of the Palace of Versailles, where "guards" were placed to protect this new and mysterious royal vegetable. At night the guards would retire, and citizens, curious about this new crop, would raid the fields, soon growing potatoes themselves.

---

* Well, just about. A bonobo named Kanzi was taught to cook by human researchers and in the early 2000s CE had advanced to the level of collecting his own sticks to make a wood pile, which he would then light with matches provided by humans. He's used fire to cook marshmallows. Similarly, in 2015 CE chimpanzees were provided with a "cooking machine"—really, a bowl with a false bottom that when food was placed into it would eject an already cooked version of the same, sidestepping the need to control fire—and they showed a preference for cooked foods over raw, even saving food in order to "cook" it later.

7.18: RICE

ORIGIN

Asia and Africa

USES

- Rice has been continuously cultivated for tens of thousands of years, so billions of humans have found it both convenient and tasty.
- Try it with a curry on top: delicious.
- A staple crop for Asian civilizations.

NOTES

- More than 1/5 of the calories consumed worldwide by humans come from rice, more than any other plant!
- Rice grows best in wet soils, so areas with high rainfall work great, but it can be cultivated to grow almost anywhere as long as you water it.
- Rice can be grown in a flooded field, which has the benefit of both deterring vermin and preventing weeds that can't survive flooding from growing.

7.19: RUBBER PLANT

ORIGIN

Different species of rubber trees are native to South America

USES

The sap from rubber trees is a flexible, sticky, and waterproof latex that has many uses:

- making erasers (from which we get the term "rubber," for rubbing out mistakes)
- pressing into sheets to make waterproof clothing
- as an adhesive or cement
- as an insulator when dealing with electrical currents (Section 10.6.1)

NOTES

• Natural rubber decays, but it can be transformed into a less sticky, more flexible, and longer-lasting material through chemistry! This process is called "vulcanization" (and while you can of course call it whatever you want, "vulcanization" already sounds pretty sweet).

• The easiest way to vulcanize rubber is to add sulfur to it while it's being heated. If you don't have pure sulfur lying around, great news! The South American rubber plant is typically found with another plant climbing on it: a vine with large, fragrant night-blooming white flowers. The juices from these "moonflower" plants contain sulfur. You've got all the ingredients you need to make vulcanized rubber side by side!

7.20: SOYBEAN

ORIGIN

East Asia

USES

• Soybean plants produce two times as much protein per square kilometer as other vegetables, five to ten times more than land used for grazing animals to make milk, and fifteen times more than land set aside for meat production. You like protein? *You came to the right place.*

• Soybeans are also a great source of many essential nutrients.

NOTES

• Like yams, soybeans are toxic to humans (and all other animals with just one stomach) when raw, and need to be cooked before being eaten. Listen: you should be cooking a lot of the mysterious food you find anyway. Plenty of foods have toxins that get destroyed by cooking, and no foods become toxic when cooked. Cooking: it's great!

7.21: SUGARCANE

ORIGIN

New Guinea

USES

• The juice pressed from sugarcane plants can be boiled down to a density at which table sugar crystallizes out! While not necessary for civilization, sugar does make life sweeter (and also contributes to diabetes and obesity *and* makes it harder for your body to process fiber, so maybe go easy on the sugar).

• If you're not in the tropics where sugarcane grows, you may be in temperate climates where sugar beets are found! Shred the plant and then boil it for hours to extract the sugars. The resulting liquid can then itself be boiled down to produce a thick molasses.

NOTES

• The dried pulp left over from sugarcane sugar extraction can be used to make paper.

• Sugarcane is one of the most efficient photosynthesizing plants: it converts more sunlight into biomass than just about any other plant! This means if you want to use plants as a fuel source, you'll get the best and most productive use of your land by growing sugarcane, drying it out, and then burning it to boil water (which can be used in your steam engine, see Section 10.5.4). You can even burn the dried pulp left over from producing sugar, making sugarcane even more efficient.

7.22: SWEET ORANGE

ORIGIN

China and Southeast Asia

USES

• Contains tons of vitamin C and comes conveniently wrapped for transit!

NOTES

• Humans require vitamin C but cannot produce it on their own, so eat an orange if you don't want scurvy (or if you already have it, because it's a cure).

• Most fresh foods contain vitamin C, but the vitamin breaks down when exposed to light, heat, and air, so most preserved foods don't contain any at all! Section 9: Basic Nutrition shows how knowing this simple fact will save *thousands if not millions of lives.*

• The sweet orange was not bred until the 1400s CE, so before then expect some very bitter oranges.

7.23: TEA

ORIGIN

China, Japan, India, Russia

USES

• Put the dried leaves in hot water and it's delicious.

• Also, it's a source of caffeine, which can be valuable if you haven't been to Africa yet to discover coffee.

NOTES

• Tea is the second-most widely consumed drink in the world (only water is more popular), so yeah, it's pretty tasty.

• Tea can be made from other plants too, but these are usually called "herbal teas." Proper tea comes from the tea plant, and you should *accept no substitutes.*

• Try it with milk and sugar, or iced with lemon!

7.24: TOBACCO

ORIGIN

Central America

USES

• Contains nicotine, a stimulant.

• You can smoke it if you want to become addicted to a plant!

NOTES

• In the twentieth century CE tobacco was the leading cause of preventable death, and 1 in every 10 people who died worldwide died because of tobacco use.

• Even secondhand exposure can be fatal, so don't smoke it or be around people who do.

• Avoid introducing tobacco to your civilization, and you will save yourself billions of dollars and millions of lives *and* prevent the invention of vaping.

7.25: WHEAT

ORIGIN

Middle East (Fertile Crescent)

USES

• This is a staple crop for European civilizations. Ground wheat (which is what it sounds like: wheat pulverized into powder by grinding it between two rocks: see Section 10.5.1), mixed with water and given a little heat, produces flatbread and biscuits that will last quite a while before rotting. You can use wheat to make ale too: see Section 10.2.5.

• Dried wheat stores well and will still grow wheat plants next spring.

• To separate the grain from the plants, just lay cut plants on the ground and hit them with sticks: this causes the grain to separate. Next, to separate the wheat from the chaff (i.e., the grains from the empty husks), just throw your grains into the wind: the chaff and straw will be carried away while the denser wheat falls down.

• Pre-domesticated wild wheat had an important feature that was soon bred away: the seed heads would open to scatter their seeds on the ground or in the wind. Humans naturally preferred to collect wheat with closed pods—since then the seeds weren't lost—and this quickly led to domesticated wheat, which keeps its seed heads closed. These closed seed heads mean that domesticated wheat is now unable to survive without humans to plant it.

NOTES

• Wheat is grown on more land than any other food and is the most popular vegetable source of protein.

• Bread is a staple food that's both simple and nutritious and has been consumed for tens of thousands of years, so it's no surprise it's inspired several sayings: depriving someone by *taking the bread out of their mouth*, being savvy and *knowing which side your bread is buttered on*, the idea that we can't *live on bread alone*, and things being *the greatest thing since sliced bread*.

• Sliced bread, incidentally, was first commercially available on July 7, 1928 CE. Before then you had to slice bread yourself, all the while whispering "the greatest thing ever would be to not have to do this anymore."

• You can invent fans if you don't want to rely on windy days when separating wheat from chaff: combine an electric motor (Section 10.6.2) with a propeller (Section 10.12.6).

• Domestication of wheat can be achieved in as little as twenty years![13]

7.26: WHITE MULBERRY

ORIGIN

China

USES

• This plant is the preferred food for silkworms (see Section 8.15), which produce silk.

NOTES

• For more than a thousand years the Chinese traded silk while also keeping the knowledge of how it was produced completely secret from the rest of the world, ensuring they'd have a monopoly on this extremely lucrative product. (The fact that anyone who exported silkworms or their eggs was condemned to death probably helped.)

• Theories of where silk came from included the petals of a rare flower, the leaves of a special tree, or even an insect that would eat until it exploded, sending silk everywhere.

• These theories were sadly incorrect: silk comes from the cocoons of silkworms and can be harvested by simply taking the cocoons off a white mulberry tree, but if you want to produce it on a larger scale you'll want to farm the silkworms yourself. Section 10.8.4 has complete instructions.

7.27: WHITE WILLOW

ORIGIN

Europe and Asia

USES

• The leaves and bark contain salicin, which metabolizes into salicylic acid in the body when eaten. Salicylic acid is the primary ingredient in aspirin, which is one of the most commonly used drugs in the world, and one you'll want to have access to.

• You can use willow to make baskets, fishing nets, fences, and walls. Humans have been using willow to make things for a really long time: willow nets date back to 8300 BCE!

NOTES

• Aspirin can treat the symptoms of fever, reduce inflammation, and provide temporary pain relief.

• Like sugarcane, willow is a good source of biomass for fuel.

• Willow, like ash, is a tree that really wants to grow: so much so, in fact, that it's possible to cut down a willow *and not kill it*. This is done through "coppicing": a process in which you cut down the tree in winter—when it's dormant—but leave the stump. In spring, the tree will use the same root system to regrow, and it can be re-harvested by undergoing coppicing again, in two to five years. Regularly coppiced trees remain at their juvenile stage, and therefore don't die of old age, which makes them a renewable source of wood fuel! Other trees can be coppiced too, but willows are particularly well suited due to their growth speed.

7.28: WILD CABBAGE

ORIGIN

Mediterranean and Adriatic coasts

USES

• Can be selectively bred into kale, Brussels sprouts, broccoli, cauliflower, and more. All are descended from the same ancestor, making this plant very versatile for selective breeding!

NOTES

• Cabbages grow readily in most climates and soil types, so they are an easy source of calories.

7.29: YAM

ORIGIN

Africa, Asia

USES

• A starchy vegetable rich in minerals, carbohydrates, and vitamins, though lacking in protein.

NOTES

• This is distinct from the sweet potato found in America, also called "yams," because humans just love being confusing.
• Many yams, especially those that haven't been domesticated, are toxic. The toxins are destroyed by boiling, baking, or roasting the yams, so be sure not to eat these raw! Plus they're better when they're roasted; you've got to try them.

Sidebar: It's Time for Some Beer

One of the oldest naturally extant recipes (i.e., one that reached the present via ancient Sumerian clay tablets from around 1800 BCE, rather than being brought here by time travelers) is a recipe for barley beer. Well, *technically* it's a hymn proclaiming the charms of one of the Sumerian gods, Ninkasi, but it actually spends most of its time describing how to make beer instead. It's as if the Christian Lord's Prayer read:

> *Our Father, who art in heaven, hallowed be thy name. Thy kingdom come, thy will be done, on Earth as it is in heaven. Give us this day our daily bread, including pizza, which is a flatbread topped with cheese, if thy pizza be plain, and with vegetables for thy vegetarians, or with meat, if thou hath put meat lovers amongst us, all of which can be prepared in thy name as follows . . .*

If you were in a religious society and wanted information to be preserved and shared for as long as possible, wrapping it in the cloak of a prayer or hymn could easily do the trick.* Here's an excerpt of that hymn to Ninkasi, translated from ancient Sumerian, that contains this ancient beer recipe:[14]

---

* This is not necessarily as crazy as it sounds! In Catholicism, the ceremony of Mass requires wine, which has meant that as long as there have been Catholics, there have also been monasteries working to ensure that wine is always available for these services—while also, incidentally, ensuring that the viticultural techniques required to produce this beverage are not lost. In the Middle Ages, Catholic monastic orders, including Cistercians, Carthusians, Templars, and Benedictine monks, were among the largest producers of wine in France and Germany, and several wines they introduced are still enjoyed today. Dom Pérignon, for example, is named after the Benedictine monk who advanced the fermentation of this sparkling wine in the late 1600s, in a French province named Champagne.

Your father is Enki, Lord Nidimmud,
Your mother is Ninti, the queen of the sacred lake.
Ninkasi, your father is Enki, Lord Nidimmud,
Your mother is Ninti, the queen of the sacred lake.

You are the one who handles the dough and with a big shovel,
Mixing in a pit, the bappir [a Sumerian unleavened barley bread] with sweet aromatics,
Ninkasi, you are the one who handles the dough and with a big shovel,
Mixing in a pit, the bappir with date-honey,

You are the one who bakes the bappir in the big oven,
Puts in order the piles of hulled grains,
Ninkasi, you are the one who bakes the bappir in the big oven,
Puts in order the piles of hulled grains,

You are the one who waters the malt set on the ground,
The noble dogs keep away even the potentates [autocratic rulers like queens or kings],
Ninkasi, you are the one who waters the malt set on the ground,
The noble dogs keep away even the potentates,

You are the one who soaks the malt in a jar,
The waves rise, the waves fall.
Ninkasi, you are the one who soaks the malt in a jar,
The waves rise, the waves fall.

You are the one who spreads the cooked mash on large reed mats,
Coolness overcomes,
Ninkasi, you are the one who spreads the cooked mash on large reed mats,
Coolness overcomes,

You are the one who holds with both hands the great sweet wort,
Brewing it with honey and wine.
Ninkasi, you are the one who holds with both hands the sweet wort to the vessel
Brewing it with honey and wine.

The filtering vat, which makes a pleasant sound,
You place appropriately on a large collector vat.
Ninkasi, the filtering vat, which makes a pleasant sound,
You place appropriately on a large collector vat.

When you pour out the filtered beer of the collector vat,
It is like the onrush of Tigris and Euphrates.
Ninkasi, you are the one who pours out the filtered beer of the collector vat,
It is like the onrush of Tigris and Euphrates.

8

THE BIRDS AND THE BEES: USEFUL ANIMALS FOR THE STRANDED TIME TRAVELER

That dog you just domesticated isn't fat. He's just . . . a little husky.

Detailed in this section are eighteen of the most useful animals on Earth, alongside three notably horrible ones. Every animal on this list predates the evolution of anatomically modern humans (excepting of course the animals created by humans, such as dogs and sheep), so the good news is that any civilization with people in it has the potential to have these animals in it too.

Before you get too excited by visions of lions plowing your fields while giraffes keep a watchful eye on your herds, you should know that only about forty different animals were ever fully domesticated before the invention of time travel, and this list includes such obvious filler as goldfish, guppies, canaries, hedgehogs, finches, and skunks: species that are generally pretty useless to humans outside of being a *reasonably* adorable pet.* Unlike plant domestication—comparatively easy—an animal candidate species for domestication must:

- be useful to humans in some way (food, labor, fur, companionship, entertainment, good at dying to let us know when coal mines are filling up with carbon monoxide, just give us *something*)
- breed in captivity
- be easily contained, or naturally stay close to humans
- reach maturity quickly
- tolerate if not enjoy the company of both humans and other animals of its species
- be calm, docile, and not flip out when panicked
- eat food that's found near humans or that humans can easily provide
- accept human presence, human captivity, and ideally human enlightened civilization-building leadership

If even one of these criteria is absent, your domestication attempt probably isn't going to work, and you'll just end up with a bunch of upset wild animals that now know *exactly* where you live. However, if all criteria are present, then you will have an animal that will accept being kept by humans, and that you can now selectively breed. Like in plant domestication, select individual animals that have the traits you want, encourage them to breed, and continue selecting for your chosen traits with each new generation. That's all it takes, and this process of artificial selection will soon be producing animals more useful

---

* The complete and alphabetical list of these fully domesticated animals is: alpaca, camel (one-hump), camel (two-hump), canary, cat, chicken, cow (no-hump), cow (one-hump), dog, donkey, dove, duck, ferret, finch, fox, goat, goldfish, goose, guinea pig, guineafowl, guppy, hedgehog, honeybee, horse, koi, llama, mouse, pig, pigeon, rabbit, rat, sheep, Siamese fighting fish, silk moth, skunk, turkey, water buffalo, and yak. Of course, this list excludes previously extinct animals that were harvested from the past and domesticated *after* the invention of time travel, which includes such specimens as the speedy dodo, the eager diprotodon, and the gentle, noble, soulful, deeply compassionate brontosaurus.

to your purposes—whatever you decide your purposes are—than those found in the wild.

What animal should you domesticate first? The most important thing your civilization can have is a large, four-legged, easily tamed, easily contained, and easily controlled vegetarian mammal, because such animals are miraculous do-it-all sources of meat, hide, milk, fur, transit, and labor. The best example found across all of human history* are horses: they can get you around, pull your plows, feed you, and provide clothing and even entertainment (i.e., watching the galloping steeds and then wagering money on the galloping steeds). If you look around yourself and see horses or protohorses (henceforth: "horsies"), great news: you and your civilization are playing on easy mode.† If you don't see any horsies, look for substitutes like camels, llamas, and alpacas. Failing that, bison, cows, oxen, and goats don't do everything horsies do, but they do at least provide meat, hide, and fur, which is better than nothing.

The bad news (both for horsie lovers and temporally stranded civilization builders) is that there are several times and locations throughout human history during which there aren't any horsies or their substitutes available. Two times to watch out for in particular are:

- The Americas between 10,000 BCE and 1492 CE (i.e., after humans first arrived and before extensive European contact)
- Australia between 46,000 BCE and 1606 CE (again, after humans first arrived but before extensive European contact)

---

* There are of course other animals available *before* humans evolved, but as you need humans to build your civilization, they won't be as much use to you. There are several vegetarian dinosaurs that can be domesticated—triceratops are especially useful once their horns are bred out—but their advantages are outweighed in your case by having to farm them in an era actually populated by *ravenous tyrannosaurs*.

† How easy? Large mammals like horses are so great at moving both humans and goods over terrain that they'll remain the gold standard option for land transport all the way until the invention of *railroads*, and the gold standard option for military transport right up until the invention of trucks and/or *tanks*. Plus, many large mammals can transform vegetables you can't digest (like grasses) into delicious milk that you can (assuming you have the right enzymes; see Section 8.5). The recurring calories these animals provide from milk alone greatly outweigh the one-time calories you get by eating their meat.

In both these cases, the arrival of humans corresponds with mass extinctions—including horsies and the horsie-adjacent—which left these continents bereft of useful pack animals until their later reintroduction from Europe.

- If you're in the Americas between 10,000 BCE and 1492 CE, while there are no horsies or camels, there are llamas and alpacas in South America. In North America there are bison, but they can't be domesticated, and good luck trying to get one to pull a plow. You're the worst off if you're in Central America during this time period, where you won't even have bison. The best you can do while stranded here is domesticate smaller animals—wolves, turkeys, ducks—and try to substitute them as best you can for the larger, more useful beasts that are available to other civilizations elsewhere and elsewhen on the planet.

- Australia—always a special case since it evolved separately from the rest of the world once it separated from Antarctica around 85,000,000 BCE—is a place where marsupials* achieved dominance over other mammals, and horsies never appeared. However, from 2,000,000 BCE until 46,000 BCE there are diprotodons available to you: these giant hippo-sized wombats provide meat, milk, hides, can be ridden, will pull a plow, and have been domesticated by other time travelers.[15] They go extinct when humans arrive on the continent, and while kangaroos and emus do survive contact with humanity, neither is well suited for transportation or for pulling a plow. As you require humans for your civilization, your best hope for easily building one in Australia is to be stranded around 46,000 BCE: after humans have begun arriving but before the diprotodon disappears. Protect the diprotodon from the humans and your civilization gets both humans *and* useful draft animals.

Examine the following pages to find animals native to your region. Like in Section 7: Putting Down Roots, each species is listed alphabetically: species that have been domesticated appear first, and those that haven't appear after. Each

---

* You know: animals that carry their young in a pouch, like kangaroos!

includes the area they first evolved in. Depending on your time period, you may get lucky and find instances of these animals—of which only a few are blood parasites—outside of their native ranges.[16]

8.1: BISON (AMERICAN BUFFALO)

NATIVE RANGE
North America, Europe

FIRST EVOLVED
7,500,000 BCE

DOMESTICATED
Water buffalo were domesticated in 3000 BCE (India) and 2000 BCE (China), but American buffalo have never been domesticated.

USES
• Every part of the buffalo can be used: meat for eating, hides for clothing, sinews for bowstrings, hooves for glue (see Section 8.9: Horses for the recipe), and bones for fertilizer. If you're throwing out buffalo parts, you're doing it wrong!

NOTES
• They can get up to speeds of around 55 km/h, so watch out.
• If you're in North America after humans show up and there are no horses or camels, you'll still have buffalo. But they'll fight you and won't pull a plow, so maybe just eat them.

8.2: CAMELS

NATIVE RANGE
The Americas, Africa

FIRST EVOLVED

50,000,000 BCE (rabbit-sized camel ancestor in North America)

35,000,000 BCE (goat-sized ancestor)

20,000,000 BCE (camel-sized ancestor)

4,000,000 BCE (modern camels)

DOMESTICATED

3000 BCE

USES

• Like cows, camels are a good source of milk, meat, hides, and labor. Plus their dung is dry enough that you can burn it for fuel.

• You can live on camel milk alone for about a month! We wouldn't recommend it, but it is an option if things are going that way.

• Two-hump camels are easy to ride: put the saddle between the humps. But what about one-hump camels? Humans messed around with saddles in front of and behind the hump before realizing around 200 BCE that building a wooden frame around it and putting the saddle on top of that worked best.

• Camels will tolerate saltier food and water than sheep or cows.

NOTES

• While mostly associated with Arabian deserts today, camels actually evolved in the New World, crossing over to Asia on a land bridge that existed around 4,000,000 BCE. Camels—along with horses, mammoths, mastodons, sloths, and saber-toothed cats—went extinct in the Americas around 10,000 BCE, shortly after humans showed up. This is a coincidence that has absolutely nothing to do with how extremely delicious these animals are.[17]

• Camels have a wobbly gait but can carry more than horses, and can go in places horses can't. They're also larger than horses, big enough to spook them in battle!

8.3: CATS

NATIVE RANGE

Eurasia

FIRST EVOLVED

15,000,000 BCE (last common ancestor with tigers and lions)

7,000,000 BCE (earliest cat-sized wildcats)

DOMESTICATED

7500 BCE (if cats can be said to be domesticated)

USES

• Cats are useful for killing vermin (mice, rats) but beyond that provide very little use to humans, except companionship, and even then only according to their capricious whims.

• Cats can be considered to be only "semi-domesticated": domestication usually involves changes between domesticated and wild specimens, and wild and domestic cats show very few genetic differences.

NOTES

• Cats, like dogs, may also have been self-domesticating: as soon as humans started keeping grain around, it would attract mice and rats, which would attract the wild cats that hunt them. As the cats provided a useful service and asked very little in return, they could enter the fabric of human society easily.

• During the Black Death in Europe (1346–1353 CE, 50 percent of the human population there died, stay away if you can) cats were thought to carry the disease, and they were slaughtered en masse in an attempt to end the plague. Ironically, fleas carried on rats were one of the major disease vectors, and with no cats, the rat population exploded. Again: stay the heck away from Europe from 1346 to 1353 CE.

8.4: CHICKENS

NATIVE RANGE
India, Southeast Asia

FIRST EVOLVED
3,600,000 BCE (common ancestor between chickens and pheasants)

DOMESTICATED
6000 BCE

USES
• Chickens are a delicious source of both meat and eggs. Plus, they're omnivorous, which makes them easier to feed than cows.

• To answer your question: the egg came first, as eggs evolved in other animals millions of years before chickens ever appeared.

• To answer your second, newly clarified question: the *chicken* egg also came first. Inside the first chicken egg was a zygote carrying a mutation that allowed it to become the first chicken. This egg, with a mutated zygote inside, was therefore laid by a protochicken. Evolution!

• Aristotle wasted a lot of time pondering this problem around 350 BCE and ended up concluding that both chickens and eggs must have always existed as two eternal constants in the cosmos. See? *These are the kinds of conclusions you reach when you don't know that evolution is a thing.*

NOTES
• After first being domesticated around 6000 BCE in China, they reached eastern Europe around 3000 BCE (possibly through another domestication), the Middle East around 2000 BCE, Egypt around 1400 BCE, and western Europe and Africa around 1000 BCE, and they were brought to the Americas with European contact.

• Chicken eggs are an extremely versatile ingredient in cooking and baking! The proteins in eggs solidify when cooked, which makes them very useful as binding agents in all sorts of foods, including in the delicious hamburgers you no doubt will one day want your civilization to produce. Eggs are also used to add moisture, to thicken sauces, to leaven, to emulsify, as a glaze, and as a way to purify liquids (see Section 10.2.6).

8.5: COWS

NATIVE RANGE
India, Turkey, Europe

FIRST EVOLVED
2,000,000 BCE (aurochs)

DOMESTICATED
8500 BCE

USES
• One of the most useful animals to humanity, cows can be seen as machines that turn indigestible (to humans) materials like grasses into delicious meat, refreshing milk, tasty proteins, and heartening fats.
• They can be used to plow fields and transport goods and people, and their skin is a great source for leather.
• The usefulness of cattle makes them one of the oldest forms of wealth: if you have a lot of cows, you are doing *pretty okay*.

NOTES
• If you're around before domestication, you won't find any cows, but you may find aurochs: the wild animal from which cows were domesticated (more than once, actually). They're bigger than cows—up to 2m tall—better muscled, *and* have giant horns, making them both the largest and most formidable animal ever domesticated. They appeared around 2,000,000 BCE in India, reached Europe around 270,000 BCE, and died out in 1627 CE. Auroch back-breeding attempts (using selective breeding to restore aurochs using genes remaining in

current-day cows) began in the 1900s CE, but they were finally restored in 2033 using DNA sequencing done in 2010.[18]

8.6: DOGS (ALSO: WOLVES)

NATIVE RANGE

EVERYWHERE (though wolves first appeared in North America and Eurasia)

FIRST EVOLVED

1,500,000 BCE (wolves and coyotes diverge from a common ancestor)

34,000 BCE (first domestication of wolves)[19]

DOMESTICATED

20,000 BCE (first domestication of wolves that led to modern dogs)

USES

• All dogs evolved from domesticated wolves: sure, wolves are clever and cunning carnivores that hunt in packs and set traps to ambush their prey, but they rarely attack humans unless rabid or starving. Plus, wolves are where you get dogs from, *so we will not hear any bad words said against wolves.*

• How great are dogs? Besides being awesome pals, dogs are a great source of labor; good for vermin control; good for hunting, herding, and livestock guarding; and can also be used for food and pelts after they die (of natural causes, hopefully, after a long life of being an *awesome dog*).

• Plus, if you point somewhere, dogs can understand your intent and look to where you're pointing. Wolves don't do this, and neither do our closest extant relatives like chimpanzees and gorillas. In some ways, domestication has made dogs more humanlike than any other animal!

NOTES

• Farming will change your relationship with wolves. Until the invention of farming, humans and wolves can be allies, working together to hunt animals and sharing in the spoils. But after farming, wolves will attack what had become valuable farm animals, and they become adversaries to humans.

• Wolves/dogs were the first species domesticated (before even farming was invented) and have actually been domesticated more than once. In some instances dogs domesticated *themselves*: wolves that were gentler, cuter, and less afraid of humans could get more food from them than those that were vicious and kept their distance, and so there was selective pressure on increasingly doglike wolves, until they could be accepted into human society as companions.[20]

• An experiment began in Russia in 1959 CE, trying to breed doglike animals from wild foxes by selectively breeding the "tamest" foxes with each other. In four generations some were wagging their tails when humans appeared, in six they licked human faces and wanted contact, and in ten about 18 percent of their foxes were doglike: calm, friendly, playful, solicitous of human touch. By twenty generations it was 35 percent, by thirty it was 49 percent, and by 2005— less than 50 years after the experiment began—100 percent of the foxes were being born tame, and the scientists now raise money to continue their research by selling the foxes as pets. You can do this with wolves in any time period. *You can make yourself a dog.*

• Wolves enter puberty around 22 months, so in a best-case scenario with ten generations, that's only 220 months, or about eighteen years, to produce a pretty decent dog. Imagine how great it'll be to have a dog after wanting one for eighteen years! It'll be *extremely great.*

8.7: GOATS

NATIVE RANGE

Turkey

FIRST EVOLVED

23,000,000 BCE (common ancestor of both sheep and goats)
3,400,000 BCE (wild goat ancestor, the bezoar ibex)

DOMESTICATED

10,500 BCE

USES

• Goats are a source of meat, milk, wool, and hide, and can be used as beasts of burden too. Like camels, their dung is dry enough to be used as fuel.

• Goat milk is closer to human milk than cow milk is, meaning we can extract more nutrition from it, plus it's lower in (sometimes-challenging) lactose. It's also more homogenous than cow's milk, which means it's great for making cheese.

• Their downy undercoat—called "cashmere"—is terrific for sweaters, but it's difficult to produce in large amounts.

NOTES

• Goats are actually extremely picky eaters—they often won't eat food if it's dirty unless they're starving—but they're very curious and will try eating basically everything.

• Goats (like all animals *except* humans and a few other primates) have no sensitivity to poison ivy. Since they eat it willingly, a few grazing goats are an easy way to get rid of this plant: just don't pet them or drink their milk for a few days afterward.

• The bezoar ibex is a species of wild goat, found in the mountains of Turkey, from which all modern goats are descended.

8.8: HONEYBEES

NATIVE RANGE

Southeast Asia

FIRST EVOLVED

Bees: 120,000,000 BCE

First honeybees: 45,000,000 BCE

Modern honeybees: 700,000 BCE

DOMESTICATED

6000 BCE

USES

• Honeybees make honey, which, until you find other sources of sugar, is one of the few ways to sweeten food! It's also energy-dense and easy to digest.

• Honey can also be used as a treatment for coughs and sore throats, and can be used to treat wounds in a pinch.

• Honeybees also make wax, which is great for candles, seals, and waterproofing clothing, and can be applied to tablets to make a reusable writing surface.

• Honey lasts almost indefinitely without rotting, so it is a very easy way to keep sugar around.

NOTES

• Finding wild beehives is pretty easy: just spot a foraging bee and follow her back to her hive.

• Botulism spores can contaminate honey, which is usually no big deal, but infants can be vulnerable to it so maybe keep your babies away from beehives (there are several reasons for doing that, actually).

• Honey has been harvested even before humans evolved, which is no surprise, since it is extremely delicious. Primates like chimpanzees and gorillas use sticks to collect honey from hives.

• Honeybees went extinct in the Americas around 10,000 BCE but were reintroduced by European colonists in 1622 CE.

8.9: HORSES

NATIVE RANGE

The Americas, Asia

FIRST EVOLVED

54,000,000 BCE (earliest dog-sized horses)
15,000,000 (horses big enough to ride)
5,600,000 BCE (modern horse ancestors)

DOMESTICATED

4000 BCE

USES

• One of the most useful animals to domesticate, horses provide meat, milk, hide, hair, bone, drugs (see Section 10.9.1: Birth Control), plus are useful in sport, transportation, war, and labor.

• The horse-drawn plow (see Section 10.2.3: Plows) greatly increases farming efficiency and is one of the fundamental inventions you'll want ASAP.

• Horse hair is used to make bows for stringed instruments like violins, and horse hooves can be boiled to produce glue, which has been used since at least 8000 BCE.

NOTES

• The horse was the basis of long-distance communication from their domestication all the way until the 1800s. Until the invention of the train, the fastest speeds horses could travel at were also the fastest speeds humans could travel at!

• Making glue is easy: when a horse dies, break up the hooves into small pieces, boil them until they dissolve, and add some acid (the stomach acid from the same horse is a convenient source). This will set into a hard resin, which can be combined with hot water to produce glue when you need it!

• The earliest horses were clever, dog-sized animals that evolved in North America around 54,000,000 BCE. So if that's where and when you're trapped, while you won't be riding any horses, you are going to have some *adorable* pets.

8.10: LLAMAS/ALPACAS

NATIVE RANGE

South America

FIRST EVOLVED

Cousins of the camel, alpacas and llamas have a similar evolutionary history, first appearing around 4,000,000 BCE.

DOMESTICATED

4000 BCE

USES

• Llamas and alpacas are a source of meat, milk, hides, and fiber, and can also be used for labor.

• After (other) humans (beyond yourself) show up in the Americas around 10,000 BCE, llamas and alpacas are the only pack animal left, and then only in South America.

NOTES

• Unlike most mammals, female llamas don't have a reproductive cycle, but rather ovulate on demand after mating. Nice! This makes breeding them slightly easier!

8.11: PIGS

NATIVE RANGE

Europe, Asia, Africa

FIRST EVOLVED

6,000,000 BCE (early ancestors)
780,000 BCE (wild boars)

DOMESTICATED

13,000 BCE

USES

• Pigs give meat, hide, and, uniquely, toothbrushes. Pig bristles make a great toothbrush, which is great because human teeth are baloney.

• Here is the thing about human teeth: they're the only tissue we've got that doesn't self-regenerate. You cut your skin, it heals, but your teeth just sit there getting covered in plaque (food particles, which are unavoidable if you eat food, *which you have to do to live*) until they decay. Ridiculous!

NOTES

• Pigs were domesticated from wild boar several times, including around 13,000 BCE in the Near East, and again in China around 6600 BCE. If you're around before that, wild boar first evolved in the Philippines around 780,000 BCE before spreading to Eurasia and North Africa.

• Careful eating pigs: their meat contains what you can call an "unusually high" number of parasites and pathogens, including *E. coli*, *Salmonella*, *Listeria*, round-worm, tapeworm, and more. You'll be fine; just make sure you cook your pork all the way through!

8.12: PIGEONS

NATIVE RANGE

Europe, Asia

FIRST EVOLVED

231,000,000 BCE (earliest ancestors)
50,000,000 BCE (earlier ancestors that are safer to encounter)

DOMESTICATED

10,000 BCE

USES

• Originally domesticated as a food animal, they became way more useful when we realized that they can find their way back to their home nest even when released from an unfamiliar location up to 1000km away, making them useful for carrying messages.

• Up to the invention of the telegraph (1816 CE) pigeons were one of the very few methods of rapid long-distance communication available.

NOTES

• Pigeons are the first domesticated birds! They're descended from the rock pigeons, which, like all birds, are descended from dinosaurs: a diverse group of animals that thousands of other temporal tourists have safely encountered between 231,000,000 BCE and 65,000,000 BCE in their FC3000™ personal time machine, and which you may also briefly and much more unsafely encounter, should you be unfortunate enough to find yourself stranded in that particular time range.

8.13: RABBITS

NATIVE RANGE

Asia

FIRST EVOLVED

40,000,000 BCE (early ancestors)
500,000 BCE (modern rabbit)

DOMESTICATED

400 CE

USES

• Rabbits are a convenient source of meat and fur: they're small, pose absolutely no threat to hunters, and reproduce rapidly: so much so that predation is generally all that keeps their population in check, so don't feel *too* bad about hunting these defenseless, adorable fur balls.

• Rabbits don't require much space or food, and so can also be raised in the home for a convenient and inexpensive source of meat.

• The ease of growing and hunting rabbits may make you want to eat them exclusively, but heads-up: rabbit flesh is super low in fat. Not getting enough fats is fatal, which makes it possible to *die of starvation* while still constantly keeping your belly full of rabbit meat. Variety in your diet: it's useful!

NOTES

• The earliest rabbit ancestors evolved in Asia around 40,000,000 BCE, but the modern rabbits you're probably familiar with (the European rabbit) appeared around 500,000 BCE in the Iberian Peninsula, where they remained until being introduced elsewhere by humans. These rabbits have since been introduced to every continent except Antarctica!

• Introducing rabbits to new ecosystems has usually gone pretty poorly: without natural predators, rabbits reproduce "like rabbits," hence the famous expression (that famous expression is "do not introduce invasive species to new continents; *what are you thinking?*").

8.14: SHEEP

NATIVE RANGE

Western Asia

FIRST EVOLVED

23,000,000 BCE (common ancestor of both sheep and goats)

3,000,000 BCE (mouflon)

DOMESTICATED

8500 BCE

USES

• Sheep are a source of meat, wool, and milk (which is great for cheese, see Section 8.7: Goats).

• Sheep were the second animal to be domesticated, after the adorable dog. Initially they were bred for meat, until 3000 BCE, when the focus shifted to their wool.

• Until silkworms and cotton plants started being used to make clothes, most people wore leather and wool, so the sheep is a very convenient animal to invent and then keep around!

NOTES

• Domestication and selective breeding produced the super-woolly sheep you're familiar with, so if you're around before 8500 BCE, you won't find any. Instead you'll find mouflon, the animal from which sheep are descended. Mouflon have short reddish-brown coats, a white belly, white legs, and giant horns.

• After first being domesticated in the Middle East, sheep spread to the Balkans in 6000 BCE, and by 3000 BCE were spread throughout Europe.

8.15: SILKWORMS

NATIVE RANGE

Northern China

FIRST EVOLVED

280,000,000 BCE (first metamorphosing insects)

100,000,000 BCE (first silk-producing metamorphosing insects)[21]

DOMESTICATED

3000 BCE

USES

• Silkworms spin cocoons made of silk thread, which you can weave with the instructions in Section 10.8.4. Silk's popularity has resulted in silkworms being one of the few insects to be domesticated!

NOTES

• Domestication didn't work out too well for the silkworms though: those that emerge from their cocoons do so without the ability for sustained flight, and will not eat unless fed by humans. They live for only a few days—during which they mate and lay eggs—before dying.

8.16: TURKEYS

NATIVE RANGE

North and Central America

FIRST EVOLVED

30,000,000 BCE (turkeys split from chickens and other birds)

11,000,000 BCE (earliest turkeys)

DOMESTICATED

2000 BCE (Central America)

100 BCE (North America)

USES

• Chickens aren't native to the Americas, but turkeys are a pretty delicious substitute.

NOTES

• Turkeys (like other birds, including chickens) can carry incredibly deadly diseases, including flus that can mutate to affect humans. See Section 3.5 for more.

8.17: BEAVERS

NATIVE RANGE

Europe, North America

FIRST EVOLVED

7,500,000 BCE (common ancestor of the North American and European beaver)

2,100,000 BCE (bear-sized cousins in North America)

DOMESTICATED

Never, don't even try, their teeth never stop growing so they're just going to chew up all your coolest stuff.

USES

• Beavers are (a) a source of meat, (b) a source of fur, and (c) a source of cutting down trees if you're willing to wait and don't particularly care about which trees get cut down.

• Beavers excrete a substance called "castoreum" (so named because humans once believed male beavers bit off their own testicles, castrating themselves, which is not true and really just tells you more about the humans who thought this than it does about the beavers) to mark their territory, and castoreum contains salicin, which is an anti-inflammatory agent in humans, and can also be used as an analgesic.

• Castoreum smells like vanilla, and this beaver juice was first added to mass-produced food in the twentieth century for this very reason—usually under the euphemism "natural flavoring."

NOTES

• If being trapped in the past has given you a headache, try consuming beaver castor sacs. That's where beaver castor comes from, and they're in cavities under the skin between the pelvis and tail. They're right beside the anal glands, so you'll know when you're close.

• Salicin can also be found in willow bark (Section 7.27), if you've got some of those trees nearby and would rather not get beaver glands all up in you.

• The bear-sized beavers that lived in North America died out around 10,000 BCE, the same time humans showed up.

• North American and European beavers can't crossbreed: they've been separated for so long that they now have a different number of chromosomes. Evolution! It's crazy!

8.18: EARTHWORMS

NATIVE RANGE

Worldwide (including Antarctica, before it became covered in ice)

FIRST EVOLVED

400,000,000 BCE[22]

DOMESTICATED

Never: we never needed to, because they already do terrific work for free.

USES

• Earthworms move by forcing themselves into crevices (baby worms can move soil 500 times their body weight!), making them incredibly useful animals to have if you're farming: they aerate and mix soil, improve drainage, and encourage plant growth. They're a handy marker for soil health: lots of worms usually means fertile soil that your plants will have an easier time growing in!

• In poor soil you can expect only a few earthworms per square meter, while in fertile soul there may be hundreds living in the same space.

• Worms work as fish bait, and can be "charmed" out of the ground by tapping it rhythmically. Seagulls will dance on soil for this reason!

NOTES

• An adult earthworm can weigh around 10g, which can put at least 1kg of worm biomass per square meter of fertile soil. Multiplied across the size of a farming field, the weight of worms underground can outweigh the animals grazing above it!

• During ice ages, glaciers scrape away topsoil and wipe out any earthworms. In most of Canada and the northeastern United States, native worms died out during the last ice age (from 110,000 BCE to 9700 BCE). As worms migrate very slowly, the worms living in these areas in the modern era are descended from non-native varieties, brought over after European contact in 1492 CE.

8.19: LEECHES

NATIVE RANGE

Europe, Western Asia

FIRST EVOLVED

201,000,000 BCE

DOMESTICATED

Why would you want to?

USES

• Leeches show up as a medical treatment in 500 BCE and *continue to be used in medicine* until the late 1800s. That's way too long to be attaching actual worms to ourselves hoping they'll cure our diseases (they won't) under the theory we have "too much blood" (we don't).

• We're mentioning them here in case you're around humans in this *colossal length of time* during which they think putting a predatory worm on your skin to feed off you will help cure disease. It's unlikely to harm you (leeches are full of parasites, but none that can survive in humans), but here is a true fact: *it's not going to do you any favors.*

NOTES

• Leeches actually made a brief comeback in medicine in the 1980s CE, when it was found the anticoagulant in their saliva could be used in reconstructive surgeries. However, we soon figured out which protein in their saliva had that property and generated it synthetically, and then we once again didn't need leeches anymore.

• So yes, when your civilization invents plastic surgery, you may have a brief use for leeches.

8.20: LICE

NATIVE RANGE

Worldwide

FIRST EVOLVED

12,100,000 BCE (hair and pubic lice, showing up when humans did)

190,000 BCE (body lice, evolving only after humans started wearing clothes)

DOMESTICATED

Again, why would you want to, what are you planning?

USES

• These parasites cover the world and were ubiquitous in human society until at least the Middle Ages. If there are humans around, you are probably going to have *insectoid parasites sucking the blood out from your skull and then laying eggs in your hair*, or "lice."

• Lice species are closely tied to their hosts, and there are three types of lice that infect humans: head lice, body lice, and pubic lice. Head and pubic lice live in hair, but body lice live on clothing.

• Lice also carry diseases like typhus, which has been responsible for several epidemics throughout history.

• You know in old-timey paintings how rich Europeans would always wear those giant fancy wigs? That's because they *shaved their heads because their lice were so horrible.* Wigs could still develop lice, but they could be more easily sterilized in boiling water.

NOTES

• Human lice evolved at the same time humans did—humans were breaking off from their chimpanzee ancestors at the same time their lice were breaking off from chimpanzee-feeding ancestors—so there's no time period where you'll find humans and not find lice. Sorry!

• Most rampant plague outbreaks on Earth occurred during winter. Why? That's when wearing the clothes of a dead person is more likely. One single human with infected lice on his clothes could be responsible for a citywide outbreak of plague. *Do not wear the clothes of dead people unless you boil them first*, especially if they died of disease.

8.21: MOSQUITOES

NATIVE RANGE

Sub-Saharan Africa, now everywhere

FIRST EVOLVED

226,000,000 BCE (earliest mosquitoes)

79,000,000 BCE (modern mosquitoes)

DOMESTICATED

Please stop asking about domesticating human parasites.

Please.

USES

• Mosquitoes are a completely useless animal that also happens to carry viruses and parasites, and can inject you with malaria while you sleep. A flying, swarming blood ectoparasite that grows in water: *terrific!*

• The mosquito is one of the few animals that, if removed, would have no lasting negative impact on the world: the activities they perform in the ecosystem (feeding birds, some light pollination) are already performed by other insects. The only legacy would be fewer human deaths from malaria.

NOTES

• Mosquitoes are found all over the world, in every region except Antarctica, Iceland, and a few small islands. *Oh well!*

• They evolved before humans or even dinosaurs, so in any era in which you have either someone to talk to or something amazing to look at / be chased by, there'll also be mosquitoes. Thanks, Earth!

• If you're in Peru, look for cinchona plants (described in Section 7.7), which can treat the malaria spread by mosquitoes.

9

BASIC NUTRITION: WHAT TO EAT SO YOU WON'T DIE FOR AT LEAST A WHILE LONGER

Remember back when you had to worry about the dangers of eating too many prepackaged and processed foods?
Great news: that's something you'll never have to worry about again!

 The story of nutrition is, yet again, one in which humanity took its sweet time to make even basic advances. It wasn't until 1816 CE that we realized proteins were important, and then only by noticing that if we forced dogs to eat nothing but sugar they still died of starvation.* In 1907 CE, four years of experiments began in which different groups of cows were fed only one of

---

* These experiments were done by one François Magendie, and if they seem cruel, then you definitely don't want to know about his vivisection lectures, during which he'd dissect animals onstage *while they were still alive*. This disgusted even his contemporaries, and here's a tip: they're entirely unnecessary! You don't need to perform either of these experiments! They're crazy!

the many grains they'd normally eat, which finally led to the conclusion that different foods have different nutritional content. We only started figuring out what vitamins were in 1910 CE, and while there were food guides created by doctors Hippocrates (around 400 BCE in Greece) and Sun Simiao (around 650 CE in China), it took until World War II (1940 CE) for most nations to introduce food guides, created as a part of wartime rationing. Labeling foods with their nutritional value began only at the end of the 1900s CE. And even today the food guides from different nations actually give different nutritional advice, so what hope do you have to eat well in the past?

Well, quite a bit, actually. While specifics differ, the central tenets of modern dietary advice—defined by an era in which food, including less-nutritious processed foods, is generally plentiful—have been stable for generations and boil down to three points:

1. Don't overeat.
2. Get some exercise already.
3. Eat foods that are better for you, like fruits and vegetables.

The first two points probably won't be an issue for you in your current circumstances, so let's add some details to that "fruits and veggies" one. Your ideal diet will have you:

- eating plenty of fruits and vegetables, because they're usually pretty good for you, even if they're not as tasty as a delicious steak.
- eating moderate amounts of fats and oils, because they're less good for you, even if they are literally as delicious as a delicious steak.
- eating a variety of food, because variety ensures you have access to the widest selection of vitamins and minerals, including the many micronutrients that you don't need much of but which you do need to gobble at least occasionally.
- eating a moderate amount of salts and sugars, because they make food tasty but too much is bad for you.

- not eating too much of any one thing, because too much of anything can kill you (humans can even die from a *water overdose*, and we're not using gangster-style euphemisms here: drink too much water and you can die).

You'll probably miss the taste of processed foods—seeing as they were *exactly engineered* to be as delicious as possible—but you won't miss the effects they have on your body. Beyond that, don't worry too much about your diet. Your circumstances will probably enforce eating in moderation and having an active lifestyle for the next several years at least, and it's not like you're going to be eating savory handfuls of breaded and cheese-stuffed fried chicken nuggets anytime soon.* It is worthwhile, however, to read the brief primer on vitamins that follows, so you'll know how to identify and correct any accidental vitamin deficiencies that crop up, in you or in others.

Until 1910 CE, all humans knew about vitamins was that certain foods gave you different perks: around 1500 BCE Egyptians knew that eating liver helped you see in the dark without knowing what vitamin A was, and as early as 1400 CE Europeans with no knowledge of vitamin C picked up on how fresh food and citrus kept you from getting scurvy. But sadly, in what can fairly be described as "a comedy of errors, only not funny," Europeans—who generally like to think of themselves as being a pretty savvy lot—managed to forget and then rediscover this fact about vitamin C at least *seven more times* over the next five hundred years, including rediscoveries in 1593 CE, 1614 CE, 1707 CE, 1734 CE, 1747 CE, and 1794 CE, until the idea finally stuck in 1907.†

---

* Actually, domesticate the chicken (Section 8.4), produce bread (Section 10.2.5), make cheese (Section 10.2.4), and use the same machines you use to print on to press oil (Section 10.11.2), and you can, in fact, summon these tasty treats again.

† How does this happen? Bad communication and bad science. Humans are one of the few animals that can't produce vitamin C on our own, but we're very good at taking it from the foods we eat and storing it in our bodies, and can rely on those stores for about four weeks before scurvy symptoms start to develop. The catch is that vitamin C is easily destroyed by heat (i.e., cooking) and exposure to air, so processed or stored foods won't have any. In the 1400s CE the fact that citrus prevented scurvy was known to Italian sailors, and some Portuguese sailors went so far as to plant orange trees on convenient islands along their way, but this

Vitamins are simply vital chemical compounds that humans need to live and can't synthesize on their own.* While they can have a large effect on your health and wellness, you need more than just vitamins for good nutrition, which is why even in our utopian era of rental time machines we all can't just pop a vitamin pill in our mouths for dinner. Specifically, you also need carbohydrates and fiber (found in grains, fruits, and vegetables); proteins (found in beans, eggs, milk, meat, and which contain amino acids); fats (found in meat, milk, eggs, and nuts); and of course water. The chart on the following pages includes a complete list of vitamins, where to get them, and what happens if you don't get enough of them. Please absorb and digest this information carefully.

---

knowledge was later lost. Rediscoveries were then ignored, as they went against the prevailing knowledge that scurvy was a disease "caused by internal putrefaction brought on through faulty digestion." Lemons as a cure did *briefly* take hold in the British fleet around 1800 CE but was lost in 1867 CE when the fleet switched from lemons to the juice of Key limes. That juice has less vitamin C, and when it was exposed to air, light, and the copper pipes on ships, it was reduced to practically none. However, as faster steam engines had come into use over the foregoing half century, the time most sailors were at sea was reduced, which when combined with better nutrition on land led to fewer instances of scurvy and obscured the (non) effect the new lime juice was having. When sailors on longer expeditions again began developing scurvy, and with the old citrus cure apparently ineffectual, new theories took hold: perhaps scurvy was caused by food poisoning in poorly tinned meat, or even by poor hygiene and morale. It was only when experiments were done on guinea pigs in 1907 CE—a lucky choice of animal, since it's one of the few nonhuman animals that can even *get* scurvy—that Europeans (again) discovered that fresh foods and citrus were a cure, and this time, it stuck. Scurvy has caused more death, created more misery, and had a greater effect on our history than any other disease that's so easily avoided and so trivially, trivially cured.

* Vitamin D is the exception here: exposure to sunlight is all your body needs to produce it. Some K vitamins are also produced inside the body, but not by you: rather, it's bacteria living in your digestive tract that are doing the work here.

Vitamin	Where to get it	What can happen if you don't eat enough
A	Liver, oranges, milk, carrots, sweet potatoes, and leafy vegetables.	Night blindness, which can lead to total blindness.
B_1	Pork, brown rice, whole grains, nuts, seeds, liver, and eggs.	Weight and appetite loss, confusion, muscle weakness, heart problems, and involuntary eye movements.
B_2	Milk, bananas, green beans, mushrooms (but the few mushrooms that are toxic can be really toxic, so be careful with those), almonds, dark chicken meat, and asparagus.	Painful red tongue; sore throat; chapped lips; oily, scaly skin rashes around your genitals; and bloodshot eyes.
B_3	Meat, fish, eggs, whole grains, and mushrooms.	The three D's: diarrhea, dermatitis, and dementia. And by "dermatitis" we mean "your skin can discolor before peeling off." Other symptoms include sensitivity to the sun, aggression, confusion, and hair loss.
B_5	Meat, broccoli, and avocados.	A chronic feeling of pins and needles, and/or a chronic feeling of bugs crawling under your skin.
B_6	Meat, potatoes (with skin), eggs, liver, vegetables, tree nuts (i.e., basically every nut except peanuts, which aren't even technically nuts in the first place), and bananas.	Anemia (your blood can't carry oxygen as well, which can make you feel confused and pass out), plus nerve damage.
B_7	(Raw) egg yolks, liver, peanuts, almonds, and leafy green vegetables.	Your hair doesn't grow properly, your skin doesn't grow properly, plus you can have abdominal pain, cramping, diarrhea, and nausea.
B_9	Leafy vegetables, beets, oranges, bread, cereals, lentils, and liver.	Your cells don't divide properly anymore. This unsurprisingly causes a bunch of problems, including fatigue, rapid breathing, and light-headedness, all of which eventually leads to nerve damage, difficulty walking, depression, and/or dementia.
B_{12}	Meat, poultry, fish, eggs, liver, and milk. That means if you're vegetarian your only natural sources are milk and eggs, and if you're vegan . . . well, maybe take a glance at what happens when you don't get enough B_{12} and decide given your current circumstances that you can maybe bend the rules a little?	All the symptoms of B_9 deficiency listed above, plus your spinal cord can degenerate.

Vitamin	Where to get it	What can happen if you don't eat enough
C	Fresh food, including oranges and other citrus fruits in particular.	You get scurvy. There's this whole giant footnote about it a little while back. When you get scurvy, your hair curls, you bruise easily, your wounds stop healing, your teeth get loose, your personality changes, and then you die.
D	Fish, eggs, liver, milk, and mushrooms.	Rickets and soft bones, neither of which can be classified as "good."
E	Leafy green vegetables, avocados, almonds, hazelnuts, and sunflower seeds.	It's actually pretty difficult to get a vitamin E deficiency, but if you do manage to, it causes sterility and nerve damage!
K	Broccoli, cabbage, dark leafy green vegetables (like kale, beet greens, or spinach), egg yolks, and liver.	Red spots on the skin and/or raccoonlike bruising around the eyes.

Table 11: Some vitamin letters and numbers are skipped because we thought certain things were vitamins, but then we realized they weren't. Even as late as 1909 CE, this chart would still blow everyone's mind!

If that column of *horrible diseases* doesn't convince you to eat your vegetables (and livers, oddly? Livers show up here more often than you might expect, on account of how they're chock-full of vitamins*) then nothing will.[23]

---

* Some, in fact, are too chock-full of vitamins! You should be using the Universal Edibility Test (Section 6) on everything you're eating that's unfamiliar, because if you don't, and you're stranded in the right time and place, you might eat some seal liver without realizing the risks of what you're doing. Seal liver has so much Vitamin A in it that you can actually cause an overdose, which results in a condition called "hypervitaminosis A" (there's hypervitaminosis for vitamins D, E, and K too). Symptoms include dizziness, bone pain, vomiting, vision changes, hair loss, and itchy skin that peels off. And it's not just seals: animals that routinely eat seals, like polar bears, have livers that are toxic to humans too—from storing all that vitamin A in the seals they ate.

10

COMMON HUMAN COMPLAINTS THAT CAN BE SOLVED BY TECHNOLOGY

... technologies that you are totally going to act like you came up with all on your own.

You are in a unique position: very few people wake up in the morning and decide that today they want to invent civilization from scratch. Historically, most people wake up in the morning, discover that they're hungry or thirsty or bored or horny, and while trying to solve those problems for themselves only end up inventing civilization by accident, *if at all*.

In this section we list the most common complaints humans have had throughout history, along with technologies you can invent that answer those very complaints, explained from first principles. Conveniently, this also doubles as the list of the most useful technologies you'll need in your civilization!

There are some obvious omissions that we assume you are already familiar with, including the wheel (because if you don't know what that is, you have no business rebuilding civilization from scratch*); cooking food with fire (this was invented before even anatomically modern humans evolved, so if you can't figure it out there's hopefully someone else around who can),† and French kissing (which, if you haven't tried yet, you should give a shot! With the right person, *it's actually not bad*).

Technologies are grouped by the human complaints they address, which allows you to peruse conceptually adjacent inventions. If there's something in particular you'd like to invent first, refer to the technology tree found in Appendix A: it illustrates which prerequisite technologies are required to unlock all the coolest stuff quickly.

---

* Okay, fine. Just in case, you can roll around on them and they look like this: O

† Creating fire in the first place can be tricky though, so here's how you create one out of nothing but trees and ambition. First, collect material that can burn easily: dry leaves, pine needles, the inside of barks, grasses, etc. This is your tinder, which you'll make into a small nest-shaped pile. Then, collect some twigs about the size of a pencil, which burns hotter and longer than tinder but doesn't light as easily: this is your kindling. Then collect fuel: dead wood that you can use to keep the fire burning indefinitely. The basic strategy is find two dry sticks that look like they can burn easily. Place one on the ground, find an indent in it that you can force the other stick into, and begin rotating that stick while at the same time forcing it downward. Your goal is to create enough friction that things start to burn. This will absolutely take a long while, and it's energetic, tiring work, but eventually the friction should cause the wood to smolder, producing a small, glowing ember. Transfer that ember to your nest of tinder, and then gently blow on it to encourage it to burn, adding more tinder, then kindling, and eventually fuel to keep it going. After you've created your first fire this way, you'll be so tired and annoyed that you will definitely never want to do it again. Instead, you'll work to keep those home fires burning, hence the expression "keep the home fires burning, because it's a lot of work to get them going again if your home fires go out, and I never want to have to do that again." Once you have fire, you unlock the technology of cooking, which for humans functions both as external teeth (cooking softens foods, meaning you don't have to chew as much) and as an external gut (cooking increases the digestibility and absorption of many nutrients in food). Not all nutrients are increased with cooking: vitamin C can easily be destroyed, so you'll always want to eat *some* raw vegetables and fruit.

Finally, all technologies in this section include a quotation that's at least tangentially related to the invention under consideration, credited both to whoever said it first in your new timeline (you), and to whoever said it first in our original and unaltered timeline. Feel free to plagiarize these ideas, taking all the credit for them when you pepper your speech with these "bons mots." It's like you (are soon to) always say: *Quotation is a serviceable substitute for wit.**

---

* W. Somerset Maugham said that in 1931 CE, but you're going to say it way before he ever gets the chance. And don't let anyone tell you you're wasting your time when you memorize a quotation that speaks to you! As Antoine de Saint-Exupéry said, "It is the time you have wasted for your rose that makes your rose so important." Hey. You should take the credit for that one too.

Water can be found over most of the planet, but it's not all safe to drink. **Charcoal** can solve that by giving you a way to filter water, but it also has many other purposes: so many, in fact, that it's easily the most useful substance you can make out of some wood and a hole in the ground. But charcoal won't turn saltwater into drinkable fresh water. For that, you need the technology of **distillation**. Besides desalinating seawater, distillation has uses in everything from civilization-building chemistry to purifying alcohol, which should put it right near the top of your List of Food-Adjacent Technologies to Invent Right Away.

Please feel free to mentally compose just such a list while reading the following section.

10.1.1: CHARCOAL

> Our spread over the earth was fueled by reducing the higher species of
> vegetation to charcoal, by incessantly burning whatever would burn.
> —You (also, W. G. Sebald)

WHAT IT IS

A lighter, more compact, and more useful form of wood that burns hot enough to *forge steel*.* Besides unlocking cool metals, charcoal binds substances to its surface, which makes it great at filtering water and gases and mitigating

---

* Wood usually only gets to around 850°C when you burn it. Charcoal can get up to *2700°C*.

any poisons you've swallowed. And if you're into writing or painting, charcoal makes a great pigment there too!

BEFORE IT WAS INVENTED

You couldn't melt glass or smelt cool metals, because your fires wouldn't get hot enough, which meant civilizations were restricted to a small subset of less-useful materials. Also, good luck enjoying water with the unwanted sediment, tastes, and odors filtered from it, because that's not happening!

ORIGINALLY INVENTED

30,000 BCE (used for cave drawings)
3500 BCE (used for fuel)

PREREQUISITES

wood

HOW TO INVENT

Fire is a reaction requiring three ingredients: fuel, heat, and oxygen.* Put any fuel (wood, say) in an area with an abundance of heat and oxygen, and you'll get fire. But if you put your wood somewhere where there's plenty of heat but *minimal* oxygen, then instead of fire, a different reaction takes place. It's called "dry distillation," and while less entertaining than watching the world burn, it's arguably more useful!

In dry distillation, the moisture and impurities in the wood get evaporated away without the wood burning, leaving behind a purer version of your fuel: purified lumps of carbon.† That's your charcoal! You might find some charcoal produced accidentally in the remains of a fire (which is where it was harvested

---

* Well: technically you need an oxidizer, but it doesn't *have* to be oxygen. But given your current circumstances, we're going to assume that when burning things you'll rely on the free and abundant oxygen all around you, instead of on more exotic oxidizers like chlorine trifluoride.

† How pure? Anywhere from 65 to 98 percent pure, depending on the skill of the charcoal maker.

from back in 30,000 BCE when it was used for drawing on walls), but if you read how useful it is at the start of this section, you'll probably want to produce it deliberately.

Since you're on Earth (a place where, in any time period in which you've survived long enough to read this far without suffocating, there is an abundance of oxygen), controlling your fire's access to oxygen is the trick. You need enough air to keep your fire going and transform whatever wood that isn't burned as fuel into charcoal, but not so much air that your fire spreads and consumes not only your wood but all your newly created charcoal as well.

One of the easiest ways to make charcoal in the modern era is to simply start a fire inside a steel container with an adjustable aperture, so only a small amount of oxygen is allowed in. Some wood gets burned as fuel, and the rest gets turned into charcoal. Easy, right? But if you need steel to make charcoal, and charcoal to make steel, you've stumbled across a chicken-and-egg problem.* No worries: we're going to sidestep it entirely by producing charcoal from first principles, using nothing more than wood, fire, leaves, and dirt.

The slacker way to produce a small amount of less-pure charcoal is to simply dig a hole and start a wood fire inside it. When the wood's burning steadily, put on some more wood (this'll be what hopefully turns to charcoal), then add a 20cm-deep layer of leaves, followed by 20cm of soil. The fire will smolder underground, and two days later you can dig up your reward. But you're going to need more charcoal than this "dig a hole and come back later" technique can provide. You're going to need a specialized kiln, a single-use *charcoal-producing machine*.

First, gather a bunch of wood logs, stripped of branches and leaves and ideally already well dried in the sun. Use hardwood if you're looking for fuel charcoal (it burns hotter) and softwood if you want to filter water (it's more porous, which allows it to absorb more impurities).† Take a 2m-long pole and

---

* Section 8.4 shows once and for all which came first.

† Hardwood generally comes from slow-growing trees with broad leaves, like oak, maple, and walnut. Softwood generally comes from fast-growing evergreen trees with needles, cones, and sap: trees like spruce, pine, and cedar. Now you know! Now

mount it in the ground—this will be the center of your pile. Using smaller logs—about 10cm in diameter—crisscross them on the ground to make a grid, and extend this grid of logs out until you've got a flat and roughly circular area about 4m in diameter around your log. This is your platform.

On your platform, place the wood that's to be turned into charcoal, packed as densely as possible. Longer pieces (up to 2m long) can be placed vertically, leaning against your central pole, with smaller pieces leaning vertically against them. Your goal is to construct a roughly circular mound, rising about 1.5m tall and filled as efficiently as possible with wood, with your 2m-tall post poking out of the middle. Once you've got your wood pile, seal it by covering it in a layer of straw or leaves and then adding sand, soil, sod, turf, clay, or other surface materials on top, building a 10-to-20cm layer of this sealant. Don't forget to leave several air holes at the base of the mound: these are areas you can seal or unseal during burning to control the fire within those particular sections of your pile.

When you're ready to burn, climb up your pile and pull up your central 2m-tall post. The empty space it leaves behind is your chimney.* Dump already-burning wood or hot ashes into your chimney to ignite the pile. When you see white smoke coming out, your fire has started to burn. Over the next several days the smoke will turn blue, then finally clear. During burning, feel the exterior of your pile and open your vents if areas seem too cool, while closing them if areas seem too hot: you should never see red burning through any of them. You're trying to get an even burn throughout the pile. Fill any cracks that form in your pile's sealant layer to keep it intact.

When is your burn finished? This depends on your wood, its moisture, the size of your pile, and how fast your burn has proceeded. Knowing the right moment is as much an art as it is a science, and it might take you a few tries. If you call it too soon, you'll make less charcoal than you could've otherwise, but if

---

you can tell other people how to broadly classify trees and not have to fake your confidence in this particular matter!

* If you don't think you'll be strong enough to pull a 2m-tall pole out of the ground, don't put it there in the first place! As long as you remember to leave a pole-shaped hole in the middle of your pile, you'll be fine.

you wait too long, your charcoal will be consumed and turn to ash. When you've decided your burn is done, seal up the vents and the chimney, cutting off all oxygen inside the mound to extinguish the fire. Let the mound cool. A few days later you can open it—with water nearby to put out any fires that start once oxygen rushes back in—and harvest your charcoal. If you called it correctly, you can expect about a 50 percent yield of charcoal (10 parts wood will give you 5 parts charcoal, by volume), but with practice and skill these yields can be easily increased from 10:5 to 10:6, and can even reach 10:8 with an exceptional fire manager.*

If this sounds like a lot of work, that's because it is! Managing these fires is easily a full-time job, and when your civilization allows for this level of specialization, you'll want to have people to devote their professional lives to it. If you want charcoal on the regular and already invented bricks (Section 10.4.2) and mortar (Section 10.10.1), you can make a permanent version of your giant single-use kiln by using bricks instead of leaves and dirt, and leaving "doors" for access that can be (temporarily) bricked up and sealed when burning.

When dry-heating wood, the sticky resin and pitch contained within the wood will separate out, forming tar. Tar's great! It's sticky and water-repellant, making it useful as both an adhesive and a sealant, and it's particularly suited to sealing boats (Section 10.12.5) and roofs against leaks and rotting. You'll probably want to build a different sort of kiln to generate and collect tar: something with a sloping floor that leads to an outlet to make collection easy. Not all woods have much resin in them, and trees that produce sap are best. Pine trees work great to make sealant tars, and birch-bark tar was used around 4000 BCE to make chewing gum. Tars also contain antiseptic compounds called "phenols" that make them a convenient sticky bandage for animal hooves and horns.

A final word of warning: once the technology to turn wood into charcoal is well known, the usefulness of the charcoal can induce people to cut down trees

---

* You'll never get a perfect 1:1 charcoal production, because some of the wood needs to burn to provide the heat that turns the rest of the wood to charcoal. Sorry! The laws of thermodynamics apply even in the distant past! (Note: the laws of thermodynamics do *not* apply in extremely distant times in the past, such as those before the Big Bang, but there is no wood there, so don't waste your time.)

like crazy, which leads to large-scale deforestation. This happened in Europe in the 1500s CE, where a lack of trees in turn induced people to exploit harder-to-harvest but less scarce sources of fuel, like coal.*

10.1.2: DISTILLATION

Civilization begins with distillation.
—*You (also, William Faulkner)*

WHAT IT IS

A way to purify or concentrate something made of two or more liquids by heating and cooling it, taking advantage of the fact that different liquids have different boiling points

BEFORE IT WAS INVENTED

It was impossible to turn a liquid into a more pure version of itself: a process that is actually extremely useful, and not just for turning alcohol into more powerful alcohol. It can also be used for desalination (i.e., turning seawater into something safe to drink) and is critical in the isolation of chemicals, many of which you'll require very soon.

ORIGINALLY INVENTED

100 CE (as part of alchemy, which is when humans thought they could purify baser metals like lead into higher metals like gold, and maybe also become

---

* If you can use coppicing rather than clear-cutting to maintain wood a supply that meets demand (see Section 7.27), you can also avoid human-induced climate change, simply by avoiding the need for fossil fuels. While it's true that both coal and charcoal are carbon-based and release carbon dioxide when burned, coal releases carbon into the air that had been stored in the ground for *millions of years*—thereby altering the composition of the modern atmosphere—while charcoal and wood are only releasing carbon that's been stored for a few decades at most. While your civilization will reach a point when charcoal does not meet your energy needs, by then you may be able to leapfrog over fossil fuels to using fuels with fewer dangerous, ecology-destroying, icecap-melting, sea-level-raising, island-drowning, and largely irreversible climate-changing downsides—such as biodiesel, vegetable oil, hydrogen, or temporal flux anti-causality inducer arrays.

immortal and cure all diseases. This was doomed for several reasons, not the least of which were: lead and gold are different elements and not pure and impure versions of the same substance; immortality isn't possible when you've got only our flawed, aging, and failing bodies to build off of; and diseases are caused by a variety of environmental, genetic, and mental sources that no one cure could possibly address. While humans spent four millennia on three continents trying to get alchemy to work in what can only be described as "a staggering waste of human ingenuity, life, and effort from which almost* nothing useful was gained except distillation, and even that took a thousand years before anyone thought to apply it to drinks," you can do far better by simply *not wasting your time.*)

1100 CE (for delicious beverages)

PREREQUISITES

fire, wood or metal to make a barrel, metal bowls (see Section 10.4.2: Kilns, Smelters, and Forges), and something to distill; alcohol (Section 10.2.5) is a good place to start

HOW TO INVENT

We said you need a barrel, but really you want to remove the top and bottom from your barrel, so it's just a tube. Put the liquid you're interested in distilling in a bowl—ideally as wide as your barrel—which will be placed over a fire. Put

---

* We have "almost" here because in 1669 CE an alchemist named Hennig Brand spent the fortunes of two wives trying to find a way to turn iron into gold. He thought that maybe, *just maybe*, you could make gold by letting more than 5500L of human pee sit until it smelled horrible, boiling that rotten urine down to a syrup, heating it until a red oil came out, letting the rest separate into two parts as it cooled (one black and spongy, and one beneath it more grainy and salty), adding the oil back into the spongy upper part of his cooled urine paste and discarding the rest, heating it again for sixteen hours, and then feeding the gases this produced through water. This does not make gold, but it does produce what Brand called "cold fire": a glow-in-the-dark compound that contains the phosphorus naturally found in urine. This made phosphorus the first new element discovered since ancient times! If you want to reproduce this process, know that letting the urine rot first is a completely disgusting and unnecessary step, and that the salty part Brand discarded actually contains most of the phosphorus. Include it in your recipe and you'll get a much greater yield!

your barrel on the bowl. Now at the top of the barrel, put another bowl, which you'll fill with cold water. As your liquid in the bottom bowl boils, the vapor will travel up, hit the bottom of the cold bowl, and condense, just like water does on the glass of a cold drink. All you need now is to add a smaller third bowl—to catch the condensed liquid that drips from the top bowl—which ideally will feed that liquid out of the barrel through a hole (this saves you from having to open up your barrel to remove the condensation every few minutes before it evaporates again). That's distillation! What you're doing is capturing steam, turning it back into liquid, and removing the liquid from the heat. And the barrel is optional—it's used to prevent steam from escaping and therefore increase the efficiency of your distillation—but really all you need are three bowls: a hot one to boil liquid in, a cold one to cool down your steam, and a regular one to catch the condensed steam and remove it.

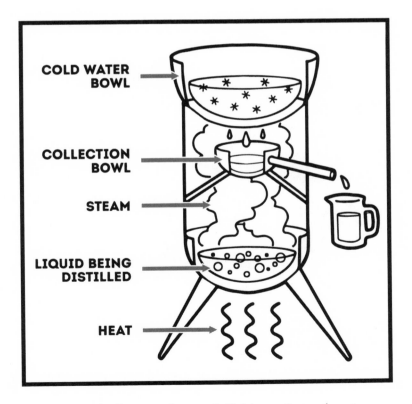

Figure 13: An illustrative schematic, distilled down to its core elements.

Distillation works because different liquids have different boiling points, and the vapor from boiling a mixture of liquids will have a different ratio of its constituent parts than the liquid it came from: it'll contain more of the liquid that has a lower boiling point.* By distilling repeatedly, you can produce a progressively more pure version of the liquid with the lower boiling point!

If you're in a cold area, you can distill liquids without even needing fire! This process is called "freeze distillation," and it's really simple: you just leave your mixture of liquids out in the cold until it begins to freeze. The liquids that freeze first have the highest freezing points, and by removing their ice as it forms, you concentrate everything else in what's left.†

Both regular and freeze distillation will separate saltwater into drinkable fresh water (which is good for staying alive) and salt (which is good for both flavoring and conserving foods: see Section 10.2.6). They're also useful in distilling alcohols (Section 10.2.5), making them the best way to correct an absence of absinthe.

---

* It's a common misconception that the liquid with the lowest boiling point boils first as the mixture gradually gets hotter, but that's not what happens: a mixture of several liquids actually has just one boiling point, rather than several. But when it boils, the liquid with the lowest boiling point adds more of itself to the vapor produced, so condensing that vapor provides a more distilled liquid.

† For example, if you have saltwater, the ice that forms first will have less salt in it than the water left behind. Collect the ice, let it thaw, and distill it again to produce even less-salty water. Similarly, if you're melting frozen sea ice, the ice that melts first will have the highest concentration of salt and will leave less-salty ice behind.

"I'M HUNGRY"

When stranded in the past, one of your first priorities will be finding food. Hunting and gathering will help you survive in the short term, but as we saw in Section 5, farming is key to producing the food that helps ensure your civilization survives in the long term.

While there are technologies in this section that make farming easier (**horseshoes**, **harnesses**, and **plows**), there are also several that allow you to transform the things you farm, like grain, into things that last longer (**preserved foods**), taste better (**bread**), or are more fun to consume with friends (**beer**). Finally, **salt production** not only makes food taste better but is necessary to keep both you and your animals alive. Knowing how to cheaply produce it will change your civilization forever.

Enjoy satisfying your hunger with the help of these tasty technologies!

10.2.1: HORSESHOES

> They went into my closets looking for skeletons, but thank God,
> all they found were shoes, beautiful shoes.
> —*You (also, Imelda Marcos)*

WHAT THEY ARE
A way to make nature's most useful animal *even better*

BEFORE THEY WERE INVENTED

Horses would need to take time off for their hooves to regrow, plus it meant that humans hadn't helped any other animals wear shoes, which honestly seems like one of our most adorable achievements

ORIGINALLY INVENTED

400 BCE (horseboots)

100 CE (metal-bottomed horseboots)

900 CE (nailed horseshoes)

PREREQUISITES

rawhide or leather (for horseboots), metalworking (for horseshoes)

HOW TO INVENT

Horses' hooves are made of keratin—just like human nails—but unlike us, if their hooves become too worn down, they won't be able to walk. This isn't a problem in the wild, but in a domesticated setting—where horses work on different terrains from what they evolved on, carry many more people and loads of cargo on their backs, and pull farmers' plows, carts, and even chariots around—horse hooves face increased wear and tear. The solution is to protect them with shoes! The earliest horseshoes were more accurately called "horseboots": leather or rawhide wrapped around the feet. These evolved into metal-bottom boots around 100 CE, and into bronze or iron shoes that were nailed* directly into the horse's hoof—there are no nerve endings in keratin, so the horses don't feel anything— a few hundred years later. Nails are hammered in from the bottom of the hoof up out the top, where they're bent back so they lie flush with the hoof. This holds the shoe in place while also preventing the nails from catching on anything. You want to nail at the edges: go in too deep and you will cut the quick, which *does* cause pain and will prevent a horse from walking until it heals.

---

* Glue can be used as an easier option instead of nails, but it's usually harder to remove.

Figure 14: A free-range horseshoe, and one attached to a horse.

Horse hooves never stop growing for as long as the horse lives. In a sense, horseshoes do their job too well, because any horse that wears shoes must have them removed every six weeks or so, so that their nails can be filed back down to an appropriate size before their shoes are reattached. If you have a lot of horses, this can become a full-time job—early on it's usually the blacksmith who specializes in both producing the iron shoes and attaching them, but when you reach a level of specialization where it's a separate vocation, that job is called "farrier." Farriers sometimes have their own furnaces so that they can heat their shoes, bending them for a better fit! Hey, remember when we talked earlier about how calorie surpluses would allow for all sorts of productive specialization? *We weren't lying.*

10.2.2: HARNESSES

Horses make a landscape look more beautiful.
—*You (also, Alice Walker)*

WHAT THEY ARE

A way to attach a load to animals, so animals can do work for you instead of you having to do it yourself with your *puny human body*

BEFORE THEY WERE INVENTED

People had to pull things with their human bodies, which we already established are *extremely puny*

ORIGINALLY INVENTED

4000 BCE (yokes)

3000 BCE (throat-and-girth harness)

400 CE (collar harness)

PREREQUISITES

wood, cloth, rope, or leather

HOW TO INVENT

Harnesses seem like a simple invention: tie a rope around your animal, ideally around its shoulders, and it'll pull your load, right? For oxen—dense, sturdy animals with heads positioned beneath their shoulders—it really is that easy. The best oxen harnesses are called "yokes," and they're made from wood. To have two oxen work side by side, lay a beam of wood on their necks in front of their shoulders, and strap it loosely in place: this will keep the animals aligned, and you can attach your load to the beam at the midpoint between the two animals. Put cloth between the wood and animal for comfort, and you're done! With just one ox, you can attach a curved piece of wood directly in front of the horns, and attach your load at the side—oxen "head yoked" can't pull as hard, but two oxen can also be head yoked individually, then held together with another rope.

But horses are trickier.

The most obvious way to harness a horse—a loop around the base of the neck connected to a loop around the chest to hold it in place—may seem like it'll do the job, but it's actually one of the worst ways you can harness these animals! This throat-and-girth harness lets a horse pull something, but as it pulls, the straps cut into the horse's windpipe, carotid arteries, and jugular veins, *all at the same time*. Any horse harnessed this way is clearly not going to be working at peak horse efficiency. To get them there, and to reap the benefits of not *accidentally choking your horses all the time*, you'll want to invent the collar harness.

A collar harness is simply a padded piece of wood or metal that fits *around* the base of the horse's neck, with points to attach the load near the bottom, on either side of the horse. This distributes the force of the load away from the neck and onto the shoulders. Instead of just pulling the load with its front, the horse can fully engage its rear end to also push *against* it. This (finally!) allows the horse to exert its full strength against a load, and your horses will go from being (accidentally) artificially constrained to operating at full efficiency, simply by changing the way they're harnessed.

This will be revolutionary. Wherever horses with collar harnesses appear, they quickly replace oxen: freed from the physical penalty of an inferior harness, a horse can do the work of an ox in half the time, with greater stamina. The increased power of a horse (or "horsepower," if you will*) allows not only more fields to be plowed in a day but for harder fields to be plowed, transforming previously unproductive terrain into productive farmland. Assuming you have horses, the simple invention of the collar harness is enough to greatly increase your farming efficiency, and therefore also the size of the population your civilization can sustain. Humans took more than three thousand years to figure out this simple piece of wood that would finally let their draft animals and civilization reach their full potential.

You did it in a paragraph.

---

* Horsepower (or "hp") is a unit of power, later standardized as the power required to lift 75 kilograms up one meter in one second. Fun fact: a healthy human can produce about 1.2hp briefly and 0.1hp almost indefinitely, while a trained human athlete can produce about 2.5hp briefly and around 0.3hp nearly indefinitely. But 0.3hp is less than a third of what a horse can do! You scoffed when we called humans puny earlier, but now all you can do is nod and whisper, "Thank you for setting this puny human straight, and I love horses forever now!"

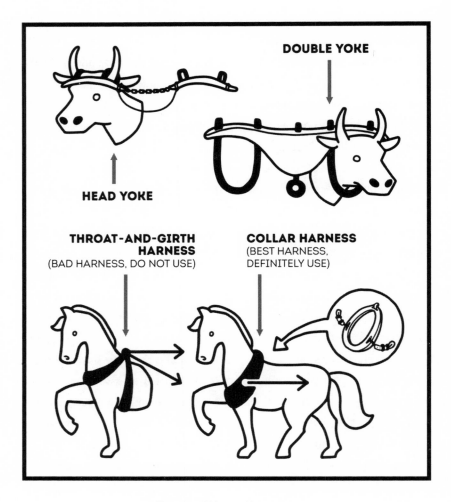

DOUBLE YOKE

HEAD YOKE

THROAT-AND-GIRTH HARNESS
(BAD HARNESS, DO NOT USE)

COLLAR HARNESS
(BEST HARNESS, DEFINITELY USE)

Figure 15: Yokes and harnesses.

10.2.3: PLOWS

It never rains roses; when we want more roses, we must plant more trees.

—*You (also, George Eliot)*

WHAT THEY ARE

One of the biggest agricultural inventions in history: a more efficient way to cut the surface of the Earth open so you can plant seeds inside it

BEFORE THEY WERE INVENTED

The only way to move large amounts of soil was through your own power, which sucks, don't try it, *it sucks so bad*. A lack of plows limits how much land your civilization can farm, which limits how much food it can produce, which limits how many valuable human brains it can support, and as we already established, human brains are your greatest resource.

ORIGINALLY INVENTED

6000 BCE (ards)

1500 BCE (Babylonian invention of seed drills)

1000 BCE (plowshares)

500 BCE (iron plowshares)

200 BCE (Chinese reinvention of seed drills)

1566 CE (European reinvention of seed drills)

PREREQUISITES

draft animals (to pull them); harnesses (to attach the animals to them); wood, or metal if you want better plows; wheelbarrows (optional, for a seed drill)

HOW TO INVENT

You can't control the weather (yet*) and you can't control the sun (again, give it time†) but you can at least *influence* your soil. You probably remember several farming tools that you can invent easily: a hoe is just a wedge on a stick used to break up topsoil, a pointy stick can be shoved into the ground to make holes for seeds, and if you put a blade on the edge of a long stick and swing it, you just invented a harvesting scythe. But these are all tools designed for human use, and while human bodies are great at many things

---

* While the foundations of even the simplest intercontinental weather control apparatus are unfortunately too complex to be included in this book, we promise that once you invent umbrellas (the recipe is "fabric plus a few sticks," and they were first invented in China around 400 BCE), you'll hardly notice. Honest.

† Specifically, give it several decades of sustained interplanetary engineering advances before your civilization can begin to think about constructing variable solar megastructures.

(carrying human brains around, surviving temporal displacement, carrying human brains around while surviving temporal displacement), they stink at manual labor. A plow is a tool that allows* animals—with their stronger, more useful, more reliable bodies—to help with the work. Plows also give you access to harder soils that humans alone couldn't reliably till, and more farmland means more calories, which means more people who can live in and contribute to your civilization.

The earliest plows were called "ards" and were an adaptation of that "pointy stick for seeds" you were inventing a paragraph ago: by dragging them across the soil—preferably under animal power, but humans can drag them too—you create long shallow trenches, or "furrows," for planting. This is better than nothing but only scratches the surface of the soil, and doesn't prevent weeds from growing around your track. What you want is a way to *reset your* soil, making it weed-free and primed for whatever you decide to plant. And you do that by cutting your top layer of soil up into slices, lifting it, and flipping it over.

This makes your topsoil less dense, which is both easier for plants and microbes to grow in and for water to penetrate. All good features! Flipping it over also cuts off weeds from their roots, covers them in dirt, and denies them sunlight—and once that kills them, it recycles their decaying bodies as fertilizer. Put some manure on top before you plow for extra fertilizer and it'll get worked right into the soil! The only problem is this flipping takes a lot of very hard manual labor, so instead of doing that, let's invent the moldboard plow.

The moldboard plow has a vertical cutting blade called a "coulter" (which can be made of wood but will work better and last longer if it's made from iron) that cuts down into the soil. Behind your coulter is a horizontal blade (called a "share" or "plowshare," which it turns out are those things hippies were always encouraging you to hammer your swords into), which cuts the soil horizontally, coming in from the furrow from your previous cut. This strip of land you've cut is then raised and flipped over on its side using a moldboard, which is just a curved wedge that you can make out of wood. The board molds the soil as the plow is drawn forward, which is a handy way to remember this plow's name.

---

* And by "allows," we mean "induces."

Add a wheel of adjustable height to help control the depth of your cut, and you're in business:

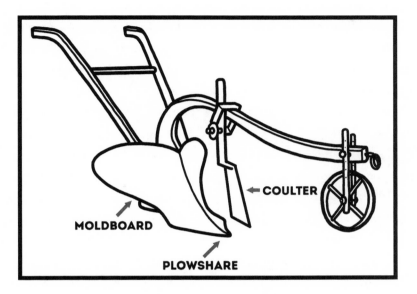

Figure 16: A plow and moldboard.

All you need to do is attach this behind an appropriate animal (as seen in Section 8: The Birds and the Bees) with an appropriate harness (Section 10.2.2). Moldboard plows cut the soil more efficiently than other plows, but they don't turn easily, so you'll want long, rectangular fields. They were invented in China, and Europeans started using them only when they were imported from there thousands of years later. The new efficiency these plows brought caused European agricultural productivity to explode, and the centuries that countless Europeans spent struggling with their inefficient, exhausting, and inferior plows—never once coming up with China's moldboard improvements—have been described as humanity's single greatest waste of time and energy.[24]

CIVILIZATION PRO TIP: You're already doing way better than the Europeans ever did.

Moldboard plowing isn't without its downsides. Go too deep and you can destroy root structures so completely that your topsoil can blow away; do it too often and you can compact the soil beneath your topsoil, resulting in a heavy "hardpan" that water can't penetrate, which then floods your fields. Crop rotation that lets fields rest can help mitigate both of these issues, and keeping soil fertilized with manure will encourage earthworms, which break up hardpan. You can also remove your plow's moldboard and replace your plowshare with a chisel. This new kind of plow (called a "chisel plow") won't flip over your soil, but it will aerate it!

After a field is plowed you'll want to smooth out your soil and break up any large clods, and for this you can use a harrow. A harrow is to the rake what a plow is to the hoe: it's a bigger version that an animal can pull instead of you. You can build one by making a large shape (triangles work well) supported by spikes, and dragging that across the ground—if you're in a pinch, even a large tree branch pulled by your draft animals can do the work.

Finally, you need to seed your land. You can just throw seeds over your soil and hope for the best, but you'll get better yields if you plant your seeds more carefully, with a set distance between each plant. This gives each seed the best chance to grow—and if you cover them in soil, you'll help protect them from being eaten by birds. To do this you'll want a seed drill, and you build one by filling a wheelbarrow with seed and adding a hole at the bottom.* Between the seed and the hole you add paddles that get rotated by the turning of your wheelbarrow's wheel—or through gears (Appendix H) attached to that wheel, if you'd like to adjust its speed. This way, as the wheelbarrow is pushed forward,

---

* Wheelbarrows are easy to invent: they're just a wheel combined with a lever. Mount the wheel at the end of two planks, and build a box on top of those planks to carry the load. Add feet to the end of the planks opposite the wheel so that when it's not being lifted the box rests flat, and there's your wheelbarrow. When lifted, the wheel carries most of the load, which allows one human to move things that would challenge two or more humans who haven't invented wheelbarrows yet. They could've been invented anytime after the wheel was invented around 4500 BCE, which *itself* could've been invented at almost any point in human history, but their earliest appearance is around 150 CE in China. In other words, it took humans hundreds of thousands of years to come up with the idea of attaching a *bucket* to a *wheel*.

a few seeds get the chance to come out too. Run an ard over your nicely plowed and harrowed field to make a furrow, then run your modified wheelbarrow over the same path. You can even attach some plates behind your wheelbarrow to push that furrow back together again, burying your seeds as you pass, which helps more plants grow from fewer seeds.

Follow these instructions, and the agricultural part of your civilization will have all the possible advantages you can give it, and we'd gladly bet the farm . . . *on you.*

10.2.4: PRESERVED FOODS

> One cannot think well, love well, sleep well, if one has not dined well.
> —*You (also, Virginia Woolf)*

WHAT THEY ARE

Preservation is a way to make food last for years or decades instead of a just few days (with meats) or weeks (with most vegetables). With preserved foods you can build up a buffer of food, which transforms any drought, animal plague, or crop failure from a civilization-ending catastrophe in which the living envy the dead into a minor inconvenience, soon forgotten!

BEFORE THEY WERE INVENTED

Food went bad really quickly, which trapped humans in precarious cycles of feast or famine, never more than a few weeks—the approximate life span of unpreserved food—away from starvation

ORIGINALLY INVENTED

12,000 BCE (drying foods)

2000 BCE (pickling)

1810 CE (canning)

1117 CE (China), 1864 CE (Europe) (pasteurization)

PREREQUISITES

none (drying, smoking, freezing, burying, fermentation), salt (salting), sugar (sugaring), bowls (pasteurization, boiling), metalworking or glass (pickling, canning), vinegar (optional for pickling), thermometers (pasteurization)

HOW TO INVENT

Food you want to eat eventually becomes food you don't want to eat because other plants and animals have started eating it too. We call this process "decay," "rotting," or "dinner goin' nasty," and while it's a natural part of life, it's also unappetizing and toxic. You'll want to delay it for as long as possible. Here is your secret weapon: all life on Earth—including the microbes that cause food spoilage—requires water to survive, and even with water, most can only live within certain temperature and acidity ranges. Once you know that (which you do now, because we just told you), you'll realize you can protect food from other life—thereby preserving it—by pushing one or more of these variables to such an extreme that life can no longer survive on your food. You don't have to limit yourself to a single preservation technique either. Foods can be both dried and salted, smoked and frozen, pickled and canned, and so on. Sometimes it's even delicious!

Drying is super simple, which is why it was invented so early. Dehydrating foods inhibits bacteria, yeasts, and molds, and you'll get the best results by cutting your food into thin strips (which maximizes their surface area) and exposing them to something that dries them out, like sun or wind. Smoking works similarly but also adds delicious flavors (and carcinogenic polycyclic aromatic hydrocarbons, so maybe use it as a sometimes treat), and was discovered when humans started drying meats in fire-heated caves.[25] Woodsmoke contains organic acids that make smoked foods slightly more acidic, which also helps preservation.

Both salt and sugar draw moisture out of food, so covering your food in salt and sugar will preserve it while *also* inhibiting the growth of a bunch of bacteria you don't want, like *Salmonella* and botulism.* Many salted meats end

---

* There are bacteria that thrive in very salty environments, and yeasts that love concentrated sugars, but odds are you won't encounter them. If you do, they can be

up tasting extremely delicious, and if you salt the back and belly meat of a pig, great news: *you have just invented bacon.**

Freezing works for preserving food because it turns the water inside food into ice, which slows both bacteria growth and any chemical reactions in the food that contribute to decay. Initially freezing will require a climate that at least occasionally gets cold enough to produce ice, but once you have it, you can make that ice last for a long time. The first refrigerator—a portable container for transporting butter—was invented in 1802 CE, but you can invent it right now: it was just a hollow square box put in a larger oval box, with the area between the oval and the square packed with ice, and the whole thing wrapped in furs and straw to insulate it. Replace the ice when it melts and you can keep food cool indefinitely. Of course, for that you'll need ice during the warm months, but that's easy: cut ice from lakes in winter, and preserve it over summer in a cave, a shaded straw-covered pit, or, if you're feeling more ambitious, in an insulated icehouse. If you don't have any ice at all, burying food can still slow down decay by providing cooler temperatures, and if the soil is dry enough, sand can dry your food too.

Boiling any liquid food—including water—kills microbes in it, making it safer to eat. If you can then prevent any new microbes from colonizing it—usually by sealing it in a can while still hot—the food inside will last much longer. If your food isn't liquid, heating it still kills microbes, which is why cooked meat lasts longer than raw meat.

Canning was invented in 1810 CE—originally by sealing food in glass jars with cork and wax, and later on with tin cans—but there's nothing preventing the invention of this technology as early as 3500 BCE, when wax was being

---

easily killed by adopting another preservation technique that they don't like in tandem with salting and sugaring. Heating and boiling work well!

* Try smoking your bacon to make it even tastier! Incidentally, tons of dishes were first invented as a way to preserve other foods, but are now enjoyed for their own merits as delicacies. Besides bacon (and all the other delicious salt-cured meats like ham, pastrami, sausages, jerky, and corned beef), you have preservation to thank for raisins (dried grapes), prunes (dried plums), jam, marmalade, sauerkraut, kimchi, smoked meats, pickles, cheese, beer, preserved fish like sardines and anchovies, *and more*.

harvested from bees and glass was being created in kilns,* or even earlier if you decide to invent kilns and glass ahead of schedule. If your cans are strong enough, you can even heat your food *after* you've canned it—this technique is called "pressure canning"—which allows the food inside to reach temperatures even higher than its usual boiling point. Botulism spores—which are basically everywhere worldwide but activate only in oxygen-poor environments like inside canned food—are killed by the higher temperatures of pressure canning: 3 minutes at 121°C usually does the trick. Pressure canning is the safest way to store food that isn't already pickled, but when it goes poorly can result in *explosive food*, so be careful with it.

Speaking of pickles: if you thought maybe it'd be nice if your foods started preserving themselves *inside* their jars rather than you having to do all that preservation work before: congrats, you just invented pickling. Pickling involves fermenting your food in brine (which is just salt and water mixed together). Cut up your food, submerge it in brine, and put a (clean) plate, board, or rock on top of your food pieces so they can't reach the surface. In the oxygen-free brine your vegetables will ferment: a process in which "good" bacteria feed on sugars in your food, producing vinegars that make them sour but also less vulnerable to "bad" bacteria that cause them to spoil.† After one to four weeks your food will be pickled, and can then be canned for longer storage.

Submerging foods in brine isn't just for pickles: you can preserve many

---

* While glass was being created in kilns as a glaze in this era, it took until 1500 CE for hollow drinking vessels—now so synonymous with glass that if you're thirsty you'll ask for "a glass of water"—to be produced. This 5,000-year delay between having the technology required to produce glass and *actually making a glass* is so embarrassing that we felt compelled to hide it in this footnote. And before you say, "Oh wait, I know about glassblowing, it actually looks complicated so that's why it took so long": these first glasses weren't blown, but were instead produced by making a glass-shaped pile of sand and pouring molten glass on top of it, which then cooled in the shape you wanted. In other words: it's at the same level of technology used by children when pouring fudge on ice cream.

† "Good" and "bad" are in quotation marks here because the bacteria are "good" or "bad" for our current purposes, but of course they are intrinsically only chaotic neutral. As an aside, if you're wondering if *anything* can be intrinsically good in a fallen world such as yours, please read Section 12: Major Schools of Philosophy Summed Up in a Few Quippy Sentences About High-Fives, ideally before your questions blossom into a full-blown philosophical crisis.

foods this way, including butters, cheeses, and meats. A quick heads-up: brine-preserved foods are often soaked again in fresh water before they're eaten, simply to leach out some of the salt they've absorbed to make them edible again. How do you know when your brine is salty enough for preservation? If you're starting with fresh water, anywhere from 0.8 to 1.5 times your food's weight in salt should give you a salty enough brine.

If you have a surplus of vinegar, you can use that to preserve foods directly (see Section 10.2.5). Cheese is really just preserved milk, and you can make it by introducing vinegar (around 120mL of it) to 1L of boiling milk.* The vinegar curdles the milk, causing delicious cheese curds to separate out, leaving a yellowy liquid called "whey" behind. Drain and press your curds (wrapping them in fabric will do the trick there; see Section 10.8.4) and you'll make a cheese that won't spoil for weeks. Salt it, or soak it in brine to preserve it, and it'll last even longer. By introducing specific bacteria to your cheese as it ages, you can control its flavor: the Camembert, Brie, Roquefort, and blue cheeses you remember were all made by introducing different strains of Penicillium mold (Section 10.3.1) to the curds—though the strains used in cheese making in the modern era aren't the same as the ones used to produce penicillin.

When you were boiling your milk in that last paragraph to make cheese, guess what? You were also inventing pasteurization! Pasteurization is a very simple application of the "boil-to-sterilize" process: take your liquid food, heat it up to not quite the boiling point, then cool it. That's really all there is to it. We say "just below the boiling point" here because milk can curdle from high temperature alone, so you'll want to stay just shy of it if you're pasteurizing milk for drinking. Without pasteurization, milk is one of the most dangerous foods to consume—tuberculosis bacteria in particular just *love* to grow in it!—but when pasteurized, milk actually becomes one of the safest. The hotter you go, the less time you need to pasteurize your liquids, and 16 seconds at 72°C is all it takes to pasteurize milk.

Fun fact about pasteurization: like anything involving heat, this process actually destroys the vitamin C in foods! The introduction of pasteurized milk sometimes coincided with outbreaks of infant scurvy until we figured that out,

---

* If you don't have vinegar: any other acid—like lemon juice—will work too!

so make sure the people in your civilization drinking pasteurized milk are also eating their oranges, bell peppers, dark leafy greens, berries, and/or potatoes. See Section 9: Basic Nutrition for more details!

A second, more ignominious, fun fact about pasteurization: it's a technology that both saves millions of lives and requires only fire to work, and as our proto-human ancestors were using fire before we ever showed up, it actually could've been invented by us at any point in human history! But we didn't figure it out until 1117 CE, and for hundreds of years after that it was used only to preserve wine. By inventing pasteurization ahead of schedule you'll be giving your civilization an easy 200,000-plus-year start on basic food safety!

A final fun fact about pasteurization: this is another technology named by the man who thought he invented it, in timeless honor of himself. Forget Mr. Pasteur and name it after yourself, and the [Your Name Here]ization process shall echo throughout history!

10.2.5: BREAD (AND ALSO: BEER) (ALSO: ALCOHOL)

> All sorrows are less with bread.
> —You (also, Miguel de Cervantes Saavedra)

WHAT IT IS

Bread is a staple food that travels well and is the foundation of many other foods, including pizza, which is *even more delicious*. And the same ingredients in bread, when turned into beer, have been used as an incentive to get people to switch to farming! While you can find food with hunting and gathering, you won't find any beer: it requires reliable agriculture to produce and is therefore one of the perks available only with civilization.[26]

BEFORE IT WAS INVENTED

People just ate raw grains, and if you've ever eaten raw grains you'll recognize it as among the very worst ways to eat grains

ORIGINALLY INVENTED

30,000 BCE

PREREQUISITES

none, but having farming makes it easier; thermometers (optional, but they make reproducing a beer you like easier); salt (to taste)

HOW TO INVENT

Bread is simple to produce: take flour (which is just pulverized grains: smash them with a rock by hand, or grind them between two rocks using the water-wheel you're going to invent in Section 10.5.1), add some water, and cook it over heat.* Ta-da! You have made bread. But the bread you just made is a flatbread, a denser loaf. Flatbreads will let you invent veggie wraps, burritos, and soft tacos, but at some point you're going to want a nice loaf of leavened bread, and for that you'll want to put some yeast to work.

Yeast are single-celled organisms found all over the world; they're floating through the air you're breathing in any period you can survive in, and you're going to farm some particularly suited to whatever grains you have. Here's how: First, blend together a mixture of your flour and water, with about twice as much flour as water. Cover it, put it someplace warm, and check it every twelve hours. You'll be looking for bubbles, which are signs of fermentation: in other words, signs some wild yeast colonized your flour mixture and is now feeding on it. Once you've found fermentation, throw away half of your culture and replace what you removed with some fresh 2:1 flour/water mixture. This gives your remaining yeast new food, and provides evolutionary pressure on yeast that can feed off your flour quickly. You, my friend, are selectively breeding yeast that go crazy for whatever type of flour you've got on hand.

After about a week you should have yeast that reliably generates a bubbly froth—not unlike the head of a beer—after each replenishment. Now you have

---

* If you've invented pans, they can make cooking your dough over heat easier, but they're not necessary. Press your dough onto a stick and you can make bread on a stick over any fire.

a yeast farm!* Keep this culture going for as long as you please by feeding it more of your flour-and-water mixture every day (or once a week if you can keep it refrigerated; see Section 10.2.4). Of course, daily feedings will soon make your culture enormous, so you'll remove some of your yeast before each feeding: ideally, to be used immediately in food. By adding some yeast to your flour/water mixture and letting it sit for a few hours before cooking, you will produce leavened bread.

This works because the yeast you've selected for are bred to feed on the sugars in your flour and water, and if there's oxygen around, they'll produce carbon dioxide as waste. This carbon dioxide is trapped by the gluten in your flour, where it causes your bread to rise. When you cook your dough, the yeast will happily keep gorging themselves in the food utopia you've given them, right up to the point where things become so hot that they all die as their entire colony is cooked to death. Congratulations! You have used the labor of microscopic animals to produce a more pleasant bread, then killed them the instant they were no longer useful. Millions of these microscopic animal corpses are baked into every slice of bread you eat.

CIVILIZATION PRO TIP: Don't let anyone tell you bread is vegetarian.

If there isn't enough oxygen around, yeast will be unable to fully break down the sugars in your grain, and they'll start producing alcohol as waste. Hey, you've invented brewing! The same ingredients and starter culture you developed for bread can also be used for beer, and vice versa. The difference in brewing is that instead of cooking your yeast and grains, you ferment them. Soak your grains in hot water to release their sugars, add your yeasts, and then

---

* Interestingly, it's more likely that rather than isolating a particular strain of yeast, you've actually got several species of yeast and bacteria in your little farm, ideally a community in perfect balance to process the food you give it. The flavor produced will be influenced by the types of yeast you have, so feel free to try capturing yeast from different areas and seeing what comes out!

sit back and relax while they feed. In bread, the yeast have all the oxygen they require, so they can perfectly convert sugars into carbon dioxide. But your fermenting liquid doesn't have oxygen in it, and under these circumstances, yeast produce two forms of waste instead: carbon dioxide (which makes beer bubbly) and alcohol (which makes beer popular).

Beer can quickly become a staple of your civilization's diet. It's got lots of useful carbohydrates, and in modern times it's the third most popular drink on the planet, after water and tea. Grains low in gluten (like barley) tend to work better than wheat for brewing, but you can brew with whatever grain is most convenient. It's only later in a civilization's development that people get picky about their beers. In the early stages you can make them happy by just saying, "Everyone settle down, for I have just invented brewskis," and they will reply, "Thank you, for that automatically makes our civilization way cooler than all others."

Fermentation doesn't just produce alcohol: yeasts can actually *add* nutrients to your beer when fermenting, particularly those in the B-vitamin family. You can transform grains—already pretty healthy—into an even more nutritionally complete food, all thanks to the free labor of microscopic yeast! While you can't survive on beer alone (at least, not for more than a few months, before the symptoms of scurvy and protein deficiency set in), you can at least make it a crisp, delicious, and sociable source of vitamin B_2! (Section 9 has more information on why that's valuable.)

One of the innovations modern brewing has brought us is germination. That's when you allow the grains to sprout a little before you use them. Remember: grains are seeds, and seeds have evolved to be *real good* at passing through animal digestive systems intact, because animals eating plants or fruit and then pooping out the seeds somewhere is how lots of plants expand their territory. Trick seeds into turning their defenses off by germinating them: soak them in water for a few hours, then let them dry for eight, then repeat. After a few rounds of this your seeds will attempt to sprout. This process transforms the starches in the grain into sugars, making the grains softer, sweeter, and easier for humans—and more to our current interests, for yeast—to digest. The more sugars you have, the further fermentation can proceed.

After the grains have germinated you'll want to stop them from growing, or else all your grain's sugars will get used up growing a stupid plant you don't even want. You can either pull the sprouts off each grain manually, or you can save time by roasting your seeds over a fire and then shaking the sprouts off. Roasting also adds flavor to your beer through the Maillard reaction,* so it's worth a try! The entire process is called "malting."

There are alternatives to malting to increase the sugar content of your grains. If you get lucky you might isolate a mold called "koji," first discovered in China around 300 BCE. This mold—which looks like a dark-gray spot on rice—miraculously converts starches to sugars while also imparting a nice flavor, no malting required. It led to the invention of several fermented foods in Asia, including soy sauce (which is fermented soy) and sake (which is beer produced from rice infected, and therefore sweetened, by koji). If you don't find koji mold and still don't want to malt, there's always the option of making chicha instead: here instead of malting you chew the food in your mouth for a while before spitting it out, relying on the enzymes in your saliva to break starches down into sugars. If you're willing to get a dry mouth and don't mind other people chewing things before you brew them, this is an ancient and viable option.

BREAD-MAKING TIPS

- Kneading the dough allows the gluten in the bread to rise too, which helps give your bread a good texture.
- The water you add your yeast to works best if it's somewhere between room and body temperature (that is, between 20°C and 37°C): cooler water makes bread that takes longer to rise, and water that's too hot—around 60°C—will kill your yeastie beasties.
- Add salt (Section 10.2.6) for flavor (this tip applies to most foods, actually).

---

* The Maillard reaction is a chemical reaction (identified in 1912 CE by an eponymous Frenchman named Louis-Camille Maillard, but you're going to beat him to *that* particular punch) between heat, amino acids, and sugars that makes hundreds of flavor compounds. It's what produces those complex, delicious flavors found in seared meat, toasted bread, roasted coffee, French fries, caramel, chocolate, roasted peanuts, and in this case: malted grains.

- Add seeds, fruit, or berries if you think you might like bread with seeds, fruit, or berries in it.

- Try bread with butter on it: it's great! To invent butter, just fill up a jar one-third of the way with milk, seal it, and start shaking. This churning will cause your milk to separate into milk solids and buttermilk. Rinse off those milk solids, press and knead them together, add some salt to preserve it, and you've got a delicious water-in-oil emulsion of spreadable fat.

- Nobody will eat your water-in-oil emulsion of spreadable fat if you call it that, though. Just call it "butter" instead.

BEER-MAKING TIPS

- Beer around 4000 BCE was drunk from a straw. These early unfiltered beers had sludge on the bottom (mostly yeast) and floating solid matter at the top (mostly stale bread, often added as starter). A straw was the best way to get at the good stuff in the middle. You can filter your beer by running it through paper (Section 10.11.1) or fabric (Section 10.8.4) to remove the need for straws. Don't be so hasty to filter it though: the sludgy non-beer parts are full of nutrition too, and were often eaten after the beer itself was gone!

- Hops—the green, fragrant, pinecone-shaped flowers of the vinelike hops plant that naturally grows in northern Europe and the Middle East—can be added to your beer to act as a preservative and flavoring. Many people will grow to like the taste of hoppy beers! Their opinion is incorrect, but popular!

- You can take the yeast sludge from the bottom of one brew and use it as a starter colony for your next batch. While early brewers didn't know what they were doing, scientifically speaking,* they did realize that doing this made that next round of beer taste similar and ferment faster.

---

* Beer was brewed for thousands of years without anyone knowing *how* they were producing that alcohol, just that it (sometimes) worked. There was even debate as to whether brewing was a chemical reaction or a biological one. It sounds silly now, but if you're in an era that doesn't know what yeast is—that doesn't know animals could even *get* that small—you can begin to see where they were coming from.

- You can distill your beers to produce other alcoholic beverages, like whiskey! Distill it enough and you can produce pure alcohol. Since alcohol kills bacteria, it's an excellent sanitizer with tons of uses in medicine.

- If you want vinegar, just take your beer (or any sugary and unpasteurized liquid) and leave it to ferment some more. New bacteria—ubiquitous airborne bacteria called "acetic acid bacteria"—will colonize your booze, feed on its alcohol, and produce (you guessed it) acetic acid. That's what vinegar is! You'll end up with a sharp, acidic liquid that you can use as an antibacterial cleaner, stain remover, or delicious pickling agent. Just like yeast, different strains of acetic acid bacteria will produce different flavors of vinegar, so try starting different colonies until you find one you like.

10.2.6: SALT PRODUCTION

Upon your saltworks all other products depend, for although there may be someone who does not seek gold, there never yet was one who does not desire salt.

—You (also, Cassiodorus)

WHAT IT IS

Salt is a substance produced by reacting an acid with a base, and the only family of rocks that humans eat! Besides being necessary for life, salts are also used in making food taste better, adjusting the boiling and freezing points of water, preserving food, smothering grease fires, and as an exfoliant, stain remover, and justice enabler.*

---

* This last one is worth an explanatory footnote. Salt has been used to facilitate (one era's idea of) justice: while laws forbidding suicide are usually pretty toothless—by the time anyone can prosecute you for successfully pulling off your master crime, you're well beyond their jurisdiction—France's Criminal Ordinance of 1670 CE changed that. Suicide became subject to prosecution, and those who were charged would absolutely get their day in court . . . as a corpse cut open and stuffed with salt. And it wasn't just for suicides: if you died in jail while awaiting trial, your body could still be preserved until such date as your trial occurred. These laws were enforced until the French Revolution (1789 CE), and

BEFORE IT WAS INVENTED

For most of human history, salt was one of the most desired and most expensive commodities on the planet! This despite it also being one of the most common substances around: the oceans are full of the stuff, and there are only a few places on the planet that don't have salt somewhere beneath the surface. In areas without efficient salt production salt has become one of the world's most expensive spices, but in the modern era it's super cheap everywhere, so we just dump it on our roads whenever we want them to be slightly less icy.

ORIGINALLY INVENTED

6000 BCE (harvesting salt from dry lakes)

800 BCE (boiling brine in clay pots)

450 BCE (boiling brine in iron pans)

252 BCE (brine wells)

1268 CE (salt mining)

PREREQUISITES

none (for solar flats), distillation and clay (for boiling brine in clay pots), iron (for boiling brine in iron pans), mining (for rock salt), eggs (for cleaning brine)

HOW TO INVENT

Humans need salt to live, and a healthy adult human carries around about 250mg of it inside their body: three saltshakers' worth. But as that salt is constantly being lost through popular human activities like sweating, peeing, and crying, it needs to be replaced. If you eat meat, you'll probably get all the salt you need from consuming that delicious animal flesh, but if you eat only vegetables you'll have to find it from other sources.* The vast majority of plants on

---

since at the time the Canadian province of Québec was a French colony, corpse-preservation laws were once found in areas of North America as well.

* One challenge you'll face is that, unlike thirst and hunger (which give humans a very distinct craving for water and food, respectively), salt deprivation doesn't result in people feeling like they could really go for some salt. Instead, humans

the planet are *killed* by salt, and only 2 percent of all plant species can tolerate high salinity.*

Fortunately, animals have the same need for salt that humans do, which means that anywhere you find animals, you'll find a source of salt nearby.† In fact, one of the easiest ways to locate a natural source of salt is to follow the trails made by vegetable-eating animals: eventually they'll lead either to a salt lick (an outcropping of rock salt), brine (fresh water that by flowing over salt has become salty), the ocean, or some other natural source of salinity.

The most common source of salt is brine: saltwater that either comes from the ocean or from a briny lake, river, or spring. Once you have brine, you'll want to transform it into a more condensed and easily transportable form: solid salt. The obvious approach is to simply boil your brine down. This is what the Chinese were doing around 800 BCE and what the Greeks and Romans were doing a few hundred years later, using cheap clay pots that they'd smash at the end to get at the salts. But with all that water to boil, it's an expensive process, and you can expect to burn a lot of wood to keep your fires going. A less destructive process is to do what the Chinese had figured out around 450 BCE and use wide, flat iron pans: heat them, then scrape off the salt once the water has boiled away.

However, if you live in a sunny area with access to saltwater (either from

---

who need salt feel nauseated, get headaches, get confused, suffer short-term memory loss, irritability, fatigue, and loss of appetite, then have seizures, fall into comas, and eventually die. Luckily, our salt needs are relatively low and we've evolved to find salt extremely delicious, so once you find a source of salt, you'll probably end up eating more of it than you need, rather than too little.

* It's pretty easy to identify salt-loving plants: just look at whatever's growing in salty areas, like seashores or saltwater swamps. Any plants growing *in* the ocean obviously tolerate saltwater too. To harvest their salt, burn them, put their ashes in water, and then boil the whole business down until all the water has evaporated.

† This does mean that when you begin domesticating plant-eating animals, you'll need to supply them with not only food and water but also with salt. Your salt source doesn't have to be fancy: in the modern era we still use salt blocks on farms, which are just giant cubes of the stuff. A horse needs about five times the salt a human does, and a cow ten times, so if you want to keep a lot of animals in pens, you'll need a steady and reliable salt supply.

the ocean or a brine spring), you can produce salt much more cheaply by letting the sun evaporate water for you. Build shallow clay channels near the source, introduce briny water into them, seal them off, and wait for it to evaporate; when it does, the salt will be left on the bottom for you to collect.*

Humans around 6000 BCE were doing this with natural lakes that dried out in summer, but it took us several thousand years until we started doing it on purpose with artificial ponds, and it was as late as 1793 CE when we invented covered ponds that worked even in less sunny areas. This technology isn't hard: it's just a cover that goes over the ponds when it rains (to protect against rainwater) and at night (to protect against dew), which stops fresh water from diluting the brine.

If you're not near the ocean, salt can be harder to come by. It can be hand-mined from the ground if you're near salt deposits, and while humans originally thought those were rare, in the modern era we know there are actually many scattered throughout the world. These are typically formed when ancient shallow seas evaporated, leaving their salt behind, which was later covered by earth. If you do find yourself above a huge collection of underground salt, be careful: the salt dust in the mine air causes miners to suffer rapid dehydration, which, along with a host of other medical issues that constant exposure to an *airborne desiccant* causes, results in salt miners historically having low life expectancies. If you've got pumps (see Section 10.5.4), you can use a process called "solution mining" to extract salt instead: just send fresh water down into the salt reserves, and when you pump it up, you'll have saltwater you can then evaporate using whichever method is easiest for you.

The weight of rocks above underground salt can sometimes push that salt

---

* A handy tip: if you add egg white to the brine and beat it, it'll form a foam on the top that traps insoluble matter suspended in your liquid. Skim this foam away before you reduce your brine and you'll produce a purer and whiter salt! This "clarification" process can be used on other foods you don't want things floating around in, like wine. Whiter salt is typically more desirable, at least until white salt becomes cheap and commonplace, at which point everyone starts paying more for exotic colors and flavors of impure "raw" salt. That's right: those expensive "red salts" you used to pay a premium for were just regular salts with some dirt in them!

upward, forming a large dome of salt visible at the surface level. If you find one, once you dig past the surface soil, these domes are terrific places to uncover literal mountains of salt.*

Before we leave this topic, fully confident that the information we provided in this section will allow you to leapfrog over the vast majority of human history to the era of cheap and abundant salt you remember, we would like to quickly mention iodine. Iodine is an element that, like salt, is necessary for human life, and whose absence causes fatigue, depression, goiters (neck swelling), and, if the absence occurs during pregnancy, intellectual disabilities in the child. It's abundant in seaweed and in both saltwater and freshwater fish, but can be more rare inland. In the modern era we spray iodine on salt, thereby ensuring adults get, on average, their 0.15mg of iodine recommended per day no matter what they eat. Salt is used as the iodine delivery vector for two reasons: it doesn't go bad, and people tend to eat a predictable amount—nobody, for example, sits down in the morning and decides to eat a 5kg pile of salt for breakfast.† Iodized salt is one of the simplest and cheapest public health measures humans have ever come up with, and it enhances both physical health *and* intelligence at the same time: when iodized salt was introduced across America in 1924 CE, intelligence scores in iodine-deficient regions went up by an average of *15 points*. And while you probably won't be producing iodized salt for a while, if you ensure the members of your civilization have access to foods high in iodine like fish, shrimp, seaweed, cow's milk, chicken eggs, and nuts, then they—and you—will do just fine.

---

* Salt under pressure compresses into an airtight substance that naturally repairs any cracks, which also makes these salt domes ideal places for organic material to transform into oil and natural gas without dissipating!

† Nobody who survives the experience, anyway.

Of all the technologies ever forged by humanity, medicine is the kindest. Combined with the knowledge in Section 14: Heal Some Body, the two innovations in this section will help humans—all humans—live the best lives they can. The impact of these technologies is astonishing: millions of humans are alive today (and will be alive in your civilization) because of these two technologies: one biological and preventative, and the other mechanical and diagnostic.

Penicillin aids you in fighting off infection at the microscopic level, while **stethoscopes** help you at the human level, in both recognizing the symptoms of individual diseases and in understanding how they affect people across humanity as a whole. Civilizations are certainly possible without these technologies—and several of Earth's "greatest" have risen and fallen without them—but these are civilizations marked by illness, infection, and fickle, premature death. Your civilization, in contrast, will be a beacon on a hill, sharing the secrets of health and wellness with the world.

And all you have to do is read the next few pages.

10.3.1: PENICILLIN

If they can make penicillin out of moldy bread,
then they can sure make something out of you.
—You (also, Muhammad Ali)

WHAT IT IS

One of the most effective antibiotics we've got, and you can get it for free on old food!

BEFORE IT WAS INVENTED

Something as pointless and stupid as a scratch could end your life, which is, it is fair to say, *less than optimal*

ORIGINALLY INVENTED

1928 CE (discovery)

1930 CE (first cure)

1940s CE (mass production)

PREREQUISITES

glass (for isolation), soap, acid, ether (for purification)

HOW TO INVENT

Famously, Alexander Fleming was growing *Staphylococcus* bacteria in 1928 CE when his samples were accidentally left exposed to air and then colonized by passing molds drifting in through an open window. Before throwing them out, one caught his eye: there was a halo around the blue-green mold infecting it, in which his *Staphylococcus* bacteria wouldn't grow. Fleming investigated why this was happening, and in doing so he isolated the antibiotic properties of that mold growing in his petri dish: a mold named *Penicillium*.

Here's the thing: humans had already known for thousands of years that some molds seemed to help prevent wounds from getting infected: the molds from rotting foods were used as a treatment for wounds in ancient India, Greece, China, the Americas, and Egypt (around 3000 BCE), though these treatments rolled the dice on whether the molds would contain helpful penicillin or other less-helpful contaminants that would only make things worse. This idea of "molds as medicine" was rediscovered in the 1600s CE in Europe, and the antibiotic nature of *Penicillium* mold was repeatedly observed by scientists in Europe in 1870 CE, 1871 CE, 1874 CE, 1877 CE, 1897 CE, and 1920 CE, and again in Costa Rica in 1923 CE, but it took until 1928 CE for anyone who noticed the antibiotic effect of mold to *also* notice, isolate, and concentrate the active agent. All humans needed to discover penicillin was glass, curiosity, and a bit of luck, and it still took us over 5,000 years to get there.*

Here's how you do it ahead of schedule.

First, you'll need a bunch of petri dishes, which are just shallow flat-bottomed glass bowls. Get your petri dishes as clean as you can with soap and water, and wash your hands too.

Next you'll need a growth medium: something that bacteria can grow in, while also feeding them. We could just mix some beef broth† and water together and call it a day, but a solid growth medium is better, because then we don't run the risk of messing up our cultures when they're moved. So we're going to add some gelatin to the mix, produced either by boiling hooves (see Section 8.9), or by boiling seaweed (try boiling different seaweeds until you find one that gels overnight). Remember that both molds and bacteria survive by growing opportunistically wherever they land, so you don't have to be too precious about setting up the perfect environment here.

Now we need bacteria to show us if penicillin has colonized our petri dishes by not growing around it. Fleming used *Staphylococcus*, so let's go with that one! They're harmless bacteria that reside inside human mucus membranes, which leads us to this Civilization Pro Tip:

CIVILIZATION PRO TIP: Sometimes all you need to do to become one of the great scientists in history is to pick your nose and wipe it on a petri dish.

---

* Actually, you don't even really need glass. You can use pottery instead; it's just easier to keep glass clean and sealed. And you really don't need much luck either! In any period you can survive in, *Penicillium* spores are all over the surface of the planet. All you really need to isolate *Penicillium* is some curiosity and some moldy food.

† Make it by boiling fresh cow bones in water, maybe with a few vegetables tossed in for flavor! What you're basically making here is *meat-flavored water*. Delicious!

Take all these petri dishes primed with your nasty boogers, and expose each to different fungus spores collected from different areas. The mold from rotten fruit or vegetables works great, moldy bread works too, and even just a sprinkling of dirt will probably contain spores of *Penicillium*. Wait a week and look for that telltale halo around your mold, separating it from those bacteria growing from the snot you picked out of your nose because a book told you to. Take samples of these molds and add them to a sealed jar of your watered-down beef broth (scientists call this a "liquid solution"), so that they can grow there in peace.* You just isolated *Penicillium!*

You can rub these molds on wounds and hope for the best, but we can purify them to do better. The penicillin produced by *Penicillium* is more soluble in ether than in water, so you'll want to strain your liquid solution to remove any solids, add some weak acid to what's left (vinegar or lemon juice works) to help keep the penicillin active, and then add some ether (see Appendix C.14). Shake it up to get it mixed, then let it sit: the ether—now containing most of your antibiotics—will rise to the top. Drain off the bottom water layer and you're good to go: you've produced purified penicillin, which can be mixed with water and injected (if you've invented needles), taken orally, or mixed with baking soda (Appendix C.6) to produce a shelf-stable "penicillin salt," which can then be removed from solution with a centrifuge (i.e., spinning it really fast, which you don't *technically* need any technology for except a wheel, but the electric motor in Section 10.6.2 doesn't hurt). Injected penicillin works best when injected on the site of tetanus, gangrene, or other wounds.

Here's the catch: it took about 2000L of Fleming's mold culture to produce enough penicillin to treat a single badly infected wound. A better source was found in 1942 CE by one Mary Hunt in the garbage of a Peoria, Illinois, grocery store: a "pretty, golden" mold on a cantaloupe that produced 200 times more

---

* *Penicillium* particularly enjoys eating leftover corn juice (also known as "corn steep liquor"), which you can easily produce as a by-product when milling corn: just soak your corn in water for about two days to soften the kernels, and then concentrate the leftover water into a viscous slurry to feed your new favorite molds. Sugars are a fun treat for *Penicillium* too!

penicillin than Fleming's strain. This mold was eventually exposed to X-rays in the hope the radiation would cause enough artificial mutations that a strain that produced even more penicillin would evolve, and serendipitously, that worked! Humanity now had a mold that produced 1,000 times the penicillin of Fleming's original strain. You probably don't have a convenient source of X-rays, but you do have lots of convenient sources of mold, so be sure to occasionally rerun this experiment until you find a better-producing penicillin source. Once you have penicillin mold isolated, you'll want to have members of your civilization working on producing penicillin full-time.

10.3.2: STETHOSCOPES

Shall we make a new rule of life from tonight:
always to try to be a little kinder than is necessary?
—*You (also, J. M. Barrie)*

WHAT THEY ARE

A medical instrument so basic that its construction can be described in a single phrase ("roll something into a tube," there, we just did it), but which still eluded all human invention for hundreds of thousands of years

BEFORE THEY WERE INVENTED

Knowing what was going on inside a living body without cutting it open and taking a look was pretty difficult

ORIGINALLY INVENTED

1816 CE

PREREQUISITES

paper

HOW TO INVENT

Like we said, all you need to do to invent a stethoscope is make a tube and hold it up to someone's chest. Done. The very first one was made of paper, but you can make more effective and durable ones out of wood or metal. You can add the flexible tube and stereo earpieces you're familiar with in modern stethoscopes if you're feeling fancy, but they're not necessary.

Despite their simplicity, stethoscopes were discovered only by chance when a (male) doctor, unwilling to get too close to his (female) patient's chest, discovered that a rolled-up paper tube let him hear what was happening inside her body much more clearly than his ears alone could. This invention at last allowed the internal anatomy and behavior of living beings to be explored both in detail and, more important, non-lethally. It was the beginning of a major change in medicine: instead of a collection of symptoms to be treated, disease began to be seen as what happens when different parts of the body start to misbehave—through age, decay, or infection.

Humans soon discovered that a lot of sound is produced inside the body, and differences in these sounds can be diagnostic, especially when dealing with the heart, the lungs, and the gastrointestinal system. This isn't a medical manual, but if you want to become a doctor, a stethoscope will quickly make apparent the differences between many healthy and diseased organs and allow you to make diagnoses based on that.

This technology is also good for hearing aids! The basic design of the tube stethoscope can be modified to form an "ear trumpet"—a giant trumpet-shaped tube you hold up to your ear as a primitive but effective way to hear things better.

10.4

"THE NATURAL RESOURCES I SEE AROUND ME SUCK; I WANT BETTER ONES"

The inventions included in this section are fundamental to any technological society.

Mining is of course necessary if you want access to any resources that don't happen to be on the surface, and **kilns, smelters, and forges** not only give you access to new materials but also unlock a host of new technologies, from metalwork to the steam engine, making them linchpins for a huge number of technologies yet to come. And finally, kilns can transform sand into one of the most useful substances humanity has ever produced: **glass**—a material with the ability to bend *light itself.* This not only helps the people in your civilization see better so they don't trip over things all the time, but also opens up whole new areas of science to explore, from microscopic life forms to the light of distant stars.

While other technologies tend to get the spotlight, many of them would not exist without the technologies in this section to support them. By being the one to invent them first, you're making yourself the most brilliant, influential, and crucially essential person in world history.

Well done on that, by the way.

10.4.1: MINING

My grandfather once told me that there were two kinds of people: those who do the work and those who take the credit. He told me to try to be in the first group; there was much less competition there.

—You (also, Indira Gandhi)

WHAT IT IS

Taking things you think might be useful out of the ground

BEFORE IT WAS INVENTED

Unless the substances you were interested in happened to be on the surface, you were *kinda out of luck*

ORIGINALLY INVENTED

41,000 BCE (earliest mining: hematite, used for red pigment for painting and makeup)

4500 BCE (fire setting)

100 BCE (hushing)

1050 CE (percussion mining)

1953 CE (landfill mining, in which it was realized that landfill sites capped before recycling became prevalent would have a higher concentration of aluminum—thanks to discarded aluminum cans—than actual aluminum mines)

2009 CE (first asteroid mining company founded)

PREREQUISITES

candles (for seeing underground), metal tools (for mining rocks and percussion mining), animal husbandry (for domestic birds like canaries)

HOW TO INVENT

Unless you get lucky—and people have throughout history*—mining is probably going to involve moving a lot of heavy rocks around, and that's going to be hard work. There's no way around it! The silver lining† is that if whatever you're

---

* For example, you might get lucky by having the material you're interested in (flint for stone tools) be covered in an easily removed rock, like chalk. Chalk is so soft it can be mined with little more than deer antlers as picks and the shoulder blades of oxen as shovels. A mine today known as "Grime's Graves" in England is an example of this, where mining began around 3000 BCE.

† Incidentally, if you are mining silver, you should know that it's barely ever found in nugget form: it's usually alloyed with other metals, which means you have to extract it. Your smelters (Section 10.4.2) can do the job!

interested in happens to be near the surface, you can try an open-pit mine: a specialized hole that, unlike most mining, has the advantage of giving miners plenty of sunshine and fresh air. But not everything you're interested in will be near the surface, and at a certain point you're going to have to dig down until you find rocks you like, hit those rocks until they break apart, and carry away the pieces.

Here are a few techniques you can use to make that job slightly easier:

Hushing: store water in a reservoir near where you want to mine, then release it all at once. The rushing water acts as instant artificial erosion, carrying away surface rocks and soil for you and leaving the bedrock exposed. If you're lucky, veins of ore in the bedrock will then be exposed for mining!

Fire setting: light fires in your mine shaft next to rocks, then dump water on the rocks once they're superheated. This sudden cooling fractures the rocks and makes them easier to break, but this process does involve lighting fires in your mine—a place where oxygen is usually treasured—so it's not *entirely* without its dangers.

Wedging: useful on its own, or after fracturing rocks with fire setting; hammer a wedge into a split in a rock, forcing it to break into pieces. If your wedge is made of wood, you can soak it in water after it's been forced into a split, which causes the wood to expand, putting even more pressure on the rock.

Percussion drilling: set up a heavy rod with a sharp iron or steel tip (see Section 10.10.2), and drop it into the same spot over and over. Levers or pulleys* can be used to raise the heavy rod up again once

---

* A lever (which is just a fulcrum that a plank pivots on) and a pulley (which is just a wheel that a rope feeds around) are two simple machines that reverse the direction of force required to move something. They're useful because pulling something up is usually harder than pushing it down: when you're pushing down, you can use your own weight (and gravity) help to do the work. Like other simple machines (including ramps, wedges, and screws), levers and pulleys can also make work easier by exchanging force for distance. Add more pulleys that your rope feeds through, and you can reduce the force required to move your load, at the cost of pulling more rope. Move the fulcrum on your lever closer

it's been dropped, and guide tubes can be made from wood or metal to ensure the rod always hits the same spot. This produces a small well shaft, better suited to extracting liquids (like brine in Section 10.2.6: Salt Production) than solids.

Some of the simplest underground mining involves "bell pits," named after the shape they take. In these, a shaft is dug (usually vertical, sometimes inclined) to the level of the ore, at which point miners start digging out and around from the shaft, forming a natural bell shape as they go. No supports are used, which means at a certain point the mine is going to start to fall in. At that point (or, ideally, several minutes *before* the collapse) the mine is abandoned, another one is started nearby, and mining continues.

Figure 17: A bell-pit mine (pictured here before its inevitable collapse).

If you don't want to go the "dig until it's too dangerous to dig anymore" route, an alternative is "room and pillar" mining: here you dig horizontally but leave vertical pillars of rock standing to hold up the roofs of the "rooms" you're extracting material from. The danger of collapse here comes from the fact that

toward your load and you do the same thing, at the cost of having to move that lever farther.

if one pillar fails, more stress is added to the remaining pillars, which can cause a chain reaction. Instead of (or in addition to) pillars, you can support mine shafts with timber, making the wood carry the load of the rock above. But even in the modern era it's very hard to guarantee that roof of rock you've exposed is up to the task of carrying its load without collapsing. Mining has an unavoidable element of risk: besides cave-ins, miners also face danger from flooding and toxic gases, the latter of which can result in you suffocating, exploding, *or both in rapid succession.*

You can mitigate the threat of suffocation by bringing along a pet bird. Many birds have a high metabolism and breathe rapidly, so they succumb to carbon monoxide and other toxic gases before humans do. Specifically, canaries faint in the presence of carbon monoxide about twenty minutes before humans. Bring a canary into the mine and keep an eye on how conscious it is, and you'll know when otherwise undetectable toxic gases are present while you still have time to escape them! This is a great improvement over the way toxic gases were usually detected for thousands of years, which was by noticing when the people around you started to die. This simple but still fundamentally lifesaving use of animal labor didn't occur to anyone until the incredibly late date of 1913 CE (late enough that we'd already invented vacuum cleaners, cellophane, *and television*), so just by bringing a canary or other conveniently sized bird pal into the mines with you during the hundreds of thousands of years before that date, you'll be doing far, far better than we ever did.

10.4.2: KILNS, SMELTERS, AND FORGES

The original lists were probably carved in stone and represented
longer periods of time. They contained things like
"Get More Clay. Make Better Oven."
—*You (also, David Viscott)*

WHAT THEY ARE

A way to get more heat from your fires, which then lets you use materials in new ways, including giving you the ability to create pottery, ceramics, and glass, and also to forge *the very metals themselves.* In tastier news, they're also a way to produce amazing pizzas.

BEFORE THEY WERE INVENTED

No metal forging, no artificial glass, no earthenware, stoneware, or porcelain, and on top of all that, pizza would be way worse!

ORIGINALLY INVENTED

30,000 BCE (campfire bisque figures)

6500 BCE (lead smelting in campfires)

6000 BCE (kilns)

5500 BCE (copper smelting in kilns)

5000 BCE (glazing)

4200 BCE (bronze production)

500 BCE (blast furnaces)

997 CE (delicious pizza)

PREREQUISITES

clay, wood, charcoal for forges, limestone for iron smelting, mortar (for better kilns), mining (for raw materials)

HOW TO INVENT

Everything you'll invent in this section starts with clay. Clays are fine-grained soils that contain aluminum silicates bound with oxygen, and luckily for you, they're found all over the world in just about every time period! They're only hard to find very early on in Earth's history: after rocks have formed but before they've had a few million years to weather into soil.* Clays will generally be under topsoils (so you'll have to dig) or near seashores or riverbanks,

---

* And great news: if this is the time period you're trapped in, this won't be a concern for you, because you'll die from a lack of food long before any lack of clay becomes inconvenient!

where erosion can expose them. Clay is easy to recognize when it's wet: it's a damp, heavy, finely grained, and easily shaped soil, but if it's dry it can look pretty rocklike. You can distinguish dry clay from rocks by scratching it: if your scratch easily brings away a fine powder, that's dry clay, and some water will fix that right up!

You may find clay mixed in with other impurities or pebbles, and there are two ways to purify it. One is to dry it out, break it into pieces, and then smash it into dust: clay particles will be the smallest, and you can sift them through a sieve. This takes a lot of energy. An easier solution is to take your dirty clay, put it in a container, and add twice its volume of water. Crush the mud with your hands to remove any large chunks, then let it sit for a few hours to get fully hydrated. Afterward, stir the mixture thoroughly. Let it settle for a few minutes, and the mixture will separate into layers: sediment at the bottom, and a lighter mixture of clay and water at the top. Pour out the "clay water" into a separate container, and let it settle again for longer, this time for a day or so. Now the clay will settle on the bottom, and you can just pour the water off. Repeat this process a few times if there are still impurities. When you're done, you'll have a wet clay that you can let dry in the sun for a few days until it's usable. The easiest way to test your clay for quality is roll it into a snake shape (arguably . . . the easiest shape to make out of clay?), and then wrap it around your finger: if it bends without cracking, you've got yourself some good clay.

The problem is that you can't just make a bowl* out of clay and wait for it to dry, because dried-out clay is brittle and crumbly, and as you just saw, it becomes flexible again when wet. It's only when clay is heated that the magic happens.†

---

* Looking for an easy way to make an attractive symmetrical bowl? Use a wheel. Put your clay in the middle of a heavy wheel and spin it, and you can pull the clay into nice bowl shapes. The wheel was actually invented for this purpose, and it was only afterward that anyone thought to turn them on their sides and use them for transportation!

† And by "magic" we mean "permanent physical and chemical changes that are, in fact, well understood by science."

At around 600 to 1000°C (depending on your clay) it transforms into bisque: a stronger material that won't become clay again no matter how wet it gets.

While bisque is better than dried clay for creating figures or bricks, it's still suboptimal for bowls: even though it won't become clay again, it's porous and readily absorbs water. To fix that, you'll want to heat your bisque a second time, up to around 950°C. At this point another transformation takes place: the clay itself begins to fuse together while the impurities within it melt, filling any gaps within the clay matrix as it cools. The result is a stronger, denser, watertight material: in other words, my friend, you have just invented earthenware ceramics. Get your bisque even hotter (1200°C) and you'll produce stoneware: a more chip-resistant ceramic than earthenware. Throw some salt in while you're firing your clays and you'll glaze your ceramics in a thin layer of glass: at these temperatures the salt breaks down into sodium gas, which reacts with your bisque to form fresh glass directly on its surface. Adding minerals can produce different colors in your glass: bone ash gives reds and oranges, and copper gives greens.

A regular campfire is hot enough to create bisque—you can get one up to around 850°C—but it's not hot enough to make earthenware or glazes. For those you need a kiln, which encloses your fire, containing and amplifying its heat. With kilns—which are really just specialized ovens—you unlock not just better pottery but also glassware, metalworking, and more: a world of different and more useful materials is available to you the second you start making fires more intelligently than just throwing wood into a pile and calling it a day.

To build a kiln, you'll use a regular campfire to make bisque bricks from clay. Bisque holds heat well, doesn't burn, and has an extremely high melting point, which makes it ideal for your first kiln. And once it's built, you can then use that kiln to make higher-quality bricks, which you can then use to make better, higher-quality kilns. A simple kiln—really just a rectangular box, with a fire on one side and a chimney on the other—looks like this:

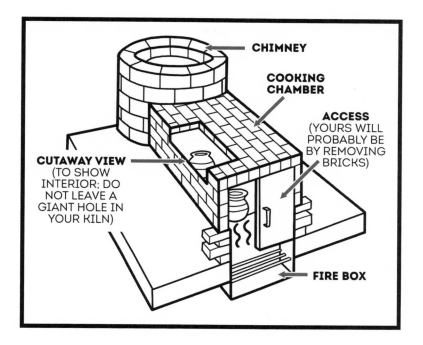

CHIMNEY

COOKING
CHAMBER

ACCESS
(YOURS WILL
PROBABLY BE
BY REMOVING
BRICKS)

CUTAWAY VIEW
(TO SHOW
INTERIOR; DO
NOT LEAVE A
GIANT HOLE IN
YOUR KILN)

FIRE BOX

Figure 18: Your new kiln.

The chimney on the far side draws the hot air from the fire across whatever you place inside. Use mortar between any bricks to ensure it's airtight, leave a few unmortared so you can control airflow and restock your fires easily, and you've just made your first kiln. Besides drying pottery, kilns can also be used to cook any thinly rolled dough that has had some tasty toppings placed on it in less than two minutes, which is how you're going to invent some *extremely* delicious wood-fired pizza.

A kiln like this can get up to around 1200°C—hot enough to melt copper, as humans eventually discovered when they noticed that some rocks would *partially melt* when put in a kiln. Adjust your design to collect this runoff and direct it outside your kiln, and you'll have convenient access to molten metals for the first time. This adjusted kiln is called a "smelter"—a device to extract base metals from ores—and you just invented it. Well—*technically*, smelters were invented the first time someone threw a tin- or lead-containing rock into a campfire and noticed the next day that there was a hardened metal mixed in with the ashes (tin melts at 231.9°C and lead at 327.5°C, well within campfire heat ranges), but yours is the first one that allows for convenient collection of molten metals.

Smelters can even be used to collect metals they *can't* melt. Iron melts at a balmy 1538°C, well beyond the range of these furnaces. But when you add limestone to iron-containing ores*—smash them up and mix them together—the limestone reacts to lower the melting point of the non-iron parts of the rock, and your smelter can work in reverse: melting the non-iron away and leaving pure iron behind, which can then be taken out and shaped in a forge. To build a specialized iron smelter, you'll build a chimney-shaped kiln with one hole at the top, and some pipes (made of clay initially, and metal later once you've got it) to draw air in from the sides. Start a charcoal fire (see Section 10.1.1) in the pit, break your ore into small pieces, and once your fire is hot, add alternating layers of iron ore and charcoal in equal parts through the top. The unmelted iron falls to the bottom and collects in a spongy lump surrounded by molten slag (which is the impurities you don't want). All that's left is to drain the slag, collect your iron, and purify it. You'll do that by hammering your metal flat while it's still warm, folding it over itself, and repeating. This process presses out any remaining slag and fuses the iron together, but you're going to need a way to keep your metal hot while you work on it. And for that, you'll need a forge.

Forges are related to kilns—you'll want to use the same bricks in their construction—but they're more open, and like smelters, they use hotter-burning charcoal. Beneath where the fire burns is an air tube that forces air directly into the fire via bellows at the other end: the more air your charcoal gets, the hotter

---

* Iron ore is just a rock that you can extract iron from! Here's the thing: most of Earth's iron sank to the center of the planet when it was molten, which formed the core of a nickel-iron alloy that also helps generate Earth's magnetic field. The iron that remained near the surface reacted with other chemicals, like oxygen, forming the iron-oxide ores you're now trying to extract pure iron from. The reason we're mentioning all this is so you know that if you ever do come across pure iron on or near Earth's surface, odds are that metal isn't actually native to Earth: instead it came here via a gosh-darned *meteorite impact*. Without smelters, meteorites are just about the only source of pure iron available, which means that the earliest iron tools, weapons, and jewelry that humanity produced were made not from Earthly materials but from meteoric iron, forged in the hearts of alien stars. Listen, we're so cool we manufacture time machines, and even *we* think that's awesome.

it burns.* Metals placed into the forge can be heated up enough to become flexible, which lets you hammer existing metals into new and more useful shapes before cooling them in water. Look at that: that clay that you first cooked in your campfire has given you access not only to better kilns but also to molten metals and the ability to shape them! It's brought you from the Stone Age, through the Bronze Age, and all the way into the Iron Age, which is pretty good for some *weird dirt that you found down by the river.*

Kilns show up around 6000 BCE, but there is nothing preventing you from constructing them at any given point in history, except knowing how to do it. And since you just learned how to do that, you've got no excuse for waiting! Do it right now![†]

10.4.3: GLASS

Don't tell me the moon is shining; show me the glint of light on broken glass.

—You (also, Anton Chekhov)

WHAT IT IS

A strong, heat-resistant, nonreactive, infinitely recyclable noncrystalline amorphous solid[‡] that you can see through, which actually makes it one of the most insanely useful substances on the planet

---

* Bellows are one of those inventions that are so simple you can get their basics just by looking at them (or, in your case, remembering them): they're two pieces of wood with an airtight sack between them and a small hole at the front. You can make them at any point in time from animal hide (Section 10.8.3), using pitch to seal any gaps (Section 10.1.1). Opening the bellows slowly sucks air in, and closing them quickly forces it out.

† If you're still not sold on kilns, just take a look at the technology tree in Appendix A. You'll see all the amazing technologies kilns unlock and realize that they're *the single most productive family of inventions in this entire book.*

‡ Glass is a solid. It's a myth that glass is technically a liquid, or a "super-cooled liquid that just takes a really long time to move." Glass, once solid, does not flow at all unless it's melted again, and you can prop up a pane of glass and come back 20 million years later and it will still be a pane and not a weird glass puddle on the floor. This has been determined both scientifically (thanks to studies of ancient natural glasses) and experimentally (thanks to your friends at Chronotix Solutions).

BEFORE IT WAS INVENTED

If you needed corrective lenses, you wouldn't get them and would instead spend your entire life not seeing things clearly. In addition to *literally* not seeing well, you'd also spend your entire life *figuratively* not seeing the benefits of the myriad technologies glass unlocks, including microscopes, test tubes, and lightbulbs.

ORIGINALLY INVENTED

700,000 BCE (natural glass used for tools)

3500 BCE (artificial glass, used mainly to make beads)

27 BCE (glassblowing)

100 CE (clear glass)

1200 CE (windowpanes)

PREREQUISITES

none (for natural glass); kilns, potash or soda ash, quicklime (for artificial glass)

HOW TO INVENT

Glass is one of the most useful substances you can produce, and to underline that, we'll describe the amazing things you can do with glass before we tell you how to make it. This way, when you do reach instructions for its production, your reaction will be "That sounds easy, I can't wait to do it!" instead of the alternative reaction of "Oh wow, glass is for losers, I'm gonna skip ahead until I find something that explodes to read about," observed with earlier editions of this text.*

Here's a short list of things you can make with glass: glazed pottery, eyeglasses, microscopes, telescopes, beakers and test tubes (useful in science because glass doesn't react with much, meaning you can even use it to safely store *sulfuric acid*), vacuum chambers, prisms, lightbulbs, thermometers, barometers, *and more*. Glass lets you bend light (through refraction), break it (through diffraction), and put it back together (by shining a bunch of lights into the same spot).

---

* If you insist on explosions, several of the chemicals in Appendix C can satisfy that need.

What are the uses of bending light? Make an outward curve and you'll cause light rays to converge, inventing the magnifying glass. Produce the opposite shape and you cause them to expand, which will correct nearsightedness. Like this:

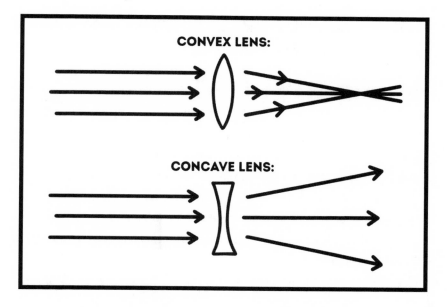

Figure 19: The shapes glass needs to be in to bend light in various useful ways.

Outward-curved lenses (called "convex lenses") are easier to make than concave lenses—since blowing a glass sphere produces convex shapes by default— so you'll find farsightedness (not being able to see close things clearly) initially easier to correct than nearsightedness.

Eyeglasses were first invented in the 1200s CE in India and brought to Europe through Italy.[27] Around the same time, sunglasses were being invented in China, and were, even then, much cooler than the nerd goggles being discussed here.* However, it took five hundred years—well into the 1700s CE—before anyone

---

* These early sunglasses didn't use glass proper, but rather flat slices of transparent smoky quartz. It doesn't matter! You still look awesome!

suggested that glasses might be conveniently held in place by arms that use our ears as anchors. Before then you'd have to either constantly hold your glasses in place, or pinch them to your nose so hard that they'd stay there themselves. Just by giving your glasses arms, your civilization will already be hundreds of years ahead of the game.

But eyeglasses are just the beginning. By taking two side-by-side convex lenses from eyeglasses and putting them in line with each other—usually held in place by an adjustable hollow cylinder—you can create a microscope, an idea that first occurred to humanity in the early 1600s CE. Combine a convex lens at one end with a concave one at the other and you've created telescopes, useful for spotting distant lands, exploring the nature of the cosmos, and spying on your freaky neighbors. Put two telescopes side by side and you've invented binoculars. Look at you: we give you a few lenses and already you're inventing like crazy!

Telescopes and microscopes are terrifically important inventions that lead to the discovery of previously unknown forms of life (bacteria), new understandings of how life exists (cells), how life reproduces (through cellular division, and with sex through microscopic sperm meeting microscopic eggs), how humans defend themselves from disease (white blood cells), not to mention the macroscopic discovery of new planets, new stars, and new galaxies—all of which will fundamentally transform science, medicine, biology, chemistry, theology, *and civilization itself*. In our timeline these innovations all had to await the invention of lenses, which required clear glass, which required hot kilns, but the instructions in this text give you all you need to invent them at any point in history.*

To invent mirrors—useful for both advanced science and advanced personal

---

* You don't even need lenses to produce a basic microscope! A simple clear glass sphere also produces magnification, and these were made well before lenses were, as far back as 100 CE in Rome—though nobody at the time treated them as anything more than a curiosity, and none realized their full potential. You can make such spheres by producing a strand of glass, then melting it—the glass will drip and form balls, which cool as they fall, producing an almost perfect sphere. The smaller the glass ball is, the greater the magnification it will produce, and a tiny sphere of glass—held close to the eye, just above the item being studied— will produce enough magnification to see both cells and bacteria. This unlocks a whole set of innovations and discoveries—including the germ theory of disease in Section 14—at any point in history: all you need to do is melt some glass.

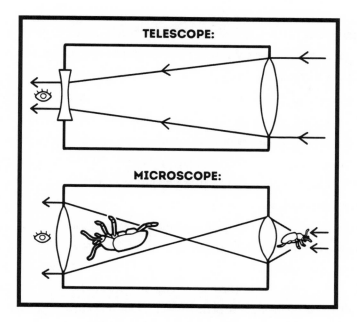

Figure 20: Schematics for telescopes and microscopes.

grooming—place a layer of reflective metal like copper, aluminum, or tin on the back of a piece of clear glass.* With the widespread adoption of mirrors comes the invention of the selfie—before the wide availability of mirrors in Europe around the 1400s CE, there was no strong tradition of self-portraits—as well as periscopes, more advanced telescopes, solar cookers that reflect and focus the rays of the sun, and the myriad body-image issues that can be unlocked only through constant, unavoidable reminders of your personal appearance at all times. Shape glass into a triangle and you'll make a prism, which breaks light out into its constituent parts (i.e., makes pretty rainbows). Put that in a dark box so that the only light that can enter your prism is from a single pinhole source, and you've invented the spectroscope. Every element in the universe, when heated, gives off distinctive bands of colored light: an incomplete rainbow that acts as a fingerprint. You can use your spectroscope to identify what's burn-

---

* Actually, you've only invented the *glass* mirror. Mirrors existed beforehand, and could be as simple as putting water into a dark vessel, waiting until the water calmed, and looking in. Polished metals also provide some measure of reflectivity, but are expensive and hard to produce!

ing right in front of you, or—when combined with a telescope—determine the elemental makeup of stars thousands and thousands of light-years away.

Not bad for some melted sand.

That's right: glass is simply melted sand—or more accurately, melted silicon dioxide, also called "silica." Silica makes up more than 10 percent of the Earth's crust by mass and is the major constituent of sand in most parts of the world, so it should be very easy to find. It melts around 1700°C—too hot for campfires, but well within the range of the kilns you just invented in Section 10.4.2. The first human-produced glass was made by accident around 3500 BCE in just such a kiln: some sand got in, melted, and cooled into an interesting substance.*[28]

You can lower the melting point of silica—and thus make glass production easier and more affordable—by adding potash or soda (Appendices C.5 and C.6, respectively) to your sand. The heat causes the potash or soda to dissolve into the sand, lowering its melting point. Include some quicklime (Appendix C.3) to increase your glass's durability and chemical resistance and to prevent it from being worn down by water, and you'll have a substance that melts around a much more achievable 580°C. You're looking for a mixture of about 60 to 75 percent silica,† 5 to 12 percent quicklime, and 12 to 18 percent soda.

When this mixture melts, it'll make a bubbling, frothy liquid—the bubbles are carbon dioxide escaping—which you'll want to let "boil" long enough to allow those bubbles to escape, and then you can pour, blow, pull, or mold your glass. Clear glass comes from white sand: brown sand usually has iron oxide in it, which makes a green glass.‡ To turn green glass clear, add manganese

---

* Natural glass predates human-produced glass: you can find it for free around volcanoes rich in silica, which, when their lava cools, can produce a dense and brittle natural glass called obsidian. Obsidian has been used since protohuman times because of a useful property: when it fractures it forms a sharp edge, which is great if you want to make blades or arrowheads. But without a way to produce new obsidian when it breaks, it's a rare material.

† You likely won't have pure silica, but that's okay: impurities will either burn off or color your glass. Bright-white sand is usually pure silica, but if you can't find any, white quartz rocks work too.

‡ Different impurities were intentionally added during the stained-glass craze of the Middle Ages, including copper oxide for greens, cobalt for blues, and gold for purples.

dioxide—which can be produced by burning certain seaweeds to ashes; experiment until you find some that work—into the mixture when molten.*

The hotter glass gets the runnier it gets, and the cooler it gets the thicker it gets. You can exploit this by cooling your molten glass down to different points, where you'll see it go from behaving like runny syrup to flexible bubblegum to thick taffy. When it's at the bubblegum point you can put a glob onto the end of an iron pipe and force air into it, and at that instant you have invented glassblowing, which is useful for making all sorts of glassware.

You'll probably want to make glass windows, an invention that (a) helps make homes feel comfortable and non-cavelike, (b) keeps them insulated, and (c) can really brighten up a room. Making a large pane of glass, however, does take a bit of doing, and there are several different techniques. And here they are, from earliest and simplest to modern and most complicated, so you can decide how fancy you want to get!

- If you've got lots of time and energy, just pour molten glass onto iron (which won't melt), let it cool, and then polish both sides until they're transparent. You'll need to start with a rough sandpaper† and gradually work your way down to a very fine grain. It takes a while.
- If you blow a balloon shape, then cut off the ends, you'll be left with a rough cylinder. Cut it in half and flatten it out on an iron plate while

---

* This discovery was originally made around 100 CE by the Romans and rediscovered in the early 1300s CE in Italy. Incidentally, these Italian glassmaking kilns sometimes caused fires—a particular danger when most buildings were made of wood—and so the government of Venice banished all glassmakers to a nearby island. While this was intended as a safety measure, concentrating all glassmakers together led to an explosion of talent and shared expertise, which allowed glassmaking to advance quickly, leading to the rediscovery of the "burn seaweed and add its ashes to make clear glass" trick. Even clearer glass can be made by a 10 percent to 30 percent concentration of lead oxide, which produces a glass (known as "crystal glass") that's very refractive and pretty but can also give you lead poisoning. The association of the upper classes of Europe and North America with gout around the 1800s CE has been traced to them drinking out of their fancy lead glasses.

† Sandpaper can be invented simply by gluing sand onto paper. You'll need paper (Section 10.11.1), glue (Section 8.9), and, of course, sand. Seeds and crushed shells can be used instead to get sandpapers with various roughnesses, or you can filter sand through fabric to get finer-grained sandpaper.

it's still flexible and you've invented broad sheet glass, a simple-to-make glass suitable for windows, but one that's rough and not often transparent. It's normally invented around the 1000s CE.

- If you blow a large globe of glass (which takes some skill), and then slowly reheat this glass up to its melting point while spinning it on something like a potter's wheel, centrifugal force will flatten your glass into a disc. This transparent "crown glass" will be thinner at the edges and thickest in the middle, where a circular "bull's-eye" will be visible, but it's largely suitable to be cut into windows. This technique didn't appear until 1320 CE in France, where it was kept as an extremely lucrative trade secret for hundreds of years. Now you know!

- If you blow large globes of glass into iron molds, you'll be able to make identically shaped glass over and over. In particular, if you blow into a cylindrical mold, you can let the glass cool, cut it lengthwise, and then slowly reheat it, and the cylinder will naturally flatten out into a sheet of glass, producing a much more uniform—and transparent—sheet of glass than you can get with the earlier broad sheet glass. This technique for cylinder glass shows up in the early 1900s CE.

- If you've produced liquid tin—a very dense metal—you can form the perfectly flat modern windows you're used to by pouring molten glass onto the tin, where it will spread across the surface in a uniform thickness before cooling. The glass solidifies around 600°C, well before the tin does, so you can just pick it off when it's cool. This technology normally shows up around 1950 CE and in less than a decade displaces all previous glass-production methods, but there's a catch: while tin won't attach to your glass, tin *dioxide* will, so you'll need to do this in a room filled with a gas that isn't oxygen so your liquid tin won't rust. If this sounds too hard, then crown or cylinder glass will be fine, we promise. Look, this is a book that in a few sections will go out of its way to explain how *buttons* work, so maybe you don't need to be messing around with molten metals just to make a smoother glass right now.

"I'M LAZY; I WANT A MACHINE TO DO WORK FOR ME"

↗ Engineering is the process of applying science, math, and other practical knowledge toward inventing new machines. You've been engineering this entire time, and you didn't even realize it! But the inventions in this section tend more toward the classical conception of engineering: the building of engines to perform different tasks for you, freeing you up to do anything else (including, but not limited to, keeping an eye on the engine you've built).

Waterwheels and windmills are the first technologies you'll invent that harness the Earth's natural processes, making them do your work so you don't have to. You'll make them even better with **Pelton turbines. Flywheels** smooth out power production and are useful in all sorts of engines, including **steam engines**, which are incredibly useful machines that require only water as a reactive agent and can run on anything that burns.

We know our more mechanically minded readers will have been anxiously awaiting this section, and we are happy to now give you everything you need to invent these technologies you've been looking forward to. In other words: time travelers . . .

. . . *start your engines.*

10.5.1: WATERWHEELS AND WINDMILLS

If we learn to feast toil-free on the fruits of the earth,
we taste again the golden age.
—You (also, Antipater of Thessalonica)

WHAT THEY ARE

A way to harness actual forces of nature to make them work for *you*, for once

BEFORE THEY WERE INVENTED

If you wanted some grain ground or wood sawed or rocks smashed or tools sharpened or ore crushed or bellows operated or iron wrought or paper pulped or water pumped you had to do it *yourself*, by *hand*, like a gosh-darned *chump*

ORIGINALLY INVENTED

300 BCE (first waterwheels and cams)

270 BCE (right-angle gears)

40 BCE (trip hammers)

100 CE (first wind-powered wheel)

400 CE (waterwheels powered by falling water)

600 CE (dams for waterwheels)

900 CE (first windmills)

1185 CE (first modern windmills)

PREREQUISITES

wheels, wood or metal, cloth (for windmills)

HOW TO INVENT

Both waterwheels and windmills are instantiations of the same idea: the planet has all these liquids and gases always moving around its surface, so wouldn't it be nice if we could make them do something for us?

A waterwheel is simple to invent: it's just a giant wheel with paddles on it so water can push it. Dip it in a stream and it'll rotate as water flows by, but this captures only 20 to 30 percent of the water's total energy. You can double that number to 60 percent by powering it with falling water instead: this way it's not just the water's *movement* that's powering your wheel but its *mass* as well. Do this by replacing the paddles on your wheel with cups, and then positioning your waterwheel beneath a waterfall. If those aren't around, you can create an artificial one: just direct a stream through a trough that exits above your wheel. While you're at it, if you dam up that stream so that the only way water can get out is by going through your waterwheel, the lake you create actually functions as a store of power ready to be used whenever you need it. That's the world's first battery, and you just invented it.

Your wheel will be connected by a shaft to the inside of your mill, and that shaft will rotate at the same speed and in the same direction as your wheel. This is useful for certain kinds of work—rotating a conveyor belt, for instance—but you can transform this rotational motion into all sorts of different movements with the help of some simple technology.

Add right-angle gears (see Appendix H) and you can make a vertical waterwheel rotate another horizontal wheel instead, which is great for grinding grain. Just set up two stone wheels: your rotating one, and one beneath it that's fixed in place. Feed your grain into a hole in the middle of your rotating wheel, and it'll grind it into flour and push it out at the edges. Change the relative sizes of the gears you use and you'll adjust how fast that grinding happens and with how much torque. Attach a crank to your waterwheel and you've changed its rotational motion into a back-and-forth one, which you can use to invent mechanical saws, pumps, or bellows. Replace your crank with a trip hammer and your waterwheel can now repeatedly smash rocks (or hammer steel) instead. All come from the same source: water pushing against a wheel!

Windmills work on the same principle, but rather than water rotating a wheel, wind pushes against a set of sails mounted like a fan around the drive shaft. This introduces a bunch of complicating factors, which we will now explore through a fictional dialogue between a waterwheel-loving critic of windmills—let's call her Dr. Waterwheels—and a well-informed and reasonable windmill apologist, whom we'll call Chompsky. We're going to imagine that Dr. Waterwheels is a human but that Chompsky is an adorable talking dog who's panting happily as he gets belly scratches from the doctor, because nobody can stop us from imagining whatever we want, least of all you:

Statement from Dr. Waterwheels (human, lover of waterwheels, windmill skeptic)	Response from Chompsky (adorable talking dog, well informed on windmills, super-scratchable belly)
Waterwheels are the best, Chompsky! They use water, whose flow is predictable and usually only changes slowly. Meanwhile, wind is famously unpredictable, wild, and capricious!	That's true, Dr. Waterwheels, but you can build windmills anywhere, while waterwheels can only be built near water, so I suppose it balances out. Can you scratch my belly a little higher up?
Is this better?	Yes, thank you. My leg is doing that adorable thing where it kicks, so you know you're scratching at just the right spot. Do you have any other issues with windmills?
I do! If my waterwheel is turning too fast, I can simply remove it from the water and it'll stop. But a windmill can't escape the wind!	While you're right in saying windmills can't escape the wind, we can easily design them to handle wind overload. The blades of a windmill can be built out of wooden frames, which then get covered by sheets. When the sheets are removed, the wind goes right through the blades instead of turning them to generate force, which allows the amount of power the windmill extracts from the wind to be controlled.
Okay, maybe . . . but water still always comes from the same direction. Wind can comes from every direction ever. What are you going to do, rotate your windmills every time the wind changes?	That's precisely what happens, yes. This can be done by hand, but if we're clever we can make the windmill self-adjusting by adding another paddle to the back of the windmill at a right angle to the drive fan, like on a weather vane. Then, any wind against the paddle turns the windmill to face it automatically. In fact, if we want to get even fancier, we can replace the paddle with a tiny windmill fan that rotates the entire structure along a circular geared track to the same effect! But it's probably worth pointing out that while it's true that wind can come from any direction, there are prevailing winds in many areas throughout the planet, which makes it possible to know the average direction wind will come from most of the time.
I guess. But water still carries way more energy than wind. For example, it's much easier to be swept away over the horizon by a river than it is to be blown away over the horizon by wind! So you have to admit a single waterwheel will often be able to do more work than a single windmill.	Yes, well, we all do the best we can with what we've got. Am I a good dog?

Statement from Dr. Waterwheels (human, lover of waterwheels, windmill skeptic)	Response from Chompsky (adorable talking dog, well informed on windmills, super-scratchable belly)
You are a good dog. Who's a good dog?	I am.
Yes you are. What a good dog. What a good dog. Good old Chompers. Look at your little face.	[belly scratches intensify; dialogue ends]

Table 12: Incidentally, the idea of education and insight happening through a dialogue between two individuals is called the "Socratic Method," and it's been a powerful teaching technique since its popularization by Socrates around 400 BCE. We used it here to discuss engineering with a talking dog!

And that's how you invent windmills and waterwheels.

10.5.2: PELTON TURBINES

Between Earth and Earth's atmosphere, the amount of water
remains constant; there is never a drop more, never a drop less.
This is a story of circular infinity, of a planet birthing itself.
—You (also, Linda Hogan)

WHAT THEY ARE

A better version of a waterwheel that's not only smaller but can also be over 90 percent efficient, which compares really well to that 60 percent we were messing around with with waterwheels

BEFORE THEY WERE INVENTED

People got by with waterwheels, but they didn't know what they were missing, and now they probably all feel like *idiots*

ORIGINALLY INVENTED

1870s CE

PREREQUISITES

waterwheels made of wood or, even better, metal

HOW TO INVENT

The waterwheels you invented in the last section (if you're reading this guide linearly) or that you might get around to eventually (if you've flipped directly to this section while muttering, "Gosh darn it, I need turbines *stat*") are powered by water in two ways: the mass of the water causing rotation, and the kinetic energy transferred by moving water when it hits the wheel. Pelton turbines work with the same mass of water but capture much more of its energy, which makes them much more efficient.*

The basic idea is to take water under pressure (the easiest way to achieve this is to run water through a pipe going downhill, with a smaller opening at the bottom of that pipe than at the top: the water's weight at the bottom forces it to pressurize) and make it hit your wheel like a super-powered hose. It didn't take any special thinking to replace the paddles on a waterwheel with cups to catch the water, but the innovation Johnny Pelton† came up with was to fire the water not directly into the middle of the cups but to put *two* cups on the wheel and to aim the water right at the wedge-shaped split between the two.‡

---

* How efficient? A Pelton turbine can be 10 to 20 times smaller than a waterwheel and still generate the same power. In scientific terms, that's referred to as "pretty friggin' good."

† Okay, his real name was actually Lester Allan Pelton, but many time travelers have found that, when it comes to naming their child, his parents are remarkably susceptible to suggestion. Names complete strangers have successfully convinced Mr. and Mrs. Pelton to call their son include "Helton Pelton," "P. P. Pelton," and "Rapmaster P, the Turbine Emcee."

‡ Cup your hands toward yourself with your fingernails touching each other, and you've approximated the shape of your turbine cups! Also, historically these are called *impulse blades* and not cups, but we're not trying to impress anyone here. Impulse blades sound like they should power starships; these are little cups you fire water into.

You can see why that made a difference simply by imagining standing beside a brick wall and spraying a hose at it. Fire that hose straight at the wall with any sort of water pressure and you're going to get wet: the water will hit the wall and splash right back toward you. That's energy in the water that's wasted instead of captured, and that's what was happening when water was fired into the middle of these waterwheel cups, where they were their flattest. But if your wall had a curve to it, and you aimed your water at an oblique angle just at the edge of that curve, you wouldn't get wet at all. Instead of being suddenly bounced off the wall, the water would be gently redirected, whipping around the curve to exit from the far edge. That's precisely how Pelton turbines work: the water imparts much more of its energy as it whips around the cups than it can when it smashes directly into them, which makes the wheel turn faster. The reason Pelton uses two cups instead of just directing the water to the edge of a single cup is for balance: this way the wheel has equal forces on both sides.

A Pelton turbine rotating at half the speed of the water hitting it will capture nearly all that water's energy. You'll be able to tell when you've got that set up properly because water coming out the far sides of your cups will be almost motionless, and that's the mark of a well-built Pelton turbine. Now you're extracting energy from moving water with over 90 percent efficiency! This not only gives you more power from the same water but also unlocks new power sources the world over, because streams and waterfalls that were too small for waterwheels will power a Pelton turbine just fine.

At this point you're probably thinking that it's pretty embarrassing humanity took more than two thousand years after inventing waterwheels to figure out that if we aimed the water to the side of our little cups instead of the middle we could almost double their efficiency! But it actually gets worse: origin stories for Pelton's turbine include "One day Pelton was hosing down some rocks and a cow got too close so he fired his hose at his cow and the water hit her between her cuplike nostrils and it knocked her head back and that's how he invented his turbine." We're not going to tell you if that story is true, because here is a fact: *there's no good way out of this.* If it's true, we're just a bunch of knuckleheads who need soggy cows to make even basic scientific advances. If it's false, we're still a bunch of knuckleheads who are apparently all more than willing to believe that no scientific advancement can happen without some soggy cow to show us the way.[29]

10.5.3: FLYWHEELS

Change does not roll in on the wheels of inevitability,
but comes through continuous struggle.
—You (also, Martin Luther King Jr.)

WHAT THEY ARE
A way to store and extract energy using nothing more than a big old wheel

BEFORE THEY WERE INVENTED
Rotational energy couldn't be stored, output from engines couldn't be smoothed, and wheels were *significantly* less fly

ORIGINALLY INVENTED
300 BCE (in pottery)
1100 CE (in machinery)

PREREQUISITES
wheel, steel (for dense and sturdy flywheels, and also for ball bearings)

HOW TO INVENT
Flywheels exploit the fact that objects in motion tend to stay in motion.* If you have a heavy wheel that takes a lot of energy to start spinning, it'll also take

---

* That comes from classical mechanics, a field of study that you're about to invent right here in this very footnote! Classical mechanics is concerned with describing how objects respond to forces that act upon them, and the three laws we're about to describe are its foundation. Before they were formalized in 1686 CE by Isaac Newton, people had a less-perfect understanding of why and how things moved, and had to make do with worse theories like "Rocks love the ground and smoke

a lot of energy to stop it. And that makes them batteries that store momentum instead of electricity! Wheels will slow down over time due to friction, so they're not perfect batteries, but a dense or large flywheel can run for a really long time before it stops. They were first used in pottery (potters' wheels are just big, heavy wheels that spin for a while once you get them moving, and that, my clay-covered friend, is a flywheel), but (no surprises here) it took humans a while to realize they could be used for anything else. Turns out, all you need to do is attach them to a rod rotated by an engine, and you're in business!

Besides storing energy, they can also be used to smooth out machinery. In piston engines (see Section 10.5.4), the pistons move intermittently, and in many uses you'll want constant power. If you're powering a tractor, for example, you'll probably want to move forward at an even pace, rather than in fits and starts every time a piston fires. If your pistons power a flywheel instead of your machinery directly, the wheel will continue to rotate even when the pistons aren't providing power, generating a much smoother force.

---

loves the sky, so that's how come smoke rises and rocks don't" (Aristotle, 300 BCE). The three laws of motion are: (1) Objects at rest remain at rest, and objects in motion remain in motion, unless acted on by a force. (2) The rate of change of an object's momentum is directly proportional to the force applied, and is in the direction of the applied force. (3) For every action, there is an equal and opposite reaction: as you push a box forward, the box also pushes you back. Keep in mind: we call these laws, but they're really an approximation that happens to work great at human scales. When you get very small (quantum scales of less than 10^{-9}m), very fast (2.99×10^8 m/s, close to speed of light), or very heavy (black holes), they tend to fall apart, and here Einstein's theories of general and special relativity describe motion with more accuracy than classical Newtonian physics can. But that's nothing you need to concern yourself about! While gravitational time dilation caused by an accelerating inertial reference frame moving through highly curved areas of space-time is a fascinating topic—and one that's also extremely useful when constructing FC3000™ rental-market time machines—unless you happened to bring your copy of *Good Grief: You Say Space and Time but Two Aspects of a Single Continuum Called Space-Time, and Furthermore That the Speed of Light in a Vacuum Is the Same for All Observers of That Light Regardless of the Motion of the Light Source? Well, Here Are 1001 Educational Comics About General and Special Relativity That Explore Those Ideas Further* back in time with you, we wouldn't worry about it just yet.

Figure 21: A "flywheel," also known by its less technical name, "wheel on a stick."

Flywheels can also release energy faster than it was originally produced. It might take you hours to get a flywheel up to speed, but attach a heavy load to your flywheel and you can direct all that energy toward doing work in an instant, giving you access to brief but powerful bursts of energy far beyond what you can normally produce. There is, of course, an upper limit on how much energy any flywheel can store: once a wheel rotates fast enough, it exceeds its own tensile strength, tears itself apart, and all those shards fly off at an incredibly fast speed. This is why a steel flywheel is safer than a cast-iron one: steel's greater strength minimizes the chance any flywheel inside your machinery transforms itself into a *surprise metal bomb*.

You can increase the energy any flywheel stores by increasing either its size or its speed. The energy stored in a rotating wheel is related to the *square* of its speed, so a smaller fast wheel will get better results than a larger slow one. And finally, while flywheels may seem old-fashioned, they aren't just used on piston-powered machinery: experimental flywheels were constructed by NASA in 2004 CE as a way to cheaply and reliably store energy in space. So *technically*, by inventing the flywheel you've also taken the first steps toward your new civilization's space program. Nice!

10.5.4: STEAM ENGINES

> The introduction of so powerful an agent as steam to a carriage
> on wheels will make a great change in the situation of humanity.
>
> —*You (also, Thomas Jefferson)*

WHAT THEY ARE

Engines that use the fact that water takes up more space when it's boiled to get things done, which is a technology *so useful* that when they finally got invented, society restructured itself around these machines in an event we call "the Industrial Revolution"

BEFORE THEY WERE INVENTED

If you wanted something done, you had to do it yourself, or get an animal to do it, or pay someone else to do it, but you certainly couldn't just boil some water and call it a day

ORIGINALLY INVENTED

100 CE (steam-powered toys that were technically steam turbines)

1606 CE (earliest steam-powered water pump)

1698 CE (first practical steam-powered pump)

1765 CE (separate condensation chamber, commercialized in 1776)

1783 CE (steam-powered boats)

1804 CE (steam-powered trains)

1884 CE (reinvention of steam turbines)

PREREQUISITES

Iron (for boilers), cast iron (for piston rings and cylinders), steel, welding

HOW TO INVENT

Steam engines probably sound old-fashioned, but even today the vast majority of electricity worldwide is generated with steam. The only real difference between old-fashioned steam engines and our newfangled ones is instead of burning wood in our boilers, we use coal, gas, or *the godlike powers of the very atoms themselves*. That's right: even with the civilization-ending power of nuclear reactors at our disposal, we still mostly just use them to boil us up some water.

The earliest steam engines were invented without much scientific theory behind how they worked, so merely by flipping through this section *before* you decide to throw a steam engine together, you're already starting off on a better foot than its original inventors ever did. It's been said that science owes more to the steam engine than the steam engine owes to science, and while that's not true (*science don't owe nobody nothin'*), it does give you a sense of how much humans were able to learn by studying these engines they'd invented, including but not limited to the second law of thermodynamics.*

A steam engine is made of two parts:

- a boiler, which uses some form of combustion to boil water, producing steam under pressure
- an engine that uses that steam to move a piston, turbine, or itself

The boiler part is easy, once you've got metal: just have airtight pipes of water that run through your fire-heated combustion chamber (this is called a "water tube boiler") or have pipes of fire-heated gas that run through your airtight and partially filled water chamber ("fire tube"). Either produces pressur-

---

* The first law of thermodynamics ("the law of conservation of energy") is that energy cannot be created or destroyed, but it can be changed. In other words, any increase of energy in a system will be equal to the energy put into that system. The second law is that the entropy (or chaos) in a closed system is always increasing. In other words, things don't just get spontaneously more organized: instead, they fall apart. It's worth noting that things have gotten more organized on Earth (life has evolved, buildings have been built, etc.), but that's because Earth isn't a closed system: it's powered by the sun. The third law of thermodynamics is that the entropy in a system approaches zero as that system approaches absolute zero: the colder things get, the less entropy you have. When you're at absolute zero (the coldest possible temperature), all physical motion ceases.

ized steam (and introduces the risk of your boilers exploding, so, uh, heads-up on that), but water tubes tend to be cheaper. Once you've got steam, you can run it through a second combustion chamber to heat it up even more, producing a superheated steam that carries more energy and can therefore do more work. Superheated steam can also be cooled a bit without it condensing back into water, which is great because then you don't have to worry about water clogging up your nice new steam engine all the time.

Putting steam to work in an engine can be done in a couple of ways, and the simplest is to inject steam into a piston. A piston is simply a mass that moves freely inside a cylinder, and it takes a little bit of precision engineering: you need a metal cylinder that's the same width all around, and a piston that's just slightly smaller so it can move freely within that tube.* To give this piston an airtight seal, you can attach a cast-iron ring to your piston: a spring-loaded piece of metal that makes constant contact with the cylinder. Before cast-iron seals, pistons would have hemp tightly wrapped around their bases as a seal. Hemp is dense, it doesn't wear out quickly under friction, and it works almost—but not quite—as well as a piston ring. Don't worry: some leakage of steam is okay, and your engine will still work, just not as efficiently.

When you take steam from your boiler and vent it into a piston, the steam expands, pushing the piston up. As the steam cools it condenses, which causes pressure in the piston to drop, and the outside air pressure to push that piston back down again. Since you generally want this steam to cool quickly, squirt cold water into the piston to speed up cooling. There's your engine. The up-down motion of the piston can drive a saw, power a pump, or be converted into circular motion via a crank (see Appendix H).

---

* Making cylinders by hand will work, but you'll probably introduce irregularities to their shape, which lets steam escape around the piston and makes your engine less efficient. The solution is to make your cylinders slightly *smaller* than what you need, run a straight metal bar through the middle, and send a drill along that bar. The drill enlarges the cylinder to a constant size, and the metal bar prevents it from following imperfections in the cylinder as it moves forward, producing a regularly shaped tube.

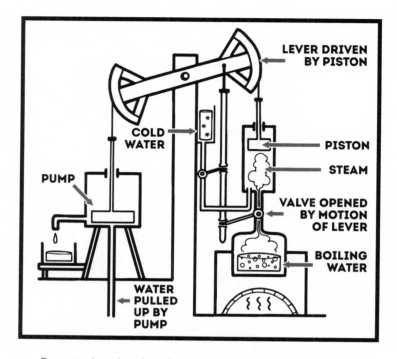

Figure 22: A machine that will power your civilization: the steam engine.

This brings your steam engine to the state of the art in 1698 CE, but if you're suspecting that heating and then cooling the same piston over and over wastes a whole bunch of energy, then put on a deerstalker* and call yourself Sherlock Holmes,[†] because your suspicions are correct! You can advance your engine almost eighty years by changing its design so that the hot parts stay hot and

---

* A hat hunters used to wear while hunting deer, hence the name. These days they're better known as "Sherlock Holmes hats" (see next footnote).

[†] Sherlock Holmes is a fake detective who never existed but who everyone agrees to imagine is the best at solving crimes. If you want to reintroduce this character to your civilization, feel free! You can even enhance him from the version you remember by dressing him up like a bat; giving him a giant bat's cave to use as his lair, similar bat-themed cars, planes, and -arangs; and having the police project his symbol into the sky whenever there's a crime nearby that needs solving. This version of Sherlock Holmes has historically been much more successful with the general public, especially when he's supplied with a murder clown to be his archnemesis.

the cool parts stay cool. Do this by introducing a connection from your piston to a separate condensation chamber, which gets opened by the piston rising. The high pressure of the piston environment pushes the steam into the condensation chamber, which you can quickly cool with a spray of cold water.

If you don't want to build a piston, there's another way to generate power with steam, and it's actually the first one humans discovered way back around 100 CE. It's called an "aeolipile," and you make one by boiling water and directing the steam into a rotating sphere, with exit nozzles positioned like this:

Figure 23: A machine that could've powered the ancient Greek civilization: the aeolipile.

The steam coming out of the nozzles acts as jets, spinning the sphere. This is a *steam-powered rocket engine*, but the Greeks who invented it never considered it to be anything more than a toy. You're about to show them up, because you're about to turn your aeolipile into a gosh-darned dynamo.

As you'll see in Section 10.6.2, dynamos transform mechanical rotation into direct current—electricity!—by exploiting the fact that wires moving through a magnetic field get a current induced in them. By putting a motionless magnet inside the sphere, and wrapping wires around the outside of your rotating sphere, you'll generate power. And if you don't want to build aeolipiles, you can always direct your jets of steam to fan blades on a turbine—similar to how you powered your Pelton turbine with water—and you'll generate rotational motion (and electricity, if you want) that way instead.*

Here's the bad news: all steam engines, whether piston-based, rocket-based, or otherwise, are fundamentally inefficient. No matter what you do, lots and lots of energy gets wasted as heat. Even with higher-pressure steam, condensers, and multiple-expansion steam engines (engines that use steam to move a piston more than once), your engines probably won't get much better than 20 percent efficiency. But even the most advanced steam engines in the modern era range from just 40 to 50 percent efficiency, so don't feel too bad: they still power our world, and they'll power yours just fine too.

The other major weakness of steam engines is their power-to-weight ratio. All that metal and water in a steam engine is heavy, and while they work great in buildings or in giant vehicles where their added weight isn't a huge factor (think trains and giant boats), they're less useful in smaller vehicles like planes and cars. For those scenarios, you'll want to invent the lighter internal combustion engine.

Steam engines are *external* combustion: you burn something outside the engine to produce steam, then pipe that into your engine. Internal combustion cuts out the middleman entirely, and instead has you blowing something up *inside* the piston itself to move it. A volatile fuel gets mixed with air so it'll burn easily, which is pushed into a piston's cylinder and compressed. An electrical spark causes ignition, pushing the piston out, and as it resets exhaust gases are expelled.† Each piston travels through this cycle of intake, compression, com-

---

* As we saw, aeolipiles were invented around 100 CE, but it wasn't until 1831 CE that the principle behind dynamos was discovered and these steam engines were *finally* put to productive use. There is one exception: they were used in the Ottoman Empire in 1551 CE to rotate meat on a spit.

† This is how gasoline-powered engines work, at least! A diesel engine works in the opposite way: diesel is introduced as the piston is compressing, and the sudden increase in pressure and heat causes the fuel to ignite. Either way, it'll probably be some time before you're dealing with gasoline or diesel as fuels, especially if

bustion, and exhaust, so that one piston is firing as the others are resetting. These engines are obviously a bit more complicated than steam engines (you're relying on a *controlled series of explosions* to drive your engine instead of good old water), but the problems involved aren't insurmountable. Four pistons in series can be arranged so one is exploding as another is refueling—this gives you more constant thrust—and cams on a rod can be used to coordinate intake and outtake valves so that the pistons work properly in series. Another rod can be bent and attached to each piston to coordinate their thrust, which itself is connected to a flywheel to smooth out its motion (see Section 10.5.3).

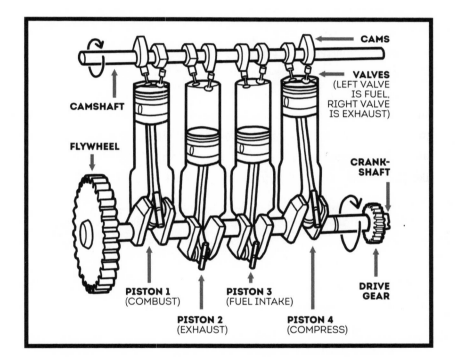

Figure 24: An internal combustion engine: piston 1 is combusting fuel, piston 2 is venting exhaust, piston 3 is intaking fuel, and piston 4 is compressing the fuel in preparation to burn it. You'll get there!

---

you haven't brought your copy of *How to Distill Crude Oil (Also Known as Petroleum) into Kerosene, Gasoline, Diesel, and Other Fuels: Chemicals That Can Be Catastrophic for the Environment but Will Let You Have Some Pretty Cool Race Cars, So Wow, Hopefully They're Worth It* with you. But you can always brew less-powerful ethanol to use as fuel in a pinch!

But before you rush toward inventing an internal combustion engine, keep in mind: they're more complicated to build, more expensive to run, and require higher-grade fuel. In an era when you have to invent everything from scratch, something like your steam engine—a machine that runs on water and is powered by *literally anything that burns*—is invaluable.

"NO, I MEAN I'M SO LAZY I JUST WANT TO FLIP A SWITCH AND HAVE MACHINES WORK AS IF BY MAGIC"

For machines that run at the flip of a switch, you want a source of power that's instantly ready, silent, and invisible. You want electricity. Electricity travels well, stores easily, and powers much of our modern world, including everything from aircraft to automobiles.

It can power your world too.

Batteries use chemical reactions to generate power on demand, and some can even be recharged by reversing the reaction. **Generators** rely on a simple law of physics to turn physical motion into electrical power, and **transformers** change electrical charges into more useful forms, all without a single moving part. The generation, storage, and transformation of electricity is the technology that allows your civilization, for the first time, to expand away from its sources of power and settle almost anywhere on the planet. Electricity allows you to conquer the world.

And you're only one page away from discovering it!

10.6.1: BATTERIES

> Magnetism, as you recall from physics class, is a powerful force
> that causes certain items to be attracted to refrigerators.
> —*You (also, Dave Barry)*

WHAT THEY ARE

A way to generate and store electrical current

BEFORE THEY WERE INVENTED

If you wanted portable electric power, you needed to carry around electric eels,*
and they're not exactly practical and predictable energy sources

ORIGINALLY INVENTED

1745 CE (first "batteries" that stored static electricity)
1800 CE (first chemical batteries that produced current)

PREREQUISITES

metal (for wires, copper is very pliable and can be stretched out into wire shapes
pretty easily), metal (for batteries, copper or silver for one end and iron or zinc
at the other work great)

HOW TO INVENT

To understand your battery you'll need to understand electricity, and to understand electricity you'll need to know about magnetism too. Humans used to think
electricity and magnetism were two different things, but then when we found out
they were so connected that you can't get one without the other, we joined them
together in our heads and called that new force "electromagnetism." The discovery and use of electromagnetism drove the second Industrial Revolution, which
(for the second time, as you may have guessed) reinvented the way humans lived.
Before electromagnetic technology, everyone had to either live right beside a
source of fuel (forests for wood, mines for coal, rivers for waterwheels, etc.), or else
pay to have those fuels delivered to them. Afterward, energy could be shipped
across the country at anywhere from 50 to 99 percent of the speed of light,† and

---

* And yes, the electric eel is actually a fish! People who know that electric eels
are actually fish tend to be really touchy about this, almost as much as those who
know peanuts are actually legumes, koala bears are actually marsupials, or guinea
pigs are actually rodents. Instead of getting mad at us, just go ahead and name
things better than we ever did, okay?

† Electrical energy always travels at the speed of light in a vacuum, but outside
of one, its speed is affected by whatever it's traveling through. Don't worry,
though: 50 percent of the speed of light is still insanely fast (it's 50 percent
of the *ultimate speed limit of the universe*, see Appendix F), and with the naked
eye you won't be able to notice the difference anyway.

people could now live comfortably anywhere we could run a wire to. In short, mastery of electromagnetism will allow your civilization to expand beyond riversides and mines to conquer the continent, the planet, and eventually *time itself.**

So here's how you do that!

Electricity is the movement of charged particles, usually electrons (and you can see "What are things made of?" in Section 11: Chemistry for more information on those bad boys). Electrons have a negative charge, so something with a lot of electrons will itself be negatively charged, and whatever material is losing those electrons will gain a positive charge. Remember: similarly charged particles will push each other apart, while opposite-charged particles will draw each other together.

Some materials let electrons move through themselves easily (these are called "conductors," and metals like copper, iron, silver, or zinc are great ones), and some keep them tightly in place and resist carrying a current (these are called "insulators," and glass, rubber, and wood are all decent examples of those). But under normal conditions, the movement of electrons in a conductor is completely random. To generate electricity, you need to somehow get those electrons all moving the same way. We call this an "electric circuit"—because the electrons move in a loop—and you're so great that you're going to invent the first electric circuit and the first battery *at the same time*.

Batteries are really easy to come up with once you've got metal, so again, it's pretty embarrassing it took us until 1800 CE to figure them out. They convert chemical reactions into electrical energy, and work by managing how two metals react with each other. Any two different metals have different affinities for electrons, so if you put them in a conducting solution, a chemical reaction will take place in which those electrons are exchanged. This conducting material is called an "electrolyte," and it can be many things: acid, saltwater, or even a delicious potato. Most salts, bases, and acids will do the job, and you can make sulfuric acid—a particularly good electrolyte—in Appendix C.12.

---

* The FC3000™ time machine is an electric, internal-combustion, and cold-fusion hybrid.

The metal that wants more electrons will draw them from the other, causing that metal (called a "terminal") to gain a negative charge while the other becomes positively charged. These negatively charged electrons gathering in the electron-hungry terminal repel each other, so if you connect a wire between that terminal and the positively charged one, the electrons will "escape" the crowds by running along that wire toward the positive end. Hey, you just induced electrons to move in the same direction along your conducting wire! My friend, you've just made electricity.*

You'll use electricity to invent electric lighting, electric heating, electric cooking, electric engines, *and more*—and in just a few paragraphs too—but for now sit tight, because we're not quite done with your new battery yet. It uses a chemical reaction to generate power, but eventually those metals will react all they can, and your battery will die. Hey, you want to keep going and invent rechargeable batteries in the same day? *Why not, right?*

Lead-acid batteries, first invented in 1859 CE, are like the two-terminal battery you just invented, but use lead-based terminals instead, which sit in an electrolyte that's a 3:1 mixture of water and sulfuric acid. One terminal is pure lead, and the other is lead dioxide. These metals react with sulfuric acid to produce lead sulfate at both terminals, but this reaction requires that the two terminals exchange electrons. So when you connect these two terminals with a wire, you'll generate electricity along that wire for as long as that reaction takes place. And critically, when you reverse the process and run electricity *into* the battery, the reaction runs in reverse: lead sulfates dissolve back into the electrolyte, and you reproduce your two terminals: pure lead at one end, and lead

---

* The first battery was made by putting silver and zinc in a stack, separated by brine-soaked cardboard. This worked, but the electrolyte is also involved in the reaction, and over time it becomes less conductive. It was improved thirty-six years later by giving each piece of metal its own different electrolyte and by joining the two "cells" by a bridge—known as a "salt bridge," since it can be as simple as a saltwater-soaked piece of paper—to allow the two electrolytes to exchange ions, keeping them electrically neutral. This battery used copper dunked in a copper-sulfate electrolyte (you can make that by adding copper to concentrated sulfuric acid) and zinc in a sulfuric-acid electrolyte. This "Daniell" battery (named of course by the Mr. John Frederic Daniell who invented it; your name is probably way better) produced more reliable power, so feel free to steal the idea!

dioxide at the other. This reversible reaction means you've effectively stored power in this battery for future use!*

So now you've got batteries that generate new power, and another kind that store existing power. And while batteries are great to experiment with, build upon, and eventually power the latest mass-market portable music players with, the fact is: *you don't build a civilization on batteries.* What you build a civilization on is a way to generate power that doesn't require someone to mine certain metals or synthesize different acids just to turn on the lights. In other words, you build them on electrical generators, otherwise known as power plants. And the best part is, if you already read up on waterwheels, windmills, or turbines, *you've basically already invented them.*

10.6.2: GENERATORS

> We will make electricity so cheap that only the rich will burn candles.
>
> —*You (also, Thomas Edison)*

WHAT THEY ARE
A way to produce energy, up to and beyond 1.21 gigawatts

BEFORE THEY WERE INVENTED
The only way to get 1.21 gigawatts of electricity was to wait for a bolt of lightning, and unfortunately, you never know when or where one's ever gonna strike

ORIGINALLY INVENTED
1819 CE (electricity and magnetism recognized as a united phenomenon called "electromagnetism")

---

* It also means you can generate the lead dioxide you need for your battery: just put pure lead in sulfuric acid and run electricity through it, and lead dioxide will form on the surface of your lead.

1821 CE (first electric motor)

1832 CE (first dynamo to generate electrical power from motion)

PREREQUISITES

metal; waterwheels, turbines, or other ways to generate rotational motion

HOW TO INVENT

We've been focused on the electricity part of electromagnetism, but here we'll take advantage of the fact that every electrical current also generates a magnetic field. You can prove it by placing a compass (Section 10.12.2) next to a wire: the instant that wire starts carrying current, that compass will move. Incidentally, by doing this you just used magnetism to turn electrical power into *physical movement*, which unlocks all sorts of inventions. Trivially, you can use it to make the world's first tools to measure and quantify electrical power. Getting a bit more complex, if you wrap your electrical wire around an iron core, you'll enhance the wire's magnetic field, thereby producing the world's first electromagnet: a powered magnet you can turn on and off. Mount a magnet so it can rotate freely, put it between electromagnets on opposite sides, and turn them on and off in sequence, and your magnet will spin for as long as you run power: this is the foundation of the electric motor.

These inventions work by using electricity to produce movement in a magnetic field, but the opposite is also true: *you can use movement in a magnetic field to induce an electrical current.* That's the foundation for generators, which you technically already invented in Section 10.5.4: Steam Engines when you turned an aeolipile into a dynamo, and it's still the basis of power production today. The core mechanic is incredibly simple: get something to rotate, wrap that rotating thing in coils of wire, put a magnet in the middle, and you'll generate electricity. The electricity generated this way is called "alternating current" (or AC), because the electrons move back and forth along the wire with each rotation. (This contrasts with the "direct current" [DC] you were making with your batteries.) That's all it takes to invent a generator. Conveniently, AC works better than DC for long-distance transmission, but that didn't stop a War of the Currents in our history, when different American corporations, backing competing AC and DC power systems, tried to convince the public that it was the *other* electrical standard that was dangerously deadly.*

You can transmit your power from your generation stations along wires, but you'll hit a limitation: any conductor has some resistance, which is the loss of electrical power to heat. This means that any wires can carry only so much power before they start heating up and, if it goes too far, melting. You can productively use this resistance to invent the toasters, stoves, ovens, electric heaters, hair dryers, and lightbulbs we teased earlier,† but to send power long distances, you'll need to adjust the nature of that power. That's where transformers come in. You should definitely invent those next!

---

* This fight included public battles over whose power would run through the first electric chair (each wanted the other to be responsible for the death, for PR reasons), public demonstrations of animals being electrocuted by *the competition's* electricity, and even the proposal of an "electric duel," in which representatives from each company would be shocked by equal but increasing amounts of electrical power from their own systems, with the first to quit being the loser. The duel never happened in our timeline, but other time travelers have proven that in this period of high tensions, it takes only a well-timed shout of "Hey, that guy just said that the *other* guy was too much of a wimp to get electrocuted for business reasons!" to set this particular event into motion.

† Incandescent lightbulbs are, at their core, just a wire overloaded to the point where it glows but doesn't quite melt. It took a lot of experiments to find a metal that worked best for this! The eventual solution, tungsten, will work for you too, but it can be a tricky metal to find and extract. Instead, do what early inventors of lightbulbs did, and use a carbon filament (which you can produce by heating bamboo or paper without burning them, using the same techniques you used to turn wood into charcoal in Section 10.1.1). These filaments won't last as long, but when you run electricity through them in a vacuum, they'll glow instead of burning. And if you're saying "I can't produce a vacuum so easily, why would you assume that I can whip up a vacuum whenever I want, that's almost as crazy as the very nature of my present circumstances," just use arc lamps instead: these are two conductors with some space between them, so the electricity jumps across and generates light.

10.6.3: TRANSFORMERS

> Let the future tell the truth and evaluate each one according to his work and accomplishments. The present is theirs; the future, for which I really worked, is mine.
>
> —*You (also, Nikola Tesla)*

WHAT THEY ARE

A way to safely manipulate electricity to make it safe for transport

BEFORE THEY WERE INVENTED

Moving electricity long distances was wasteful and dangerous, but to be honest most civilizations invent transformers pretty quickly after figuring out electricity, so you probably should too

ORIGINALLY INVENTED

1831 CE (principles of magnetic induction discovered)
1836 CE (first transformer invented)

PREREQUISITES

electricity, metal

HOW TO INVENT

We've been discussing electricity without using a lot of units (mostly because they're named after people who from your perspective likely haven't even been born yet, and who wants to give *them* all the credit), but we'll introduce one here: the volt. The volt measures a difference in electrical potential energy between two points in a circuit. If you think of electricity like water, then your wires are pipes, current is the amount of water moving through those pipes, and voltage is the pressure driving the water along. If you want more water from a pipe, you can either increase its size, increase its pressure, or both.*

---

* We haven't given you a precise definition for a volt here, because in the mod-

The same holds true for electricity: the power you get is a factor of current times voltage. The catch is, the more current you have, the more heat a wire is going to generate, and the closer you are to it melting. This leaves you, like in water pipes, with two options: you can either make your pipes bigger (thickening your wires to increase the amount of current they can carry before melting) or increase their pressure (i.e., increasing your voltage). High-voltage wires are more dangerous to be around,* but if you could transform your electricity up to high voltage for cross-country transportation—away from people and their curious, grabby, wire-touching hands—and then transform it down to safer lower voltages for use, you'd be set.

Transformers are simple, since they have no moving parts (except, of course, the electrons moving through their wires). Make a large square ring of iron. Coil an insulated wire connected to your incoming alternating current around one side. On the opposite side, make another coil to transmit your outgoing current. The two coils of wire aren't connected electrically, but when current runs through your incoming coil, it'll create an electromagnetic field (just like we saw before) that'll induce the electrons in your outgoing coil to move too. At this point your latest invention doesn't transform electricity (yet), but it does use a magnetic field to wirelessly transmit electricity over a short distance.

The real magic comes when you change the number of coils in your outgoing circuit. If the two circuits have the same number of coils, currents and voltages

---

ern era it's more than a bit messy. It's based on the ampere, which is defined either as "approximately equivalent to 6.2415093×10^{18} elementary charges moving past a boundary in one second" or "the constant current which, if maintained in two straight parallel conductors of infinite length, of negligible circular cross-section, and placed one meter apart in a vacuum, would produce between these conductors a force equal to 2×10^{-7} newtons per meter of length." These definitions? *Completely useless.* Feel free to come up with your own measurements of volts (pressure) and amps (current) whatever way you want.

* Well, technically, high voltage combined with sustained current is what's so dangerous. A sustained charge of 50 volts is enough to conduct through your skin, interrupt your heartbeat, and start cooking your organs—and yet a simple static-electricity charge carries up to 20,000 volts. So what gives? The answer is that, yes, while touching a doorknob after rubbing your feet on the carpet does produce a high-voltage shock, it's one with very little current and discharges within a nanosecond. A nanosecond of high voltage is fine; it's the fact that high-voltage wires carry sustained current that makes them so deadly.

in both wires will be identical. But if there are *more* coils in your outgoing wire, then the charge induced there will have decreased current and increased voltage, making it ideal for long-distance transmission. If there are fewer, you'll decrease your voltage and increase your current, making your electricity ready for local use. Voltage is directly proportional to the number of coils, so a 3:1 ratio of coils in and out will produce an outgoing current at one-third the voltage as incoming. Turns out, iron and some coiled-up wire is all it takes to transform electricity, and it's all possible because electricity and magnetism are two sides of the same coin! *Thanks, electromagnetism.*

So with the other inventions in this section, you can now produce, transmit, store, and transform electricity. It's worth noting that just as batteries could've been invented at any point in history once basic metals were discovered, so too could've power plants and transformers. Even when humans had invented waterwheels and windmills, they still used them to generate direct force—turning wheels, moving cranks—for more than *two thousand years* before anyone thought of inventing the dynamo and producing a much more versatile and transmittable electric current. With your knowledge of the steam engine and the dynamo, you now have the ability to produce *two* distinct industrial revolutions in your society at any point in history you choose.

"IT'S LATE AND I'M COLD, AND I'D LIKE TO KNOW HOW LATE AND HOW COLD IT IS"

Clocks are the first inventions that allow you to precisely quantify time. This, it turns out, is a surprisingly deep subject—even in a world that predates the invention of rental-market time machines. And once you have glass, you just need a little cleverness and a little water to invent **thermometers and barometers**, which let you quantify heat and pressure for the first time too.

Wanting machines to tell you the temperature and time may, given your current circumstances, seem like superficial and even useless desires, but they aren't. The technologies in this section unlock leaps forward in fields as disparate as manufacturing, chemistry, medicine, and even weather prediction, which you will definitely want to have sooner rather than later. And while the digital clocks you remember from the world you left behind might seem more advanced than the ones you're about to invent, don't worry.

You'll soon make up for lost time.

10.7.1: CLOCKS

Being with you and not being with you is the only way I have to measure time.

—*You (also, Jorge Luis Borges)*

WHAT THEY ARE

An actual time machine—though only in the "What time is it?" sense, not in the sense of "Finally, at long last, instructions for a device that will return me to my native time." Sorry.

BEFORE THEY WERE INVENTED

The passage of time was not quantified, which meant it was measured in more qualitative ways, like "from sunrise to sundown." On the plus side, if someone asked you what time it was and you lied to them, *they could never prove you wrong.*

ORIGINALLY INVENTED

1600 BCE (water clocks)

1500 BCE (sundials)

350s CE (hourglasses in ancient Greece)

700s CE (hourglasses rediscovered in Europe)

1300s CE (hourglasses common in Europe)

1656 CE (pendulum clocks)

1927 CE (quartz clocks)

PREREQUISITES

pottery (for water clocks), glass (for hourglasses), latitude and compasses (for sundials)

HOW TO INVENT

Modern wristwatches use tiny pieces of quartz to keep time: it's the second-most abundant mineral on Earth, and it has a useful property called "piezo-electricity." When you squeeze a quartz crystal, a small amount of electricity is generated—and when you do the reverse and run a small amount of electricity through quartz, the crystal vibrates at a predictable rate. This allows the construction of cheap electronic clocks, and in the modern era tiny pieces of rock vibrating 32,768 times per second are the world's most widely used timekeeping technology. But since you don't have modern electronics or quartz crystals, you'll be relying on simpler inventions to duplicate modern clocks.

A clock actually has two functions. A properly set clock can tell you what time it is, but even a watch set to the wrong time can measure how much time has *passed* since a given moment. If you're just interested in tracking the passage of time, then much simpler inventions—like water clocks—can solve your problem.

Water clocks were the first clocks invented, and the simplest versions are dead easy: just poke a tiny hole in a container of water. You're done! Water will drip out at a (reasonably) constant rate, and so by marking the fill line, and then measuring how much water drains from your bucket over different units of time, you can measure minutes, hours, and with a gigantic-enough bucket, even days. Until the invention of pendulum clocks in the 1600s CE, water clocks were the most accurate and commonly used devices to measure time, so you're doing great.

Hourglasses work on the same principle as water clocks, but use sand instead of water, and recycle the sand every time you turn them over. A few handfuls of sand, with a hole small enough to limit sand consumption but large enough to avoid jamming, will track—you guessed it—about an hour, and you can add or remove sand to get the exact unit of time you're interested in. If you have an hourglass you can technically measure as many hours as you want— just flip it over when the sand runs out and mark down how many times you've flipped it—but this requires constant vigilance, and errors are going to creep in.

To avoid having to flip over hourglasses or refill water clocks, you'll want to invent the sundial, which (during sunny days, at least) indicates the time of day. Sundials are easy to build—just shove a stick in some flat ground and mark its shadow over a day and you've built one—but they're a bit complicated to get right, especially if you want to know precisely what hour it is (which we imagine you do, because otherwise you could just glance up at the sun, say, "Looks about quittin' time," and leave it at that).

First, rather than shoving the sundial stick straight into the ground, you'll stick it in at an angle equal to your latitude (which you can determine in Section 10.12.3: Latitude and Longitude), and positioned so it's pointing toward true north (which you won't know, but magnetic north is a good approximation in most time periods: see Section 10.12.2). If you've done this accurately, then the stick's shadow at noon will always be directly underneath it, and 6 a.m. and 6 p.m. will always be at 90-degree angles to it on either side. To get the angles for the remaining hours, use the following formula, where l is your latitude and h is your hour:

$$\text{angle} = \tan^{-1}(\sin l \times \tan h)$$

Didn't memorize your trig tables before you went back in time? No worries: *it's crazy to do that*, and full tables are included in Appendix E to make this calculation easy.

There is a catch, though: even after all this measurement and math, your sundial still isn't going to be accurate. If you have a watch to compare it to (and hopefully you do, because staring at your watch in horror is de rigueur in the time traveler aesthetic), you'd see your sundial varies in inaccuracy throughout the year, becoming up to around fifteen minutes fast or slow. Great news: for once this isn't your fault, and these errors aren't happening because you made your sundial poorly!

They're happening because *the sun is lying to you.*

Or, more precisely: the Earth is causing the sun to lie to you. The planet's conspiring to mess up your sundial in two different ways. The first happens because Earth's yearly orbit around the sun isn't a perfect circle the way we like to imagine it: it's actually very slightly oval, with the sun a little off to one side. This ovalness is called "eccentricity":

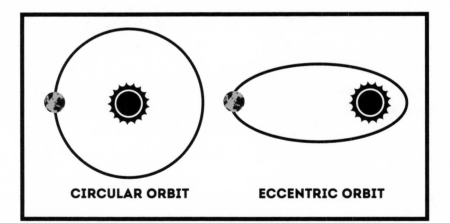

Figure 25: Circular and eccentric orbits. These images should not be taken literally, as they are exaggerated for illustrative purposes. Earth's eccentric orbit is not nearly so extreme, for example, and usually the Earth itself is larger than a few millimeters tall.

In eccentric orbits, planets don't always orbit at the same speed: rather, they speed up as they orbit close to the sun, and slow down as they orbit farther away.* Earth's eccentric orbit causes the sun to appear at the same spot in the sky up to eight minutes earlier or later than it would with a perfectly circular orbit, which means your sundial is going to be up to eight minutes off true time at different points in the year.

The other factor is the Earth's angled rotation. Rather than rotating straight up, like a top, it actually rotates on an angle, about 23.5 degrees off vertical.† This is called "axial tilt," and it causes the sun to appear higher and lower in the sky at the same time of day, adding up to ten minutes of error to your sundial. Eccentricity alters the apparent time of the sun on a one-year cycle, while axial tilt does it on a half-year cycle, like so:

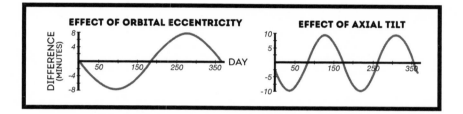

Figure 26: The separate effects of orbital eccentricity and axial tilt.

---

* This eccentricity itself varies slightly over time, forming a rough 100,000-year cycle. These changes in Earth's orbital speed lengthen summers or winters, depending on which hemisphere you're in.

† And just to make it more complicated, this axial tilt changes over time too, varying from 22.1 to 24.5 degrees on a rough 41,000-year cycle. When axial tilt is higher, the difference between seasons becomes more pronounced, with colder winters and warmer summers. To measure the current axial tilt of Earth in your time, wait until the June solstice, when the Earth's rotational axis is most inclined toward the sun (see Section 10.12.3) and put a stick in the ground. Make sure it's vertical, then measure the length of its shadow when the sun is at the highest point in the sky. Take the inverse tangent of the length of the shadow (with those trig tables in Appendix E), divide that by the length of the stick, and you'll get an angle. You're almost done! Now you just need to measure your latitude (again refer to Section 10.12.3). If you're north of the Northern Tropic (the northernmost point where the sun reaches a point directly overhead at least once a year; its latitude is identical to Earth's axial tilt and will be anywhere from 22.1 to 24.5 degrees), subtract your measurement from your latitude. If you live south of the equator, subtract your latitude from your measurement. And if you're between the equator and the Northern Tropic, add your latitude and measurement together instead. The result is Earth's current axial tilt!

So at one point in the year the eccentric orbit might be subtracting a certain number of minutes, while the axial tilt might be adding some. By combining these two charts together we can see their cumulative effect, showing what adjustments you need to make to the apparent time on your sundial to actually get accurate time!*

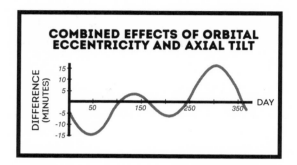

Figure 27: The combined effects of orbital eccentricity and axial tilt.

There is, of course, one more catch: both orbital eccentricity and axial tilt slowly change over time, and the Earth itself is precessing too (which means it's very slowly wobbling like a top: see Section 10.12.3 for more). If you apply this chart without adjusting for your current time period, you'll introduce a few seconds of error for every century you are away from the present, which quickly adds up. You can attempt to alter this chart (built for Earth's current eccentricity, axial tilt, and precession) by factoring in the following measurements[30] of how they've changed over the past 1 million years . . .

---

* The chart is historically known by the very impressive name of "The Equation of Time," with the word "equation" there being used in its medieval sense of "reconciling a difference." The Equation of Time is, sadly, unrelated to the actual temporal equations that brought you here and will therefore be of no use in returning you back home again, so you should give up on that fantasy already. Believe us: we wouldn't have gotten this far into writing this book if there were any easier option.

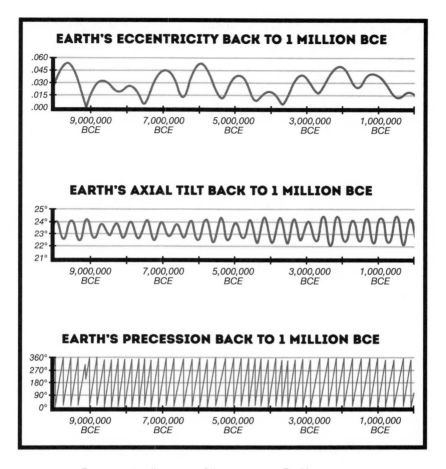

Figure 28: A million years of data to measure Earth's movement, here reduced to a few squiggly lines in a chart.

... but at this point it's probably worthwhile to tell you that you don't actually need to know exactly what time it really is anyway, and that the fifteen minutes of error we're going to such lengths to correct *honestly don't matter.*

It's only in the last few centuries that accurate time measurement became at all important, and then initially only to calculate longitude at sea: a requirement you'll be sidestepping entirely (see Section 10.12.3: Latitude and Longitude). Even today, the vast majority of humans live their lives not according to what time the sun actually says it is but rather to a mere *approximation* of that time. That's what time zones are: groups of people agreeing to pretend it's a single time across large areas of the planet, thereby doing away with the confusion that comes from each town keeping its own slightly different times by the sun. Time zones were first used within nations in 1847 CE, and worldwide time zones were proposed a few decades later. Invent time zones now and not only will the approximate time your unadjusted sundial produces do just fine, but you'll also save yourself a bunch of math homework.

10.7.2: THERMOMETERS AND BAROMETERS

> You will hear thunder and remember me, and think: "she wanted storms."
>
> —*You (also, Anna Akhmatova)*

WHAT THEY ARE
A way to measure heat (thermometers) and pressure (barometers)

BEFORE THEY WERE INVENTED
People just kind of guessed at how hot things were, what temperatures they were cooking at,* and what the weather was going to be like

ORIGINALLY INVENTED
1593 CE (water thermoscope)

1643 CE (barometer)

1654 CE (alcohol thermoscope)

1701 CE (the idea of temperature scale)

1714 CE (mercury thermometer)

---

* Before thermometers, various methods were used for measuring temperature in kitchens. The crudest was to simply stick your hand inside your oven, fireplace, or hearth and see how much it hurt, but this has (hopefully obvious) downsides. In France in the 1800s CE, pieces of white paper were used to measure temperature, with the color they turned in an oven after a set period of time—assuming they didn't instantly combust—giving a measure of how hot things were. "Dark-brown heat" was good for glazing pastries, slightly cooler "light-brown heat" was good for pie crusts, "dark-yellow heat" was best for larger pastries, and the coolest "light-yellow heat" for meringues.

PREREQUISITES

glass, a liquid (water, alcohol, oil, wine, mercury, urine—all these and more have been used)

HOW TO INVENT

Both thermometers and barometers measure things that would otherwise be invisible, and you can invent both of them with just water and glass. For thermometers, the property we'll be exploiting is the fact that (most) liquids and gases expand when they're heated and contract as they freeze: measure this expansion and contraction, and you can measure temperature!

The first thermometers weren't that far from what you'd recognize today: long glass tubes with a ball at one end, with the open end of the tube placed in a bucket or lake when the ball was hot. The air in the ball would contract when cooled—pulling water up the tube—and expand when it was warmed, pushing water down. The problem was, there was no scale: all this device could tell you was whether things were getting warmer or colder. These were thermo-*scopes* (a way to see temperature), rather than thermo-*meters* (a way to measure it).

Knowing humanity's track record as you now do, you will not be surprised that it took more than a hundred years after thermoscopes were invented for anyone to think of applying a fixed scale to them. Two men (Isaac Newton and Ole Rømer) finally came up with this idea independently in 1701 CE, but Rømer's scale was better: Newton used a lot of subjective temperature references ("the heat at midday about the month of July": *what the heck, Newton?*) while Rømer at least used constants like the freezing and boiling points of water as the basis for his temperature scale.*

There's another problem: since the water in a thermoscope is open to the air, it's also susceptible to changes in pressure, which actually makes this invention a *combination* thermoscope and barometer, or "thermobaroscope." Sealing the

---

* You can build a temperature scale around anything (as some of Newton's whimsically useless temperature references illustrate), but physical constants are better. They'll make your measurements consistent and reproducible, even if for some reason you don't have access to Isaac Newton's backyard, in summer, on an Earth whose climate approximates what it was like in England in 1701 CE. See Section 4 to reproduce the centigrade scale of measurement used in this book.

glass solves this by removing air pressure as a factor. With a sealed glass ball at the bottom filled with liquid, and a glass ball at the top filled with air, you can produce a thermoscope immune to changes in pressure. Put a scale on the outside, and you've made a thermometer. Ta-da!

The problem is this: water is weird and doesn't expand and contract linearly. Like most things it gets denser as it cools, but as you drop below 4°C it actually *expands*, which is why ice floats on water instead of sinking: ice is less dense than water.* This makes water suboptimal for thermometers: measurements between 4°C and 0°C will be off, and you won't be able to measure anything below 0 degrees because your thermometer will be frozen. Modern thermometers use mercury—which expands dramatically with heat, boils at a distant 357°C, and doesn't freeze until 38 degrees below zero—but you probably won't have any of that for a while.† Alcohol (see Section 10.2.5) expands more linearly and freezes way down at -173°C, but it boils at a mere 78°C, which is less convenient. You can always use more than one thermometer: alcohol thermometers for cold temperatures and water thermometers for warmer ones. Alternatively, wine—a delicious mixture of water and alcohol—has been used to mitigate the effects of alcohol's low boiling point and water's weirdness.

So that's thermometers! Barometers are basically the same, as you saw earlier when we invented a combination barometer/thermometer by accident.

---

* It's also why you can use lakes to cool your buildings, when your civilization gets to that point! When water reaches around 4°C (3.98 degrees, if we're being exact) it's at its maximum density. This means that all water in a lake that isn't at 4 degrees will float above the water that is, which makes the water at the bottom of a large enough lake all but guaranteed to be a constant 4 degrees. You can pump this water through your buildings in summer as a renewable and efficient way to invent air conditioning. Lakes that are 50m or more deep and far enough away from the equator that they don't get fully heated above 4 degrees work best!

† Mercury is the only metal that's liquid at room temperature. If you do want some, you can extract it from a bright-red mineral named cinnabar, veins of which are typically found near hot springs or sites of recent volcanic activity. It's toxic to humans, so be careful with it! To extract the mercury from cinnabar, crush the rock as small as possible and roast it. Use distillation (Section 10.1.2) to collect and condense what evaporates off the rocks (mercury's 357°C boiling point is achievable even with just a campfire) and there's your mercury! Despite its toxicity, humans have been mining cinnabar since 8000 BCE, although back then it was used as a pigment: crushed cinnabar produces a vibrant shade of red now called vermillion.

Take a hollow tube, fill it with liquid, seal one end, and put the open end in a body of the same liquid: that's a barometer. The weight of air pushes down on the liquid outside the tube, which prevents all the water inside the tube from draining out.* Higher densities of outside air cause water to rise in the tube—which is how the barometer measures air pressure—and the vacuum at the top allows the water to expand into the tube easily. And while this barometer works well with liquid mercury, if you use liquid water—a more easily available but much less dense substance—you'll need a tube that's around 10.4m high for it to work: any shorter and all the water will run out of the tube before the outside water holds it in place.† A more clever approach for using water to measure air pressure is the following design, called a "Goethe barometer." It was invented by a human named Johann Wolfgang von Goethe in the early 1800s CE, but is now going to be invented by a human named your name in whatever time period you're in:

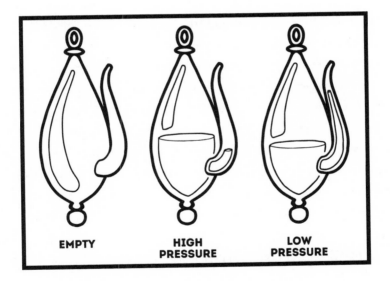

EMPTY   HIGH PRESSURE   LOW PRESSURE

Figure 29: Behold, your latest invention: a good barometer.

* It wasn't until barometers were invented that humans realized air even had weight—up till then it was assumed it was weightless. After all: it floats, right? But all matter has mass, and air is just a layer of gas pressed to the Earth's surface by the same gravity that's holding you there too.

† With mercury your column has to be only around 76cm high: much more achievable, assuming you've got lots of mercury splashing around nearby!

This is just a glass container with a spout, open to the air, that reaches up and above the container itself. Lay the barometer on its side and begin to fill it with water: you're getting water into it while allowing whatever air that water is displacing to escape, so the air left inside will still be at your current atmospheric pressure. Once it's halfway full, turn it upright. Water will fill the bottom of your barometer, trapping the current-pressure air inside. The spout indicates pressure: if the outside air pressure is *lower* than when you filled your barometer, the water in the spout will rise, because the air *inside* the barometer is under comparatively higher pressure. Similarly, if current air pressure is higher, the water in the spout will drop. Fill your barometer with water on a calm, average-pressure day and you'll have a terrific barometer that'll work for as long as you keep it filled with water: you'll need to top it up through the spout occasionally to account for evaporation.*

Your main use for barometers will be in predicting the weather, and you don't even need units for that: a rapid drop in pressure is associated with clouds, winds, and storms, and a rapid increase in pressure indicates that bad weather is about to be pushed away. Hey, you just invented short-term weather prediction! For longer-term weather prediction you'll need more complicated technology, but don't worry about it too much: predicting Earth's weather long-term is not just difficult but *actually impossible*. Even with the entire atmosphere filled with a perfect lattice of point-sized sensors from the surface all the way up to 100km above land, each positioned 1mm apart, and even with all the data they gather computationally processed in an instant, long-term predictions still quickly become inaccurate. Errors grow in scale from 1mm to 10km in less than one day, and from that to planet-sized inaccuracies within only a few weeks.[31] You will likely discover that it's much cheaper, and of approximately equal predictive power, to answer all long-term weather questions with a simple "Sunny, with a chance of clouds."

---

* Draw a line on your barometer and you'll know what the water level should be kept at. It helps if you color your water to make it easier to read (pigments can be made in Section 13). Since we're dealing with expanding liquids, your barometer will give different measurements in different temperatures, so you'll need to either control the temperature of your barometer or take that into account if temperatures swing wildly.

10.8

"I WANT PEOPLE TO THINK I'M ATTRACTIVE"

While there are no grooming or fashion tips included here per se (though we will say that however you present yourself and whatever you choose to wear, confidence is always attractive), we have included instructions on inventing the technologies needed for looking good. This gives you the opportunity to reinvent good taste from the ground up in whatever style you choose.

Soap is an easy way to keep yourself looking (and smelling!) your best and has the side benefit of curtailing the spread of illness in your civilization and dramatically reducing the risk of infection too, so that's nice. **Buttons** are a trivial way to produce form-fitting clothes that still took humans thousands of years to figure out. You can use **tanning** to convert animal hides into sturdy, protective leather, which is useful in manufacturing clothing, boots, water bottles, and more. And finally **spinning wheels** convert natural fibers into thread, which can be sewn to produce clothing as humble as a potato sack and as fancy as a kimono made from the finest silks. Your civilization can really have it all.

After all . . . while you may be trapped in the past, that's no excuse for not looking *fabulous*.

10.8.1: SOAP

Things are beautiful if you love them.
—You (also, Jean Anouilh)

WHAT IT IS

A substance that keeps you clean, in both the "get that dirt off you" sense and the "thanks to the germ theory of disease we know that even superficially clean skin can still carry harmful microbes so wash your dang hands with soap and water before you stick them in your mouth" sense

BEFORE IT WAS INVENTED

Washing, bathing, bacteria-avoidance, and general cleanliness were more difficult, because there was no substance that would lift up oils in water. On the plus side, you could visit your grandparents and say whatever swears you wanted all day long and they couldn't wash your mouth out with *anything*.

ORIGINALLY INVENTED

2800 BCE

PREREQUISITES

for lousy soap: olive oil and lime (see Appendix C.3); for better soap: potash or soda ash, salt; for great soap: lye

HOW TO INVENT

The olive-oil-and-lime "soap" (the "lousy soap" mentioned above) is the easiest to make: just mix olive oil and lime together (or sand, if you don't have any lime), rub it all over, then scrape it off. This isn't so much a soap as it is a "lubricating grease," but it has been used in ancient cultures to help clean skin. It's obviously less useful elsewhere: cleaning clothing by rubbing a mixture of sand and oil into it can, at best, be described as "minimally efficient."

To make actual soap, you'll need potash, soda ash, or lye: these are alkalines you can produce easily with Appendix C. An alkaline is a substance that at the atomic level accepts protons from any chemical donor: they're the opposite of acids, which are substances that donate them.* A neat thing happens when

---

* And just as acids can be extremely acidic, alkalines can be extremely basic. Substances that are extremely basic or acidic can be dangerous to be around, because both of them can react with your (more chemically neutral) flesh. Acids taste sour and tend to feel like they're burning the skin, while bases taste bitter and feel slippery. But as it's really dangerous to identify acids and bases by rubbing them all over yourself and tasting them, instead test for acids by placing

you combine alkalines with oils or fats: you induce a chemical reaction called "saponification." During saponification, the fats chemically combine with the alkalines to form new molecules: long and skinny hydrocarbon chains.* These chains have a cool (and for you, very useful) property: one end loves water and hates oil, while the other end hates water and loves oil.†

You probably already know that oil and water repel each other. Put water inside a greasy pot (or on your greasy, greasy skin) and you can see what happens: the oil and fats compress at the bottom, or float to the top, but they don't mix. That's why water doesn't give you much help in removing grease, which is what drove us (and now, you) to invent soap in the first place.

When your saponified substance (i.e., your soap) meets fats and oils, your hydrocarbon chains surround the grease with their oil-loving ends, forming a tiny sphere around all the grease they can find. Since their oil-loving ends are attached to the grease, that means their water-loving ends point out, which effectively coats grease in a microscopic water-loving shell. Your grease, now water-soluble, lifts up from whatever surface it was attached to and is ready to be rinsed away. These shells (called "micelles") looks like this:

---

a few drops on any carbonate (see Appendix C): acids react to produce bubbling carbon dioxide gas. Bases can be tested for by mixing them with fats and looking for a reaction: in other words, by seeing if you can use them to make soap, just like you're doing right now.

* If you don't know what a hydrocarbon chain looks like, it's not important: just imagine a tiny caterpillar. If you don't know what a caterpillar looks like, just imagine a cute fuzzy worm. And if you have no idea what a worm looks like, we are sorry to inform you that you may have problems that extend beyond the scope of this book.

† While hydrocarbons cannot actually experience the emotions of love and hate, the phrases "loves water" and "hates oil" are easier to understand than the more technically correct terms of "hydrophilic" (attracts water) and "lipophobic" (repels oil).

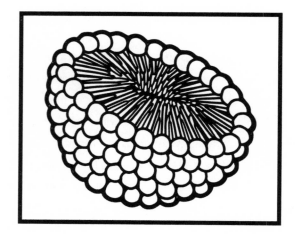

Figure 30: Micelles, which are what make soap possible.

And now that you can explain in detail why soap works—which puts you well ahead of the rest of humanity that made and used it for thousands of years before anyone knew what it was doing—here's how you make it!

The easiest soaps are made with potash and soda ash (see Appendices C.5 and C.6, respectively): just mix them in a pot of boiling fat. Use the fat and grease from whatever animals you're eating, but don't forget to purify them first in a simple but stinky process called "rendering." Take your collected fats and grease, chop them up, put them in a pot, add an equal amount of water, and boil. The fats will melt into the water: once they're all melted away, add more water (about the same amount as the first time), and let your pot cool overnight. The fats will rise to the top of the water—we're using the fact that oil and water don't mix to our advantage here—and the impurities will sink to the bottom. That top layer of purified fats is what you're interested in.

Scoop that off, put it in a new pot, bring it to a boil again, and add in your potash or soda ash. Stir until it's well mixed, which can actually take several hours. Now you've got two options: you can let it cool, which produces a soft, jelly-like brown soap, or you can throw some salt in and then let it cool. This causes little pieces of soap to solidify out of the liquid and congeal on top, which will give you a harder, more pure soap that's also easier to store. You can optionally purify your hard soap even further by boiling it in water and doing that salt trick again to precipitate it out.

You'll make the best soaps with lye instead of potash or soda ash. Lye is more strongly basic, so it makes more effective soaps. It can be tricky to know when you've got lye in the right concentration, but historically soap makers have used the test of "will an egg or a potato float in it" as a rough metric for the proper concentration. Add more water to a lye to weaken it, and boil it down to strengthen it. And that's it! You're done! You're a soap master!

Soap will allow you and your civilization to keep infection and disease at bay with much greater effectiveness than you could without. You also have the advantage of inventing this stuff in the right order: usually *surgery* is invented before humans figure out that washing their bacteria-covered hands in soap and water before shoving them inside another human is a good idea,* so you're doing great. For an extra-tough clean, use alcohol (Section 10.2.5)—it's also an antiseptic and can be used after washing with soap to get that surgery-level clean doctors crave.

---

* This idea was first proposed by a Dr. Ignaz Semmelweis in 1847 CE. He worked at a hospital with two maternity clinics: one staffed with midwife students, and the other with medical students who performed autopsies before assisting in births, all while never washing their hands. After noticing the mothers at the med student clinic would get vaginal infections *so horrible it killed them* as much as 30 percent of the time (compared to around 5 percent at the midwife clinic), Doc Semmelweis introduced a hand-washing regimen. Death rates from infection dropped to 1 percent in both clinics. At the time, the causes of disease were considered to be unique to each patient, and the notion that disease could be prevented simply by washing hands was thought extreme. After being dismissed from the hospital, Semmelweis wrote letters to other doctors urging them to wash their hands, and when that failed, he wrote new letters denouncing them as murderers. For his efforts he was committed to an insane asylum in 1865, where he died fourteen days later— from an infected wound he contracted after being beaten by guards. The idea that cleanliness could stop infection didn't gain acceptance until twenty years *after* he died—when we finally realized germs were a thing—and today, the way humans can quickly and almost reflexively reject information that contradicts their established beliefs is named the Semmelweis reflex.

10.8.2: BUTTONS

It is ever so much easier to be good if your clothes are fashionable.

—*You (also, L. M. Montgomery)*

WHAT THEY ARE

A way to fasten clothes shut, as well as temporarily seal things together. They're also used for fashion purposes!

BEFORE THEY WERE INVENTED

Clothing was either held closed by rope, or if not it was really baggy and shapeless because you needed to be able to get it on and off over your head

ORIGINALLY INVENTED

2800 BCE (for decoration)

1200 CE (as fasteners)

PREREQUISITES

thread

HOW TO INVENT

Buttons are created by taking any rigid material (wood and seashells work well) and loosely fastening them to a piece of clothing. Then—*and this part is critical*—you simply weave, knit, or cut a hole into another part of the clothing that you want to seal, and then the button can slide through it and clasp it shut.

Look, you know how a button works. We don't need to explain this. They're one of the simplest practical inventions we have . . . but figuring out how buttons work still took humans more than *four thousand years*.

Since 2800 BCE, buttons were used as "nice shells that we put on our clothes to look good" and it was only around 1200 CE in Germany when someone finally realized they also had a practical purpose. That's multiple millennia in which humans walked around with buttons sewn on their shirts, thinking they all looked pretty sharp, when actually they looked like big idiots *who didn't even know how a button works.*

Buttons could've been invented at just about any point in human history. Save humanity from doing the cultural equivalent of walking around with our fly down for four thousand years straight. Invent buttons already.

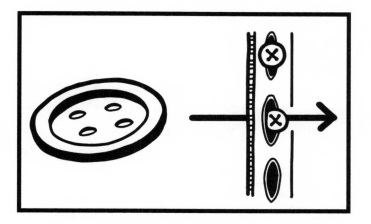

Figure 31: A button. There. Now you have no excuse.

10.8.3: TANNING

I remember a very important lesson that my father gave me when I was twelve or thirteen. He said, "You know, today I welded a perfect seam and I signed my name to it." And I said, "But, Daddy, no one's going to see it!" And he said, "Yeah, but I know it's there."
—*You (also, Toni Morrison)*

WHAT IT IS

a way to transform animal skins from rotting flesh into *rich Corinthian leather*

BEFORE IT WAS INVENTED

Animal skins broke, stank, rotted, and were less comfortable to wear. Plus: no cool leather jackets.

ORIGINALLY INVENTED

7000 BCE

PREREQUISITES

trees (for tannins), animal husbandry (optional, but it gives you a steady source of hides), salt (optional)

HOW TO INVENT

You might think, "Oh, I'm trapped in the past, time to kill a lion and do that thing where you skin it and then put its head on top of your head so you can wear a lion's head like a hat." This is a bad idea. Without tanning, animal skins quickly rot, and even dried ones become hard, inflexible, and brittle. Tanning transforms these skins into leather: a substance so resistant to rot that leather shoes from 3500 BCE have survived into the modern era. It's something you'll definitely want to do, but you should keep in mind that preparing animal skins for tanning involves not only *skin fermentation* but also soaking them in urine *and massaging them in poop slurries*, so maybe set up your tanneries downwind.

Immediately after slaughtering the animals, lay the skins flat and cover the fleshy side in salt or sand, which will dry them out and delay decomposition. In a few days the hides will become hard and almost crispy, and they can then be transported to your tanning area. Once there, you'll soak the hides: this cleans off dirt and gore and softens them up again. Scour the skins to remove any remaining flesh, then soak them in urine—this loosens the hair, which can then be scraped off. You'll make the poop slurry we advertised earlier by mixing poo and water,* and then soak your skins in that: enzymes in the poop will cause the skins to ferment, softening them and making them more flexible. You can help this process by standing in your poop slurry and kneading the skins with your feet—just keep telling yourself you're crushing

---

* As we saw in Section 5, you'll want to use animal poops, and not human poops, to minimize the transmission of disease.

grapes—but be sure to wash up with soap and water afterward, or better yet use nonhuman power (like a waterwheel) to knead the skins for you.

After all this, two things will have happened: your animal skins will be soft, flexible, and ready for tanning, and nobody will want to get anywhere near you.

To tan these skins, you'll need to collect some appropriately named "tannins," which come from trees containing the equally appropriately named "tannic acid." The bark from oak, chestnut, hemlock, and mangrove trees is high in tannin, as is the wood of cedar trees, redwood trees, and more. Tannins are brown, so if you're using wood instead of bark, look for red- or brown-colored wood, and keep in mind that hardwoods generally contain more tannin than softwoods. To extract tannins, shred your wood or bark and boil it in water for several hours. If you've produced it already, adding baking soda (Appendix C.6) to your water will make it more basic, which will draw out the tannins more effectively. You can repeat this process several times with the same bark to produce more dilute tannic solutions.

After all this, the tanning process is actually the easiest part: just stretch the animal skins out and immerse them in your tannic solutions of gradually increasing concentrations for a few weeks. During this process, the stretched-out skins trade their moisture for tannins, altering the hide's protein structure to make it more flexible, more resistant to rotting, *and* water resistant. And that's it: you've produced leather! Leather's useful not just for cool jackets; it's also for shoes, boots (you can make both entirely out of leather), harnesses, boats, canteens (leather can hold water without leaking, and it won't break when you drop it like pottery does), whips (fun fact: a whip-crack is actually the sound of a small sonic boom when the tip of the whip exceeds the speed of sound, so technically you're already inventing supersonic technology), and protective armor.

If you prepare your skin but don't tan it, you'll produce rawhide, which softens when wet but which has the useful property of hardening and contracting when dry. You can exploit that for binding: to attach a blade to a stick and make an ax, just wrap a strip of wet rawhide securely around both and let it dry. Besides making a delicious treat for your dogs (which we can only *assume* you've already started breeding; please see Section 8.6 right away), rawhide is useful in making drum skins, lampshades, and primitive horseshoes, and has even been used for casts. If you're doing that, though, be sure to leave some room for it to contract: wrapping limbs in too-tight rawhide has been used for torture. That's right: while rawhide contracting around hands and feet won't produce enough pressure to break bones, it will produce enough pressure to *move those bones to places they don't normally go.*

Anyway, enjoy the fun and practical technologies of leather and rawhide and try to forget the part where you splashed around in watered-down poop.

10.8.4: SPINNING WHEELS

> The spinning wheel is in itself an exquisite piece of machinery.
> My head daily bows in reverence to its unknown inventor.
> —*You (also, Mahatma Gandhi)*

WHAT THEY ARE

A machine that uses *physics* to transform natural fibers (wool, cotton, hemp, flax, silk) into thread at 10 to 100 times the efficiency of doing it by hand

BEFORE THEY WERE INVENTED

Drop spindles were used (a stick with a weight on the bottom and a hook at the top): you'd attach the hook to the wool and spin your stick in the air as you gently pull wool out, letting it drop as it formed a thread, but this took *forever*. Before drop spindles, wool was spun by twisting it by hand, but this took even longer!

ORIGINALLY INVENTED

8000 BCE (drop spindle)

500 CE (spinning wheel)

1500s CE (treadle aka foot pump and flyer)

PREREQUISITES

wheels, wood, natural fibers (so, you'll want farming, to give you easy access to plant breeding or animal husbandry)

HOW TO INVENT

We're not going to mess around here. Your civilization is going to skip over the thousands and thousands of years of spinning wool by hand or with drop spindles and instead jump right to the endgame: fully modern spinning wheels. With them you can make thread much more efficiently. Abundant thread not only lets you do obvious things like "make fabrics completely out of thread so you don't have to wear dead animal skins all the time"; it also unlocks less obvious perks like:

- giving you the ability to stitch shut wounds without it being a big deal
- making candles by dipping thread in warmed wax or fat
- inventing fishing line, so now you can easily catch fish
- inventing nets, so now you can easily catch fish in a different way, plus birds too
- inventing quilted armor, which works great for cushioning impacts from clubs but does become completely useless when swords get invented
- flying (see Section 10.12.6)

We'll assume you're using wool here—it's the easiest source of natural fibers—but the same principles apply to any other fibers. To get wool ready for spinning, first clean it in soapy water to remove any grease, then comb it. Combing aligns your wool fibers in the same direction, while also breaking up any clumps and puffing it up into a fluffy ball that's ready to be spun.*

---

* Silk doesn't need to be combed, since it's already in thread form! If you're got spinning wheels, silkworms, and white mulberry trees (see Section 7.26), you've got all you need to produce silk on an industrial scale. First, grow white mulberry trees, and harvest leaves from trees five years or older. Spread them on a bed of straw, and allow silkworm caterpillars to go to town on the leaves for the next thirty-five days. When the caterpillars have spun cocoons for themselves, harvest the cocoons, then dip them in boiling water to kill the silkworm. The cocoons can now be unraveled: each is made from a single piece of silk thread up to 1,300m long. These threads can now be spun together on your spinning wheel. It takes about 630 cocoons to produce enough silk for a single blouse, so this will never be a cheap way to not be naked, but it is a nice one! Before the secret of silk production leaks from China around 200 BCE, the information contained in this footnote alone is worth billions of dollars.

The basic idea is to make a cylinder (called a "bobbin") spin, which will then draw wool forward and turn it into thread. A bobbin needs to rotate very quickly to effectively draw wool forward, so we'll attach it to a large wheel—called a "drive wheel"—via a belt. When a larger wheel's rotation is connected to a smaller one, that smaller wheel must rotate faster than the larger one to keep up. By adding areas of different thickness to your bobbin, you can make different-sized "wheels" for your belt to connect to, which will make it spin slower or faster.

To spin your large drive wheel, you can just use your hands—people did for thousands of years—but you'll probably want to skip ahead to inventing the treadle. A treadle is a simple plank that lets you power the drive wheel with one foot, freeing up both your hands. Put a plank at foot level, with a rod beneath it, and use another wooden rod to attach that plank to one of the spokes of your drive wheel. Now when the drive wheel spins the board will rock up and down—and conversely, when you rhythmically pump that board with your foot, you'll turn the wheel. If you want to get fancy, you can add two treadles so that both of your feet can be put to work: just connect each at opposite parts of the rotation, and they'll work like pedals on a bike.

Your spinning wheel, at this point, is the minimum viable product that humans used for a thousand years. To start, pull some of your wool into a thread by hand, and attach that thread to the bobbin. Grip your thread with one hand, and with the other gently pull the wool behind it, stretching it into a thin line of fibers. As the bobbin spins it'll pull those fibers toward itself, winding them into thread. As great as it is, it's not perfect: twisted thread is stronger, but the biggest twist you can get with this setup is the very slight one you can achieve by holding your wool at an angle as it's fed into your spinning wheel. To get better twisting, you need to add one last innovation: the flyer.

The flyer is a simple U-shaped piece of wood that can rotate freely around the bobbin, with hooks on its wings so you can adjust the point on the bobbin that the thread is wound around. As the bobbin rotates and pulls wool in, it forces the flyer to rotate too, at exactly the same speed. All you need to do is change the speed of the flyer: you can do this by adding a brake to it (a belt around the flyer's shaft can be tightened or loosened to adjust its tension), or by adding a separate drive belt to the flyer, spinning it at a different rate. When the flyer and bobbin no longer rotate at identical speeds, a twist is introduced to your wool as it's spun. Flyers were one of the few inventions Leonardo da Vinci came up with that were actually constructed in his lifetime, and now you're doing him one better by inventing them before he was even born!*

Once you have two lengths of thread, you can twist them together to form a stronger twine, using the exact same wheel. Just twist them in the *opposite* direction those threads were spun in, and they'll naturally lock together. You can repeat this process indefinitely, going from twine to rope to industrial cable that can support your entire civilization . . . all thanks to your little spinning wheel.

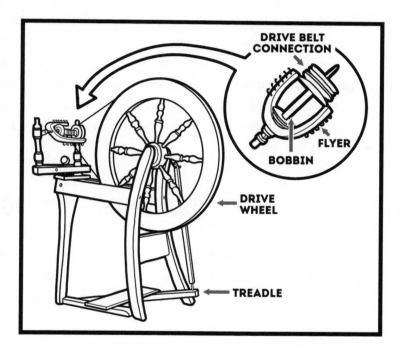

Figure 32: A spinning wheel and flyer.

---

* Here we are assuming you're in a time before Leonardo was around, from 1452 to 1519 CE. However, if you are trapped in the Renaissance, temporal experiments have shown remarkable success in simply slipping Leo a copy of this guide and letting him run with the information inside. He's *super* down for it.

10.9

"I WOULD LIKE TO HAVE SOME COOL SEX"

↗ For many people, sex is a pretty neat part of life, and as it's the only way you've got to produce more humans, sex has a huge effect on both the lives of individuals and your civilization as a whole. **Birth control** helps people in your civilization plan both their families and their lives. Once they decide to have children, **birthing forceps** and **incubators** will help your civilization's newest, youngest, and most precious members survive precisely at the moment when they're at their most vulnerable: during and immediately after their birth.

10.9.1: BIRTH CONTROL

No love-story has ever been told twice. I never heard any tale of lovers
that did not seem to me as new as the world on its first morning.
—*You (also, Eleanor Farjeon)*

WHAT IT IS

A way for families to be planned, instead of them happening accidentally, which lets both women and men decide the course of their lives and freeing them from the tyranny of unwanted and unplanned parenthood

BEFORE IT WAS INVENTED

If you had sex, you might get a kid and then that was your life, you're parents now, congratulations, *sorry about all your plans*

ORIGINALLY INVENTED

1500s BCE (physical barriers)

1855 CE (first rubber condoms)

1950s CE (birth control pills)

PREREQUISITES

HOW TO INVENT

The earliest attempts at physical forms of birth control—which is not to say the earliest *effective* birth control—were pretty basic: women* in ancient Egypt[†] would combine honey, acacia leaves, and lint to use as a physical barrier to sperm that they'd insert before sex. This is actually more effective than it might sound: acacia plants produce lactic acid, which is spermicidal.[‡] If that wasn't available, women were recommended to use crocodile poop as a substitute barrier: in contrast to acacia leaves, this is just about as ineffective as it sounds. In Asia, paper discs were oiled and inserted as early diaphragms—which were at least more effective than other wildly incorrect beliefs about reproduction floating around, including "If women are passive and just lie there without moving, they can't get pregnant" (China, 1100 BCE) and "If women wear cat testicles and/or asparagus as jewelry they can't get pregnant" (Greece, as late as 200 CE), and "If women drink a man's urine or spit three times into a frog's mouth they can't get pregnant" (Europe, 1200 CE). The Roman Empire used female-inserted

---

* Here and throughout this guide, we use "women" as shorthand for "people with vaginas" and men as shorthand for "people with penises." Of course, not all women have vaginas, and not all people with vaginas are women. Language! It's not as good as it could be!

† In many early cultures—including ancient Egypt, Greece, and Rome—birth control was usually solely the woman's responsibility. Do better!

‡ Acacia plants are trees or bushes with small, bright-yellow flowers and multiple leaves sprouting from a single stem, like ferns. They evolved around 20,000,000 BCE and are native to Australia and Africa. Not sure if you've found acacia? Human sperm are big enough to show up under microscopes (Section 10.4.3), so you can test different plants until you find one with spermicidal properties: sperm will stop wiggling their tails only when they die.

physical barriers too, but when their empire fell in the 500s CE, knowledge of this technology was lost (much like concrete in Section 10.10.1) until it was reinvented in the 1400s CE.

Early attempts at male birth control were made and included dipping the penis in lemon or onion juice or covering it in tar (Europe, 1000 CE): an idea that was actually reinvented in the 2010s CE in the form of "Put a sticker on the tip of the penis to temporarily seal it closed" technology. Just to be clear: none of these are effective either. You can make condoms from linens, silk, or animal intestines, but they will be less effective than what you're used to: unlike latex, these substances are more porous and still let some sperm through.

Unfortunately, the effective birth control you remember relies either on chemical changes (like the pill) or strong, flexible, and impervious barriers (like latex condoms), neither of which you're likely to have for a long while.* Various herbs have been used throughout history as a way to reduce pregnancies, but many are poisonous, and others cause birth defects if fertilization has already taken place. Wouldn't it be nice if, somewhere on the planet, a plant evolved that could prevent pregnancies with 100 percent accuracy and with no unwanted side effects?

If you are muttering to yourself, "Yes, absolutely, that would be very nice," then good news: this plant exists! It's called "silphium" and it grows naturally along the coast of what's now Libya. You're looking for something with a thick stalk, a rounded head of flowers at the top, and distinctive heart-shaped fruit

---

* If you're near rubber plants (Section 7.19), you can at least have condoms a bit sooner. The pill works by introducing (synthetic) pregnancy hormones into the female body, making it behave like it's already pregnant, and thereby preventing the release of an egg. The hormones you need are estrogen and progesterone. Estrogen can actually be farmed from the urine of pregnant horses—and is also useful for treating the symptoms of menopause—but progesterone is much more difficult to synthesize. But just in case you get there, the chemical formula you're aiming to produce is $C_{21}H_{30}O_2$. Even knowing something like the fact women have eggs puts you ahead of the game throughout most of history: in Greece around 350 BCE, Aristotle believed that men provided the "seed" and women only the "nourishment." In Europe around 1200 CE, these ideas were still being debated, with the opposing side at least giving women a *little* more credit: they believed that male seeds and "weaker" female seeds combined somehow to produce new people. It was only in 1827 CE (!) that the fact that human women carry eggs inside them was confirmed.

pods. It resists cultivation, but it's *so effective* at birth control that it was valued more than silver and considered to be a gift from the god Apollo by the ancient Romans.

And then they ate it to extinction by 200 CE.

Nice going, ancient Romans. If you're around before they are, silphium is your perfect choice for birth control.[32] But if you're not, you still have options. They're not as good as what you're used to in the modern era, but they're better than nothing:*

Technique	What this is	Effectiveness
Withdrawal	Removing the penis from the vagina before ejaculation. It doesn't work great because it relies on timing, judgment, and some sperm make it out before that happens anyway.	78 percent effective, which is to say that if 100 women use this technique every time they have sex, on average 22 of them will end up pregnant.
Rhythm	Having sex only when the woman isn't fertile. This is more effective today than it used to be, because at least now we know women are most fertile around ovulation, 12–16 days before menstruation! Before this was confirmed in the 1930s CE, theories of peak fertility included an idea that women were most fertile during and just after their periods. Rhythm method practitioners would therefore have sex in the weeks before menstruation, which is of course when women are at their most fertile. Needless to say: as a birth control technique, this wasn't very successful.	76 percent effective. But remember, as with all these techniques, you can use more than one at the same time to better your odds.

* That said, don't make the mistake of thinking that it's only us in the modern era who were smart enough to figure out contraception! In many times and places there's folk tradition of these techniques, along with knowledge of plants like the silphium already mentioned, passed down through a female oral tradition: mothers telling their daughters what they needed to know to control their reproduction. A salad made from these antifertility plants and herbs could give women control over their own reproduction, while men—eating from the same bowl—would suffer no ill effects and may not even realize why those particular plants were being served by the women in their lives.

Technique	What this is	Effectiveness
Breastfeeding	Maintaining breastfeeding after childbirth for as long as possible, because hormones prevent women from ovulating while they're breastfeeding.	This is only effective for six months after childbirth, but it is 98 percent effective during that time. Also the baby must be exclusively breastfed, which means at least every 4 hours during the day and every 6 hours at night; otherwise your body stops producing the required hormones.
Don't put any penises in any vaginas	This option is like abstinence, but you can still do other fun stuff!	100 percent, assuming you can stick to it, and you should keep in mind that historically, humans who are interested in doing penis-in-vagina stuff are really bad at not doing any penis-in-vagina stuff whenever they get the chance.

Table 13: A list of birth-control techniques you can use if you don't have the *markedly superior ones* we already invented in the modern era. Hey, did you bring back some condoms or an IUD with you? Use them!

Finally, we should stress that none of the techniques in this table protect against sexually transmitted infections, which you'll need to watch out for. Syphilis, in particular, should be avoided: it has much more awful strains that died out before the modern era. When it first appeared, syphilis sufferers would be horrified as their entire bodies became covered in pustules, and that was before the flesh would *fall from their faces*.* Penicillin (Section 10.3.1) is an effective syphilis cure, though in our timeline we only found that out centuries *after* the "face-fall-off" variant had died out.

And on that note, we'll end this section.

We hope your civilization enjoys some really cool sex!

---

* There are many diseases that are much more deadly in the past (which is to say, your present) than they are today (which is to say, your distant future)! The reason is simple: strains that are too deadly tend to kill their hosts before they can spread and therefore die out, leaving only the less-fatal strains to survive. And it's not just syphilis: some diseases, like sweating sickness, had forms so infectious and deadly that death would occur only hours after symptoms first appeared. We know that this is probably not the sort of thing you want to read given your current circumstances, which is why we hid this bad news here in a footnote in an apparently unrelated section. If it's any relief, sweating sickness first appeared in 1485 CE and had died out entirely by 1552 CE, and syphilis doesn't appear until the 1400s CE. The diseases you're likely to come across will be different from these ones, and therefore surprising!

10.9.2 BIRTHING FORCEPS

> No act of kindness, no matter how small, is ever wasted.
>
> —*You (also, Aesop)*

WHAT IT IS

A pair of tongs that can be used to grab things inside a body, which is particularly useful during difficult births

BEFORE IT WAS INVENTED

Mothers and children suffered from what would otherwise have been easily preventable deaths

ORIGINALLY INVENTED

1500s CE, but kept secret for more than 150 years because multiple generations of monstrous men in the inventor's family wanted to bring the *entire profession* of midwifery under their control

PREREQUISITES

alcohol, soap, metal (wood can be used but is much harder to clean, which can cause infection)

HOW TO INVENT

Birthing forceps are a simple invention: detachable tongs with curved edges that can be positioned around a baby's head, used to rotate and then gently remove a baby from the birth canal. Birthing forceps allowed difficult or obstructed births to be successful, saving the lives of both mother and child. Even though this technology was invented remarkably late—they could've shown up at basically any time after humans were using tools—birthing forceps were still kept secret for generations after their invention so that the family of the inventor could personally profit. All that was publicly known was that the Chamberlen family had a secret device that could help in childbirth, and the Chamberlen men went so far as to carry these forceps into birthing rooms in a sealed box, only using them once everyone else had been kicked out of the room—except of course for the mother, who was *blindfolded*. It was only after the secret of this invention leaked that forceps were commonly used to help in difficult births, and they were a standard tool until cesarean sections were made less deadly in the 1900s CE.*

Birthing forceps should be used when the cervix is fully dilated and the head of the baby is in the lower birth canal. The mother should be on her back (stirrups can help keep her legs supported). Each half of the forceps is individually inserted and then joined together. The baby's head is then rotated to an optimal birth position (head down, chin tucked into chest, facing the mother's spine, so that the smallest part of the head emerges first), and then pulled from the birth canal with gradual, gentle force.

10.9.3: INCUBATORS

Hello, babies. Welcome to Earth. It's hot in the summer and cold in the winter.
It's round and wet and crowded. On the outside, babies, you've got a
hundred years here. There's only one rule that I know of, babies—
"God damn it, you've got to be kind."
—*You (also, Kurt Vonnegut)*

WHAT THEY ARE
A warm box you put babies born too early in that reduces the chances of them dying by *almost one-third*

---

* C-sections had been practiced for thousands of years before that time, of course—but only as a last-ditch effort due to their insanely high maternal mortality rate (85 percent and above in England in 1865 CE, approaching close to 100 percent a few centuries before there and elsewhere). This was due to a lack of medical knowledge, antibiotics, poor or no anesthesia, and horrible surgical cleanliness. Once those problems were handled, C-sections could become almost routine, and by the early twenty-first century CE were used in over one-third of all births.

BEFORE THEY WERE INVENTED

People looked at the exact same thing being used in chicken incubators and thought, "Nah, it'd never work"

ORIGINALLY INVENTED

2000 BCE (for chickens)

1857 CE (for humans)

PREREQUISITES

glass, wood (for construction), soap (for cleaning them between babies), leather (for a warm water bottle), thermometer (optional)

HOW TO INVENT

Incubators were first invented around 2000 BCE in the form of houses and caves that were kept warm to help eggs hatch. By this point humans had noticed two things: chickens were delicious, and eggs kept at a warm temperature by hens tended to hatch more often, which produced more delicious chickens. Incubating houses were a way to scale that process up.

However, it wasn't until almost four thousand years later that anyone noticed that human babies, if born prematurely, would *also* benefit from a consistently warm environment that emulated their mothers' wombs. Before this point, premature babies were simply handed off to parents and midwives, and everyone just kinda hoped for the best. And yes, while modern incubators are complicated machines, supplying oxygen, heat, moisture, and intravenous nutrition while simultaneously keeping a constant eye on baby's heartbeat, respiration, and brain activity, you don't need all that complexity to make a real difference in any time period you're stranded in. The first baby incubator was just a double-walled tub that was periodically refilled with warm water to generate heat.

By 1860 CE the design had evolved to be heated by a water bottle, with one more critical innovation: a glass lid. This reduced random airflow while still allowing babies to breathe, which helped protect them from airborne infection, draft, noise, and the excess handling from nurses that can also be a disease vector. Something as simple as a glass box with a warm water bottle inside had an astounding effect: at the hospital where it was invented, infant mortality dropped by *28 percent*. If you've got thermometers (Section 10.7.2), you can quantify their temperature: human baby incubators are typically kept at 35°C, but if you're raising chickens, 37.5°C tends to be ideal.

If you consider the goal of health care to be giving another person more years of life than they'd otherwise have, then just helping premature babies survive is the most effective and efficient health care you can provide. You won't be giving an adult a few more years: you'll be giving a newborn baby their *entire life*.

And all you need is a small bed, nestled in a warm box.

"I WANT THINGS THAT WON'T CATCH ON FIRE"

↗ While the inventions in this section have many uses outside of building fireproof buildings, they can help with that problem too. In fact, **cement and concrete** are building materials that, despite being inexpensive, still allow you to construct buildings that stand for more than a thousand years. Even more useful is **steel**, an incredibly strong and versatile substance that gives your civilization the ability to construct everything from bridges to ball bearings. Finally, **welding** allows things larger than can be contained in any kiln to be built and for those constructs to be as strong as if they were made from a single piece of metal.

It is with these technologies that the modern era begins to be restored, so we're really glad you're inventing them.

10.10.1: CEMENT AND CONCRETE

The ideal building has three elements:
it is sturdy, useful, and beautiful.
—*You (also, Marcus Vitruvius Pollio)*

WHAT THEY ARE
Building materials you might think of as boring until you realize they can be described as *liquid rock*

BEFORE THEY WERE INVENTED
Rocks had to be laboriously cut into whatever shapes you wanted, rather than just pouring liquid into a mold, waiting for it to cure, and calling it a day

ORIGINALLY INVENTED

7200 BCE (lime plaster)

5600 BCE (early concrete, used for flooring in Serbia)

600 BCE (hydraulic cement)

1414 CE (rediscovery of cement and concrete)

1793 CE (modern concrete)

PREREQUISITES

kilns (for heating limestone), volcanic ash or pottery (for cement)

HOW TO INVENT

By following the instructions in Appendices C.3 and C.4, you can convert lime-stone into quicklime, and quicklime into slaked lime—which reacts with carbon dioxide in the air to harden on its own. Add some clay (or sand and water) to your slaked lime, and you've just invented mortar: an easily spreadable paste that dries like stone. Replace some of that sand and water with straw or horse-hair to increase its tensile strength and you've invented plaster: a substance durable enough to be used for exterior coverings that is also waterproof once it's cured. This makes plaster a great way to build underground food storage: food stays cool, and the plaster keeps any water out.

But all these technologies require air and time to fully cure: plaster can take months! The solution is to add aluminum silicates to your mortar. This creates hydraulic cement: a mortar that not only cures faster and is water resistant but can also cure *underwater*, which is obviously extremely useful when you want to build lighthouses, breakwaters, and other water-adjacent buildings. Aluminum silicates are found in volcanic ash and clay, so if you've got volcanic ash lying around, you can just mix it in with your mortar. If not, take old pottery, crush it up, and add that instead. Horsehair can be added to prevent cracks (just as in plaster), and you can add animal blood too, which will produce tiny bubbles in the cement that make it more resistant to the stresses of freeze-thaw cycles.*

---

* It sounds crazy, but it works! Because cement is alkaline (see Section 10.8.1), when it cures, it reacts with the fats in the blood to make what are effectively

Cement's great, but you can make it even better simply by mixing gravel, stones, or rubble into it. That's concrete! This simple addition of *literal garbage rocks* actually makes the cement much stronger: the rocks carry more of the load, allowing greater and larger structures.* Besides buildings, concrete can also be used to create paved roads. Remember to give your roads a slight slope on each edge (like a roof) and water will drain off, which helps prevent puddles and icing.

Cement and concrete reached an early peak in the Roman Empire, but after that empire fell around 476 CE, the technology was all but lost for a millennium. There were some cement structures built after that date, but the knowledge required was kept within guilds, rarely written down and never disseminated. It was only when an obscure Roman manuscript from 30 BCE (written by the architect and engineer Vitruvius, whose quotation graces this section) was rediscovered in a Swiss library in 1414 CE that the secrets of cement and concrete were recovered.[†] It took a few hundred years more—until 1793 CE—for that "heat limestone to produce quicklime" discovery to be made, which made cement and concrete simpler to produce. You can easily improve on humanity's actual history by *not forgetting how to make concrete for a thousand years.*

You may, for example, choose to store the recipe in a more popular library.

---

tiny flakes of soap, leaving the bubbles behind. Technically, *any* blood will work, but use animal blood, okay? You don't need to use human blood here, we promise.

* While concrete is very strong under compression (forces that squeeze it together), it's weak under tension (forces that stretch it apart). This makes it great for load-bearing walls (where the load acts like a compressive force) but less so for beams or aboveground flooring (where the floor's own weight works to bend and finally break the concrete in two, causing collapse). You can fix that by adding reinforcements to your concrete before it sets! Lateral bars will add tensile strength to your structures: if you've got the metal, steel reinforcing bars (or "rebar") work great, and bamboo can work in a pinch too. Nobody thought to do this until 1853 CE!

† The illustrations to Vitruvius's text had been lost, however, so artists created new ones: including, eventually, Leonardo da Vinci. His gorgeous and striking "Vitruvian Man"—that picture you've probably seen of a full-frontal naked guy with extra limbs superimposed and a circle and square drawn around the whole business—was intended to illustrate that humanity's proportions were as perfect as those two ideal shapes (they're not) and that, in a larger sense, the way the human body works is analogous to how the universe works (it isn't).

10.10.2: STEEL

> The solutions all are simple—after you have arrived at them.
> But they're simple only when you know already what they are.
> —*You (also, Robert M. Pirsig)*

WHAT IT IS

An alloy of iron and carbon that's sturdier than either of those two elements alone, with an incredible tensile strength: the ability to withstand heavy loads without snapping or being pulled apart. Need awesome buildings, tools, vehicles, machines, or anything else? *Maybe consider steel.*

BEFORE IT WAS INVENTED

Everyone had to "steel" themselves for much more disappointing building materials

ORIGINALLY INVENTED

3000 BCE (iron smelting)

1800 BCE (earliest steel)

800 BCE (blast furnaces)

500 BCE (cast iron)

1000s CE (earliest Bessemer process)

1856 CE (Bessemer process rediscovered by Europeans, which a European then named after a European)

PREREQUISITES

smelters and forges, charcoal or coke

HOW TO INVENT

In Section 10.4.2, we saw how with a smelter you can melt off non-iron metals from ore to extract iron and how you can then hammer that iron in a forge to purify it. But what happens when you add carbon to it? We'll tell you what happens: carbon interacts with the iron to form an alloy with great tensile strength that also holds an edge. We call it "steel," and it's great for making all sorts of things, such as:

- bridges
- railways*
- reinforced concrete
- wires and steel cables
- nails, screws, bolts, hammers, nuts
- needles
- canned foods
- ball bearings[†]
- saws and plows
- turbines
- forks, spoons, knives
- scissors
- wheel spokes
- strings for musical instruments
- swords
- barbed wire[‡]

---

* Okay, *technically* you don't need steel for railroads, and you can build them out of iron. *Technically.* But know this: when humans tried it, busy railroad tracks made from iron would sometimes need replacing as often as every six to eight weeks. Once we invented steel, the same tracks had lifetimes measured in *years*.

[†] You probably know what these look like enough to invent them, but just in case: ball bearings are tiny spheres that run along a groove between two concentric wheels. They're useful in many machines, like engines and wheeled vehicles (such as bikes, cars, and cool skateboards), because they greatly reduce the friction between moving parts—think of them as the circular equivalent of moving a heavy rock by rolling it over logs instead of just shoving it along the ground. Put each ball in a cage that prevents it from rubbing up against the other balls, and you've reduced friction even more! Ball bearings like this normally get invented around 1740 CE, but da Vinci was messing around with them as early as the 1500s CE.

[‡] Barbed wire is the first wire capable of keeping cattle fenced in—they get poked by it once and learn to keep their distance forevermore—and it's way cheaper than building a full fence or planting kilometers of hedges. As advertisements put it at the time, barbed wire "takes no room, exhausts no soil, shades no vegetation,

- two swords hinged together so you can use them like a giant pair of scissors
- and more??

Different amounts of carbon give different alloys, and only alloys with carbon levels between 0.2 percent and 2.1 percent get the "steel" label. Even within steels, different carbon levels give different hardnesses and tensile strengths, so you can experiment to find the kinds you like. Kitchen knives—that can hold a tough edge and won't break easily—have around 0.75 percent carbon.

To introduce carbon to iron to make some sweet, sweet steel, you could pack your iron into boxes of powdered charcoal and heat them to 700°C for about a week. The charcoal's carbon will react with your heat-softened iron, producing a thin layer of steel. However, only the *exterior* of your iron will be steel now, so you'll have to fold and flatten your metal on the anvil again, thereby "stirring" the metal to produce a uniform material. This is obviously a slow and expensive process that requires you to have already hammered and flattened metal to make iron, and then do it *again* just to get some steel. It may not surprise you to learn that hitting metal with a hammer for hours on end is a long, hot, difficult, labor-intensive, and tedious process that sucks, so you're going to invent a better way to do it riiiight . . . *now.*

Hey, congratulations on inventing the blast furnace!

As we're sure you already know, the blast furnace is basically the intensified version of your forge. Instead of your smelter sucking in air, you now force it in through your materials from the bottom up. And instead of alternating layers of iron ore and charcoal, you're layering iron ore, limestone, and hotter-burning coke.* You're producing a more intense combustion that smelts iron ore just like

---

is proof against high winds, makes no snowdrifts, and is both durable and cheap." This simple idea of "make the wire pokey every foot or so" revolutionized farming and allowed animal husbandry to scale up to sizes previously not affordable. While it could've been created at any point after humans began working with metal, it was only invented in the mid-1800s CE.

* Coke is just dry distilled coal, which you can produce (once you've mined it out of the ground) using the same process you used to dry distill wood into charcoal,

your smelter did, but this goes further: the iron reacts with the carbon in the stack, forming a new alloy with a melting point down near 1200°C: low enough to melt in your furnace! The high-carbon liquid iron runs out the bottom and cools, and you've got your metal.

Buuuuuut it's not quite steel. The problem now is you've got too *much* carbon in your iron: you wanted between 0.2 percent and 2.1 percent, and the output of a blast furnace can be as high as 4.5 percent. This high-carbon iron (also called "pig iron") is brittle: too easily broken if bent or stretched to be useful in bridges or buildings, but its low melting point does mean you can pour it into molds to cast frying pans, pipes, and so on. This "cast" "pig" "iron" is called "cast iron," and you just invented it.

To reduce the carbon level of pig iron enough to make steel, you'll be using the "Bessemer process," whose basics were discovered in East Asia in the 1000s CE. The idea then was to blow cold air across the molten metal, and the more modern version (patented in 1856 CE by, you guessed it, *some guy named Bessemer*) is to force air through the liquid pig iron instead, with bellows or air pumps. The air introduces oxygen to the mixture, and the oxygen reacts with the molten carbon to form carbon dioxide. This either burns off or bubbles out, leaving a purer iron behind. And as a bonus, these reactions also generate heat, which heats up the molten metal even more, allowing the reaction to continue even as the melting point of your liquid metal rises.* It's very hard to know when precisely to stop the bubbling air to get just the right amount of carbon remaining, so don't bother: just burn off all the carbon you can—producing a pure iron—and then mix whatever carbon you want back in.

---

back in Section 10.1.1. If you don't have coke, you can still use charcoal—it's what the earliest blast furnaces used—but coke burns hotter!

* Other impurities, like silicon, also form oxides, and these will sink to the bottom as slag. Here's a tip: if your iron ore contains phosphorus (and a lot of the iron ores on Earth do, so . . . *maybe?*), your steel won't be as strong as it could be. To solve this problem, toss in something chemically basic (hey, there's limestone again!) with your ore. It'll react with the phosphorus to form more slag at the bottom, which not only gives you a better steel, but when the phosphorus-rich slag cools, you can grind it up and use it as fertilizer!

Iron is the sixth most abundant element in the universe and the fourth most common element in the Earth's crust, but until humans invented blast furnaces and the Bessemer process, it was impossible to turn it into steel cheaply or efficiently. But you've just figured that out, and now one of Earth's strongest metals is also one of its cheapest. Nicely done! Once your civilization has engineers in it, *they'll definitely thank you for that one.*

A final note on steel: you can produce high-quality steel wire by taking advantage of steel's high tensile strength and using a technique called "wire drawing." All you do is make a rough wire out of steel, and then pull it through a cone-shaped hole, as so:

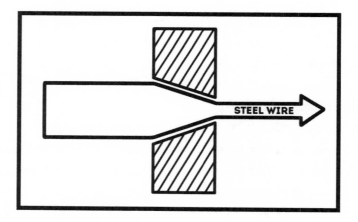

Figure 33: An apparatus to draw wire, as seen from the side.

This produces a wire of consistent area and volume, and that unused mass goes into lengthening your wire. By using several progressively smaller holes, you can produce wires much thinner than you can make by hand. A ratchet (see Appendix H) can be used to pull the steel forward, and conveniently, this can all be done at room temperature: you just need some lubricant.

Here's where it gets embarrassing for us. In the early 1600s CE grease or oil was used, but this required softer steels, and too much friction would cause the wires to break. By 1650 CE, one Johann Gerdes "accidentally" discovered that if the steel was soaked in urine for long enough, a soft coating would eventually develop (we now call this process "corrosion"), which worked to reduce friction when drawing wire. This process—named "sull-coating"—was used for *150 years* until someone noticed that diluted beer actually worked perfectly fine as a substitute, and it was only around 1850 CE that anyone thought to check if water works too. It does. It works perfectly.

Do better than we did. Don't soak your steel in pee for over a hundred years for no reason.

10.10.3: WELDING

When I told my father I was going to be an actor, he said,
"Fine, but study welding just in case."
—*You (also, Robin Williams)*

WHAT IT IS

A way to fuse two metals together in a way that can actually be stronger than the base metals

BEFORE IT WAS INVENTED

Any metal item had to be forged as a single piece, because once it existed, the only way to join it to another one was with bolts and screws, which are much weaker than a good weld

ORIGINALLY INVENTED

4000 BCE (forge welding)
1881 CE (arc welding)
1903 CE (torch welding)

PREREQUISITES

metal, forges, electricity (for arc welding), acetylene (for torch welding)

HOW TO INVENT

Forge welding is easy: just heat the two metals you want to weld to about 50 to 90 percent of their melting point in your forge, at which they're flexible but still solid. The challenge is when metals reach this point, their surfaces tend to oxidize, which prevents a good weld. By sprinkling sand (or ammonium chloride, or saltpeter, or a mixture of all three; see Appendix C) on top of your metal, you solve this problem: they lower the melting point of the oxides, allowing them to flow out from between the two metals as you beat them together. "Beat them together," you say? Yes. This is not a fancy form of welding, hotshot. This is the form of welding where you heat two metals up and hammer them together until they stick. If your arms get tired, you can use a waterwheel (Section 10.5.1) to produce a mechanical hammer that will strike your metals repeatedly.

If you have electricity (Section 10.6.1), you can invent electric arc welding: a less labor-intensive version that also lets you weld items too big to fit in a forge. Arc welding uses the heat generated by electricity arcing from an electrified piece of metal called an "electrode" to the metals you wish to weld. The electrode is placed near the point on the metals you wish to weld, and the arc jumping from it causes them to melt and fuse together. A rod of a filler metal can also be used to join your two metals together, which can make the weld stronger than the base metals themselves. Just ground your metals,* bring an electrode close enough to arc, and weld away. Try to keep the distance your arc needs to jump consistent: otherwise the current it carries will fluctuate, which alters the heat and therefore quality of your weld.

Needless to say, this option can be insanely dangerous, especially if you're stranded in the past and have never worked with electricity before.† You'll probably want to stick with "heating up metals, dumping some sand on them, and hammering them until they stick" for the time being.

---

* To ground something, run a wire that travels from it to a conductive piece of metal inserted in the ground. The earth conducts electricity, and this gives electric current a path to dissipate safely. Without a grounding wire, electric current may reach the ground by traveling through your body instead, which you'll want to avoid, because that's what being electrocuted is.

† There are other forms of welding too, but they're even more dangerous, and probably a bit beyond your current means. Torch welding—melting metals with a flame—lets you not only weld but also cut right through metals, but it requires an extremely hot torch. Burning acetylene gas in pure oxygen is hot enough (3100°C), but producing acetylene requires dry distilling coal to produce coke, combining it with lime at 2200°C (hotter than conventional fires can reach, but which can be achieved by building an electric arc furnace, which is exactly what it sounds

Regardless of whether they're made from paper or electrons, books are critical to civilization. Obviously the guide you're currently reading makes that case literally, but even books of fiction are vital: they are, after all, the stories humans write about themselves.

Paper is the technology that will turn trees into the thin, flexible, easily burned substance upon which you will commit all your discoveries and achievements to posterity. It's also good for wiping your bum. Once you've got it, **printing presses** are the machines on which the knowledge of your civilization will be distributed, debated, shared, and stored. They are an absolutely transformative technology, and critical for any civilization that wants its ideas to have wide and affordable distribution, reliable reproduction, and the ability to survive outside the limits of fragile mortal bodies, which we regret to inform you are precisely the kind of bodies you're stuck with.

---

like), and then combining the output of that process—a powder called "calcium carbide"—with water. That reaction produces acetylene gas, but also heat, and since acetylene gas is explosive, it can be *a pretty delicate operation*.

10.11.1: PAPER

> There have been great societies that did not use the wheel,
> but there have been no societies that did not tell stories.
> —*You (also, Ursula K. Le Guin)*

WHAT IT IS

A cheap thing to write on

BEFORE IT WAS INVENTED

People wrote on the skin of animals (aka "parchment"), which meant that if you were alone and wanted to write a book, you first had to either raise or hunt down an animal and then slaughter it, which obviously slowed the creative process down just a little bit

ORIGINALLY INVENTED

2500s BCE (parchment)

300s BCE (paper in China)[33]

500s CE (toilet paper in China)

1100s CE (paper in Europe)

PREREQUISITES

fabric or metal (to produce a fine-mesh screen), wood, rags, or other natural fibers, waterwheels (for grinding pulp), sodium bicarbonate or sodium hydroxide (optional, speeds up pulping), pigments (optional, but you'll probably want ink once you have all this paper lying around; see Section 10.1.1: Charcoal)

HOW TO INVENT

Before you invent paper, you can jot down notes on animal bones, strips of bamboo sewn together into scrolls, parchment (if you've got the time and inclination to de-hair skins and stretch them until they've dried; see Section 10.8.3), silk (if you've domesticated silkworms), wax tablets (you can get the wax from bees, or by boiling fat in water, letting it cool, and using the waxy stuff—"tallow"—that solidifies on top), clay tablets (which you then fire if you want the information to stick around; see Section 10.4.2), or on papyrus (see Section 7.16). But these media are all heavy, awkward, expensive, hard to transport, or some combination of the above. What you really want is something light, convenient, cheap, and ubiquitous enough that even if it doesn't literally grow on trees, it's at least made out of their ground-up bodies. What you really want is paper, which gives your civilization not just books, magazines, and newspapers with the printing press but also unlocks playing cards, paper money, toilet paper, paper filters, kites, party hats, *and more.*

The basics of paper making are pretty simple: you'll be taking plant fibers, breaking them up, and then re-forming them into thin sheets. Anything with cellulose in it will work, and as all plants that photosynthesize produce cellulose as part of that process, it's one of the most common organic compounds in the world. One large tree can be transformed into upward of 15,000 sheets of paper, but there are lots of other sources for cellulose: old clothing and rags, for example, will also make great paper, either on their own or as a way to bulk out wood fibers. Heck, you could've made paper from the lint you collected from your dryer, back in the future where dryers were a thing!

The first step in making paper is to produce a pulp, which is done by breaking your raw material into small pieces (i.e., turning your wood to chips or tearing your rags into shreds). You can let them soak in water for a few days to get the fibers loosened, before grinding or beating your plant fibers down into, well, a pulp. To speed this process up you can add sodium bicarbonate or sodium hydroxide (see Appendices C.6 and C.8, respectively) to water and simmer your wood chips or rags in there, which chemically separates the plant fibers.* Once you've got a watery pulp, stir it to get the fibers moving, then drag a mesh screen up through it—you can make it with either metal or threads (see Section 10.8.4)—which will collect some of the fibers in a flat layer. Flip your screen

---

* This process breaks down the lignin in your plant fiber. Lignin is an organic polymer that binds plant fibers together, but it's also what causes paper to turn yellow with age. Less lignin in your pulp means you may have to add some glue to make your paper stick together better, but you'll get a whiter and stronger paper as a result!

upside down to remove the pulp, press it to remove the water and force the fibers together, and let it dry. You just made paper! And once you've used your paper, you can recycle it by repeating the same process: just tear your paper up again, break it down into fibers, and press a new sheet.

> **CIVILIZATION PRO TIP:** The basic papermaking process (breaking down plant fibers, layering them on a screen, drying them out) hasn't really changed in the thousands of years since paper was invented. Even though you're hopelessly stranded in time, the paper you're making will still share a connection with all other paper on the Earth you left behind, which has a nonzero chance of making you feel at least marginally better!

While paper was invented in China around 300 BCE, the means of its production were kept a closely guarded secret to prevent other civilizations from benefiting from it. By the 500s CE, paper in China had become so routine that people were wiping their bums with it (thanks, toilet paper), but it would still take more than half a millennium before Europeans would learn of its invention, let alone rub it on their dirty, dirty poo bums. It wasn't until 1857 CE that toilet paper was first produced commercially in the United States (before, any old paper could be used, and tearing pages from books was not uncommon), and it was only in 1890 CE that toilet paper was sold in rolls instead of in stacks. To make going to the bathroom more comfortable for members of your civilization—and to prevent them from resorting to cleaning themselves with wool, rags, leaves, seaweed, animal furs, grass, moss, snow, sand, seashells, corncobs, their own hands, or *a communal sponge on a stick**—you'll want to consider inventing toilet paper ahead of schedule.

---

* All these have been used at different points in history, but the Romans did that last one, inserting a sponge on a stick through a hole in the front of the toilet between their legs, which let them wipe without having to stand up. Give the sponge a quick rinse before you use it and you're good to go, assuming you have no idea that germs are a thing!

10.11.2: PRINTING PRESSES

The preaching of sermons is speaking to a few . . .

printing books is talking to the whole world.

—*You (also, Daniel Defoe)*

WHAT THEY ARE

A way to disseminate information en masse both quickly and cheaply, which is great if you want to get into the dissemination-of-information-en-masse business

BEFORE THEY WERE INVENTED

Books were extremely expensive, so only rich people read them, which meant all the non-rich people who might've come up with amazing ideas if only they could metaphorically stand on the shoulders of giants found that they couldn't,* and so civilization wasn't becoming nearly as great as it could if it were harnessing the full potential of every human brain within it, which is *complete baloney*

ORIGINALLY INVENTED

33,000 BCE (stencil paintings of hands)

200 CE (woodblock printing)

1040 CE (moveable type in China)

1440 CE (moveable type in Europe)

1790 CE (rotary press)

PREREQUISITES

pigments (for ink, see Section 10.1.1: Charcoal), paper (for printing), pottery (optional, for building letters), metalworking (to build the press, though this is

---

* This metaphor dates back to 1159 CE, where one Bernard of Chartres expressed the idea in a bit more loquacious form: "We [modern humans] are like dwarves perched on the shoulders of giants [the ancients], and thus we are able to see more and farther than the latter. And this is not at all because of the acuteness of our sight or the stature of our body, but because we are carried aloft and elevated by the magnitude of the giants." This image of accumulated knowledge raising everyone up is so powerful and expressive that humans haven't stopped talking about it for a thousand years.

technically optional and they can be built out of wood), glass (for eyeglasses, so that everyone can read papers, even the farsighted, who might not even realize they're farsighted until you ask them to read tiny letters on paper held in their hands)

HOW TO INVENT

If you've got pigment (which you can get from charcoal in Section 10.1.1), and you've got something you can cut (like paper, but even large leaves work), then you can make stencils of words—and therefore mass produce books—in any time period you care to name.* The earliest stencils humans ever made were of their hands, and some of them survived on cave walls into the modern era. If only someone at the time had thought to invent writing, those same ancient humans 35,000 years ago could've used stencils to carry their ideas, their beliefs, their hopes, their dreams, their successes, their failures, their stories, and their legends into the modern era, instead of just making a record of what their hands looked like. And in case you're wondering what human hands looked like 35,000 years ago, we can tell you with absolute certainty: they looked like hands.

We didn't even need a time machine for that one.

Stencils work okay for printing, but it's hard for fine shapes (which means your books are going to be large-print), plus you need some form of spray-paint (early humans used their mouths; you can use pigment blown through a tube with a nozzle on the end using the bellows described in Section 10.4.2). To avoid these problems, you may want to skip ahead a few tens of thousands of years to woodblock printing, first invented in China around 200 CE. This involves carving an entire image in reverse onto a single block of wood, which is then coated in ink and pressed onto papers, silks, or anything else you care to print on.† Woodblock printing works great for art, but for language it has several

---

* It involves making a different stencil for each page of your book, and cutting in each letter of each word by hand, but it works. And once you have the stencil you can easily make copies until your stencil wears out.

† Printing like this actually could've been invented much earlier, around 500 BCE! Certain maps in Ancient Greece were carved into metal plates: this was for pres-

downsides, not the least of which is how hard it is to correct a mistake. Mess up on a single letter and you may have to re-carve your entire page out of a new block of wood! Nobody's got time for that, and this slow, labor-intensive process means producing a book takes years. And even once you've carved out every page for your book, you still face the challenge of storage: a 2.5cm-thick piece of wood means you need over 404,128.224 cubic centimeters of storage space for this book alone!

For writing you'll probably want to jump ahead straight to moveable type. Here, rather than carving an entire page, you instead produce stamps of individual letters, which you put together on frames to create a stamp for a whole page. Besides solving the problem of storage—you'd only need to hold on to tiny letters, instead of giant wood pages—it also rewrites the economics of printing. Arranging the letters into a page takes minutes, compared to the weeks or months required to carve a page out of wood, so books can be produced more cheaply, and a much greater variety of books could be printed. Before moveable type, most of the texts being printed were religious texts: things that didn't change and that had a huge, enthusiastic, and sometimes legally mandated audience. After moveable type, anyone (with enough money to pay for it) could print anything, which set off one of the largest cultural changes civilization had seen until the invention of the Internet hundreds of years later.

Moveable type existed in China around 1040 CE, but it really only took off when the technology reached Europe a few centuries later. That was due to another innovation: the alphabet. Chinese writing used not a small set of letters representing sounds like phonetic languages do but rather a large set of characters representing ideas, with more than 60,000 different characters found in a single book. Each system of writing has advantages and disadvantages, but Chinese's disadvantage when it came to moveable type was signifi-

---

tige or to make them sturdy enough to survive long journeys. But if more than one copy of a map was needed, then a duplicate was carved into a second metal plate. In other words, the Greeks had everything they needed to invent printing (including the presses, which they used to make olive oil) and could've easily done so, had but a single person thought to duplicate their metal maps by spreading ink on them before pressing them to papyrus. None did.

cant: it's a lot cheaper and easier to keep and sort through a set of 26 different characters than it is of 60,000.*

The letters you'll be printing—your type—can be carved from wood, but this has downsides: wood wears down with regular printing, its grain can sometimes show in the final result, and wood distorts when it absorbs printing ink. Fired clay was used in China and produces strong, durable letters. You can print with either wood or clay letters, but you can also use them as prototypes for new metal type by pressing the letters into either fine sand or a soft metal (copper works well), and then pouring liquid metal into that impression. Printers eventually settled on a standard metal to forge type with: an alloy of lead, tin, and antimony called "type metal" that produces strong, long-lasting letters.†

To typeset, letters are put together into a wooden frame.‡ Once that's completed, printing is done by coating them in ink and pressing them onto paper. To mechanize that, and to get equal pressure applied across the large flat surface, you'll want to invent the screw press. The screw press is simply a giant vertical screw§ connected to a large flat surface at its bottom. Handles are attached to

---

* Of course, no printer would have only 26 different characters. Printers would store multiple copies of each character in compartmentalized wooden boxes—"type cases"—where they'd be kept alongside punctuation, spacers, and other characters. Capital letters would traditionally be stored in a separate case on the top: the origin of calling them "uppercase" and "lowercase" letters.

† Lead shrinks when it cools, which can distort your letters, but adding some tin and antimony produces an alloy that shrinks less and is harder once set. Different printers use different ratios, but a mixture of 54 percent lead, 28 percent antimony, and 18 percent tin was used traditionally, with a mixture of 78 percent lead, 15 percent antimony, and 7 percent tin being used for more durable type intended for longer print runs.

‡ If you think you may be reprinting a book in the future, you can make a metal casting of your page, quickly producing unalterable page-level duplicates that echo the woodcuts we discussed earlier.

§ Screws turn rotational force (moving in a circle) into linear movement (digging in in a straight line). They're easy to invent in theory, but in practice they need an even thread to work. Rather than eyeballing it, make a right triangle (Appendix E) out of paper (Section 10.11.1), and wrap it around a pointed cylinder, starting with the triangle's thinnest point. The shape made by the upper edge of the paper as it wraps is a helix, which coincidentally is also the shape of a screw's thread! Engrave ridges around that helix line, and you'll produce a perfect screw.

the screw at the top, allowing the screw—and therefore the pressing surface—to be raised or lowered by rotating those handles, which transforms that easy rotational force into stronger downward force. They look like this:

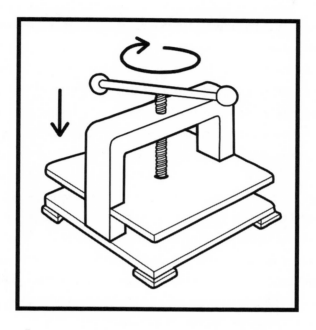

Figure 34: A screw press: an apparatus used in both printing and wine-making, though usually not at the same time.

And as a bonus, once you've got a screw press (which is ordinarily invented around 100 CE), you can use it for all sorts of other things. It works for pressing wood pulp to remove moisture—useful when making paper—but it can also be used in tastier pursuits to press grapes when making wine, or to press olives when making olive oil. Attach it to something smaller than a large flat pressing surface and you can use it to punch holes in metal too.

An innovation that helped make printing presses as successful as they were was the substitution of water-based ink—typically made from soot, glue, and water—with oil-based ones, usually made from soot, turpentine (which you can get by distilling pine resin), and walnut oil (which you can get by pressing walnuts in that screw press you invented two paragraphs ago). Oil-based inks better adhere to metal type, and they don't soak into paper as deeply, which prevents words from becoming blurry. To ink your letters, you can dab them with a flattened ball of ink-soaked leather on a stick (the number of dabs controls how much ink your letters get: an improvement over simply dipping the letters in ink, which douses them in the maximum amount of ink every time), but if you're smart, you'll invent the ink roller, which is a cylinder that can be rolled over your type to distribute ink.*

The faster your press can operate, the more books you can produce. Multiple people working together can operate a simple press at peak efficiency: typesetters set up pages in advance, one printer coats the letters in ink while another feeds in the paper, while yet another lowers the press onto paper to produce the image. Hey, you just invented the assembly line! Your screw press will initially be powered by hand, but it's easy enough to adapt them to steam or electrical power when you have that technology. And when you have the engineering to build it, you'll be able to adjust your press to be a rotary press, an invention that first showed up in 1790 CE. Here, instead of flat type being pressed onto paper, gently curved type is attached to a giant wheel, which rotates onto a strip of paper, pressing letters as it goes.† While a standard printing press requires breaks as each new sheet of paper is inserted, a rotary press can operate indefinitely, as long as there's enough paper and ink to feed it.

The easiest thing to print at first will be posters: notices that can be "posted" publicly (hence the name), allowing fast, cheap, and accurate mass communication across your civilization. Fold, cut, and bind poster pages together and you'll produce books, and the more copies of a book that are printed, the more likely

---

* Despite ink rollers being invented around the 1810s CE, paint rollers—which are the exact same idea but scaled up for walls instead of type—were only invented around the 1920s CE, which is an incredibly late arrival for an invention that's as simple as saying, "You guys, what if we put a furry cylinder on a stick?"

† We say that rotary presses were invented in 1790 CE, but this was actually a European reinvention of Mesopotamian technology that dates all the way back to 3500 BCE. The Mesopotamian invention, now called "cylinder seals," were small cylinders with figures engraved in them. By rolling these cylinders over wet clay, their images were quickly reproduced. Cylinder seals were used in Mesopotamia for everything from decoration to signatures, but unfortunately were never scaled up to mass produce texts.

that information will survive across time. When printing becomes cheaper, it'll be possible to bind a smaller number of pages together on a regular basis to produce a disposable book, or "magazine." These can take the form of scholarly journals that allow scientists to collaborate and share discoveries no matter where they happen to live, or as news and entertainment periodicals to help people to stay informed of current events and celebrity bloopers. Printing will eventually become so cheap, in fact, that it'll one day be profitable to use the lowest-grade paper to print single-use, disposable documents on a weekly or even daily basis, and these will be your world's first newspapers.

The printing press will allow your civilization, and the people within it, to become their best selves: entertained, educated, informed, and up-to-date, so it's really great you thought to invent it just now.

— **10.12**

"IT SUCKS HERE AND I WANT TO GO LITERALLY ANYWHERE ELSE"

 Without transportation, civilizations are small and constrained, unable to fully explore, or benefit from, the larger world around them.

With transportation, however, they can expand, stabilize, and incorporate disparate geographic areas into a cohesive whole. **Bikes** are one such mode of transport: exquisite machines better suited to moving humans around under their own power than even their own legs. **Compasses** let anyone determine the direction they're traveling in, which pairs well with **latitude and longitude**, wherein each location on the Earth is given its own coordinates. With them, anyone can determine their precise location anywhere on the planet. In the absence of clocks that work at sea, **radio** is the technology that makes longitude possible. Finally, **boats** open up the oceans to explorers from your civilization, and **human flight** does the same to the *very skies themselves*.

Invent these technologies, and the people in your civilization will be able to go anywhere they want . . . and find their way back home again.

10.12.1: BIKES

> Let me tell you what I think of bicycling. I think it has done more to emancipate women than anything else in the world. I stand and rejoice every time I see a woman ride by on a wheel. It gives a woman a feeling of freedom and self-reliance. It makes her feel as if she were independent. The moment she takes her seat she knows she can't get into harm unless she gets off her bicycle, and away she goes, the picture of free, untrammelled womanhood.
>
> —*You (also, Susan B. Anthony)*

WHAT THEY ARE

A way for human bodies to move themselves around with three times more efficiency than walking. We'll say that again: *humans invented a way to get around that's actually better than walking around on their own two legs.* We've been dunking on humanity a lot in this book, mainly for taking a really long time to figure out some very simple stuff, but bicycles are a beautiful piece of technology no matter where and when you invent them.*

BEFORE THEY WERE INVENTED

We don't even want to talk about it

ORIGINALLY INVENTED

1817 CE (earliest self-propelled two-wheeled tandem vehicles: you pushed them with your feet)

1860s CE (bicycles with pedals attached to the front wheel)

1880s CE (penny-farthing bicycles with the giant front wheel and the tiny rear wheel)

---

* Even if, yes, it did still take us thousands and thousands of years to dream them up even after we had their basic prerequisites of roads, wood, and wheels.

1885 CE (the so-called "safety bicycle" that had two wheels of the same size, and therefore much reduced the danger of flying off the giant front wheel of a penny-farthing)

1885 CE (the first time an engine was attached to a bike, aka the first motorcycle)

1887 CE (first bike with a chain to power the rear wheels)

PREREQUISITES

wheels, metal (optional, for chains and gears), fabric (optional, for a drive belt), or a basket (optional, for a nice picnic)

HOW TO INVENT

Attach two wheels to a frame you can sit on, one in front of the other. Put pedals on one of the wheels so you can move this whole contraption with your feet, add a seat in the middle, and make sure that front wheel can swivel freely so you can choose what direction you go in. Guess what, you just invented the bike! It may not be the shiny metal one you're used to, but it doesn't matter: some of the earliest bikes were built almost entirely out of wood. Bikes will fundamentally transform society, allowing ordinary people to travel long distances quickly and easily, all under their own power!

We're cheating a bit here: bikes can get a little more complicated. The bike we've described above has the pedals connected directly to the wheels, so each rotation of your pedals corresponds to one rotation of the wheel. Unless your wheel is giant, you'll be pedaling a lot to move only a little, and there are two ways to solve this problem. The easiest option is to make your drive wheel bigger—that's where those old-timey penny-farthings came from, with that colossal wheel up front—but this results in a (ridiculous but arguably awesome-looking) bike with a very high center of gravity. You'll fall a lot, and you'll have a lot farther to fall when you do.

A better solution is to add one more wheel to your bike: a small pedal wheel that goes in the middle of your bike at foot level, connected to a rear drive wheel. This way you can add gears around that rear wheel, which changes the amount of work one rotation of your pedals does. These days this connection is done through chains that fit around teeth in gears, but if you don't have metal-working yet, you can use a drive belt instead: a loop of fabric wrapped tightly around both the pedal wheel and the rear wheel.*

If you go with chains and gears, you'll want to invent the "derailleur," which is simply a moveable chain guide that rests between the pedals and rear gears. When it's moved horizontally, it induces the chain up or down from one gear to the other, allowing you to shift gears while the bike is in motion. Without a derailleur, you'll need to stop your bike to adjust your gears manually—which you shouldn't feel bad about, since it's what everyone did before the French came up with the derailleur in 1905 CE. Add some brakes (a clamp that closes around the wheel to slow it down) and you've got the basics of the bicycle, the fundamentals of which haven't significantly changed since their invention. The bicycle is one of the few technologies that humans got almost perfect right off the bat! Improvements since then—like rubber tires filled with air for comfort and spoked tires for lighter wheels—have been evolutionary rather than revolutionary, and while they're nice (pneumatic tires helped bicycles lose their early nickname of "boneshakers"), they're not necessary.

We said earlier that bicycles are more efficient than walking, and you can prove it yourself by going for a walk and watching what your body actually does. Many of the movements you make don't actually result in forward motion! The forward swing of your foot is what moves you forward, sure, but that's only part of what you're doing: your legs are moving up and down (*wasting energy*), your arms are swinging back and forth for balance (*wasting energy*), the entire mass of your body is bobbing up and down (which, you guessed it, *is wasting energy*). On a bike, most of the energy you're producing goes toward pumping the pedals, and the vast majority of that goes toward producing forward

---

* Friction causes the movement of one wheel to induce movement in the other, but drive belts are inefficient (a lot of energy is lost to friction) and they can slip easily. Another early alternative to chains and gears were "treadles": rods connected to the rear wheel that you push up and down to induce movement, similar to how you move a spinning wheel in Section 10.8.4. If you've got the metal to produce them, chains and gears are more efficient and reliable!

motion.* And on hills bikes are even *more* efficient, since you can coast your way downhill.

Bikes enhance civilization in lots of other ways too: they allow people to cheaply commute, which means crowding in cities can be reduced. They allow moderately sized goods to be easily transported under human power, which helps not only farmers bring wares to market but craftspeople and service workers to operate in a much wider area than they could reach on foot. And while you won't be achieving human-powered flight for a while, when we finally pulled that off in 1961 CE, it was with a human pumping the pedals of what was, in effect, an extensively modified bicycle.

In our own history, bikes have also been fundamental to early women's liberation. While this will hopefully not be an issue in your civilization—you're starting on a better foot than we ever did, seeing as you don't have to labor under the hangover of thousands of years of patriarchy—it's worth noting how something as simple as giving people the ability to cheaply transport themselves under their own power changed European society in the late 1800s CE. This newfound mobility not only allowed women to participate in civilization in ways they couldn't before, but actually changed the way women saw themselves. They were no longer observers moved around by society: instead, they were active *participants* who could—and would—move themselves. The clothing women wore also changed in response to the bicycle, as demands for a new "rational dress" that allowed for a modicum of physical activity meant the end of the restrictive corsets, starched petticoats, and ankle-length skirts that had previously been worn.

Besides their simplicity, affordability, civilization-altering utility, and

---

* It was once thought one of the reasons humans evolved to walk upright was because of how much more efficient it was for getting around, but that's not true. Human walking isn't even particularly efficient! Once you take their different weights into account, the efficiency of our walk is within 95 percent of what you'd expect when evaluating the performance of other mammals, including horses, dogs, mice, bears, platypuses, elephants, monkeys, and our nearest cousins, chimpanzees. In fact, the efficiency difference between humans (walking on two limbs) and chimps (walking on all four) is so unremarkable that you'll find larger differences between foxes and dogs, kangaroos and wallabies, and even within closely related species of mice, chipmunks, and squirrels!

virtuoso-like efficiency when paired with the human body, bikes are also just a heck of lot of fun to ride. You can even put a little basket on the front and fill it with a bottle of wine (Section 7.13), some nice breads (Section 10.2.5), maybe a cozy blanket (Section 10.8.4), and even some tasty pickles (Section 10.2.4). Is it any coincidence that a guide to reinventing civilization also functions *pretty well* as a guide to having a really delightful picnic? Picnics are objectively one of the crowning achievements of humanity, and don't worry: by following our instructions, you'll get there eventually . . .

. . . on the wheels of your bike.

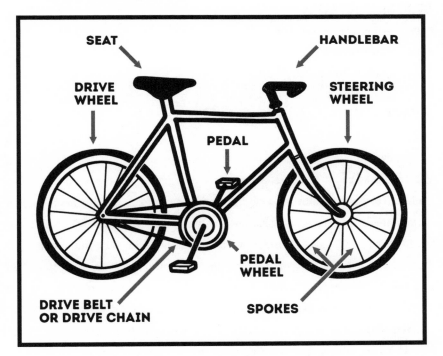

Figure 35: A beautiful machine.

10.12.2: COMPASSES

The winds and waves are always on the side of the ablest navigators.
—*You (also, Edward Gibbon)*

WHAT THEY ARE

A way to know what direction north is in, which also tells you what direction you're facing

BEFORE THEY WERE INVENTED

It was very easy to get lost, because it turns out an invisible global way-finding signal that you don't even need electricity to access is actually extremely useful

ORIGINALLY INVENTED

200s BCE (for fortune-telling)

1000s CE (for navigation)

1200s CE (for European navigation)

PREREQUISITES

rope (optional)

HOW TO INVENT

The first compasses were invented in China around 200 BCE. To make them, all you need is some magnetic rock, which conveniently naturally occurs near the Earth's surface. Look for rocks that stick together and you'll find magnets!* Once you've got one, you can use it to find more, and can even use it to make new magnets from scratch by magnetizing other things made of iron.†

These early compasses weren't the "needle on a pin encased in plastic" ones you remember: instead, they were as trivial and basic as tying a rock to a string. The string lets the rock rotate freely, the rock moves to point north, *and you*

---

* If you can't find any magnets, don't worry: if you've got a small amount of metal that can be magnetized, like iron, you can induce a magnetic field in it—turning it into a magnet—by repeatedly rubbing it through your hair in the same direction. A needle-sized piece of iron will work best here. And of course if you've got electricity (see Section 10.6.1), you can create magnets *whenever you gosh-darned feel like it.*

† Repeatedly rubbing a magnet against iron in the same direction will transform that iron into a magnet even faster than rubbing it through your hair.

just invented the compass. If you don't have string, just put a tiny piece of your magnetic rock on a leaf and float it in a pool of water, and *you just invented the compass twice in one paragraph.*

The catch is that these early compasses were used for fortune-telling, not way-finding, and it took more than one thousand years—until the 1000s CE—for anyone to think of using them for navigation. Europeans took even longer to figure this out, which means there's plenty of room for you to blow some minds here.

A warning: the Earth's magnetic field reverses occasionally, flipping "north" and "south." These reversals occur unpredictably every 100,000 to 1,000,000 years and can take between 1,000 to 10,000 years to complete. While "north" and "south" are just labels you'll apply to arbitrary poles of the planet's magnetic field, during these times of reversal, the strength of this magnetic field can drop to only 5 percent of normal. This, needless to say, will make compasses more difficult to use, so if you find yourself in the middle of one of these reversals, *maybe don't depart on any transoceanic voyages for a bit.* Here's when they happen over (what for us are) the past 5 million years or so: dark represents the poles you're used to, and white when they're reversed:

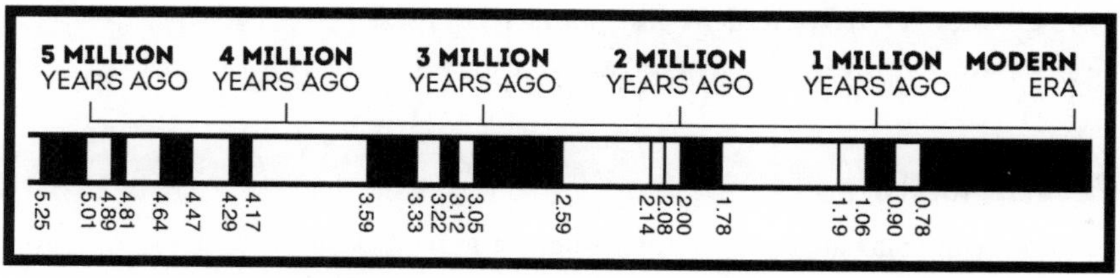

5 MILLION YEARS AGO	4 MILLION YEARS AGO	3 MILLION YEARS AGO	2 MILLION YEARS AGO	1 MILLION YEARS AGO	MODERN ERA

5.25 · 5.01 · 4.89 · 4.81 · 4.64 · 4.47 · 4.29 · 4.17 · 3.59 · 3.33 · 3.22 · 3.12 · 3.05 · 2.59 · 2.14 · 2.08 · 2.00 · 1.78 · 1.19 · 1.06 · 0.90 · 0.78

Figure 36: Magnetic pole reversals over the past 5 million years.

As you can see, we are actually overdue for a magnetic field reversal in the modern era! This was a minor concern until pole-stabilization technology was invented in the early 2040s CE, at which point nobody ever had to worry about it ever again. Except, now, for you.

Sorry about that.

10.12.3: LATITUDE AND LONGITUDE

I think you travel to search, and come back home to find yourself there.

—*You (also, Chimamanda Ngozi Adichie)*

WHAT THEY ARE

A way to precisely define every location in the planet with just two numbers, reducing the question of "Where am I?" to the mere task of figuring out what those two numbers are

BEFORE THEY WERE INVENTED

Directions were local rather than universal, more "turn right at the big tree" and "sail west until you hit land" than "these coordinates describe a location to an accuracy of 10cm"

ORIGINALLY INVENTED

300 BCE (first geographic coordinate system)

220 BCE (quadrants and astrolabes)

1675 CE (ineffective marine chronometers)

1761 CE (more effective marine chronometer)

1904 CE (time signals transmitted over radio)

PREREQUISITES

calendar (for solar latitude), radio (for longitude)

HOW TO INVENT

If you assume the Earth is a sphere (it's not*), you can cover that sphere in hori-

---

* The Earth's rotation causes the planet to bulge out around the middle, which transforms it from a perfect sphere into what's technically an "oblate spheroid"— but the bulge isn't that big, and assuming Earth is a sphere makes the math easier! And we're all about making things easy for *you*, the involuntarily stranded time traveler who once tore through time at their whimsy but who now is restrained like everyone else to traveling through time in only one direction: forward, and even then at the fixed rate of one second per second.

zontal and vertical lines. Arbitrarily label the horizontal ones "latitude" and the vertical ones "longitude,"* and you've invented the planet's first geographic coordinate system. Hey, that was easy!

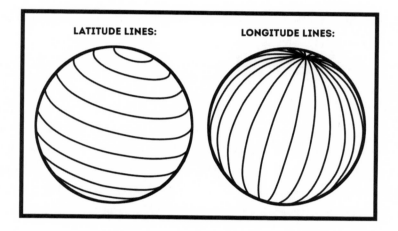

Figure 37: Two spherical approximations of Earth that someone drew all over.

Since the Earth rotates, it has a natural top, bottom, and middle.† We'll call the line around the middle of the planet "the equator," define that as 0 degrees, and label every other latitude line by whatever angle a line drawn from those latitudes to the center of the Earth makes with the equator. So latitude lines will start at 0 at the equator, and they'll increase as you go north until they reach 90 degrees at the north pole, and decrease as you go south until they reach -90 degrees at the south pole.

For longitude lines—also called "meridians"—there's no obvious "vertical

---

* Again, as in all things, you can name these lines whatever you want. If you'd like to outperform our own history, we'd recommend *not* giving these two confusingly similar lines the same confusingly similar names. Call one "latitude" and the other "Gary."

† Where "top" is on a map—or the entire planet—is actually entirely arbitrary. We're used to "north means up" and that's what we use in this text, but an upside-down map is just as accurate, so feel free to choose whichever orientation you think is prettier. Fun fact: the appearance of maps used thousands of generations from now may well be determined by whatever choice you make in the next few seconds!

equator" to make your zero point, so you're going to do what every other human throughout time has done when faced with this problem: shrug and pick one at random. In the modern era everyone uses an imaginary line passing through Greenwich, England, as the zero point (also called the prime meridian), since at the time that was the path of least resistance: England had already printed a lot of maps with that feature. But different nations have used different prime meridians passing through their favorite cities instead, and it really doesn't matter what you choose!

We'll label longitude lines slightly differently from latitude. Latitude lines are treated as rings—each latitude line wrapping around the entire planet like a belt—but lines of longitude are treated as half-circles: each meridian defined by the curve you get when drawing a line between the north and south poles. This means that instead of ranging from –90 to +90 degrees like latitude does, longitude will range from –180 to 0 (for places east of the prime meridian) and from 0 to 180 degrees (for west), and your final coordinate system will look like this:

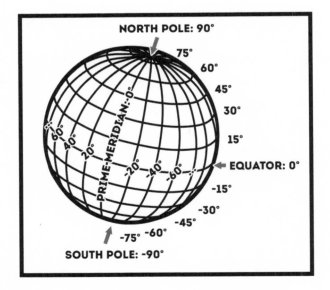

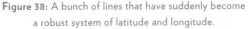

Figure 38: A bunch of lines that have suddenly become a robust system of latitude and longitude.

Now that you've given each location on the planet a precise coordinate,* all that's left is to figure out some way to know the coordinates you're standing at.

For latitude, you'll use the stars to measure your position. You may be thinking, "Oh yes! I have heard of a 'north star' used for navigation by sailors, so I'll just use that!," but when you consider the matter, your next thought will be, "Oh no! I just remembered that while the Earth spins like a top, which causes 'days,' it also wobbles like a top too, which causes a rough 25,700-year cycle we call 'axial precession'! Because of axial precession, any imaginary lines drawn straight up from the north pole or straight down from the south pole will appear to sweep across the cosmos in a giant circle, meaning any star used for navigation in the modern era likely won't appear to be in the right place in the sky in whatever crazy time period I'm stuck in, and that's a huge problem, made worse by the fact that the stars are themselves all slowly moving over time!"

And yes, this thought you're having is correct. While you might recall in your pre-time-travel life looking up to the sky at night and finding a north star (if you're in the northern hemisphere) or a south star (if you're in the southern), the position of these stars is not guaranteed, and without knowing precisely what time period you're in, relying on them is impossible. However, there is one star visible from Earth that *is* always going to be where we want it no matter what time period you're surviving in: our star, the sun. Measure the angle between you and the sun at noon—when it's highest in the sky—and you'll still be calculating latitude using the stars.

For this you'll need a quadrant, which is just a quarter of a circle with

* For example, time travel was first successfully demonstrated at 43.660155 degrees latitude and -79.395196 degrees longitude when a small mass was sent back in time 250 years for three seconds before being retrieved. For such a momentous achievement the experiment itself is famously calm and restrained, except of course for one moment near its end. When the mass returns, researcher Bennett suddenly realizes that the first successful instance *of* time travel is, by its very nature, also going to be one of the most famous moments *for* time travelers. Almost despite herself, she glances around the room to see if any future time travelers really have come back to witness it. Her reaction, if you then choose to step out from the shadows, makes this one of the most popular choices in our *Earth's Most Wild and Wacky Moments in History* tour program.

angles marked on it: in other words, it's just half the protractor you already invented in Section 4. Here's a template; make yours out of wood or metal.

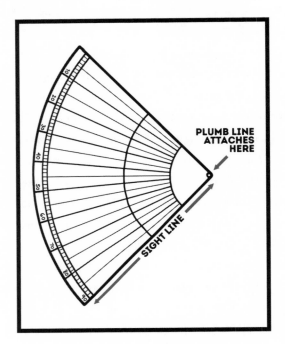

Figure 39: A quadrant template.

Attach a rock on a string to the corner: this is your plumb line, and it will always point straight down (assuming you can protect it from the wind). At each edge of the sight line make two loops. When you stare through the sight line, lining up your quadrant so the sun appears in the middle of both loops, your plumb line indicates the angle you've got the quadrant at—which also happens to be your latitude. Take several readings and average them for more accuracy, and latitude is now a solved problem!

Or at least it would be, if staring at the sun long-term didn't cause *irreversible blindness.*

The quadrant you've just invented is great for navigating by the stars, but if you're using the sun, you'll want to adjust it to allow for indirect solar observation. Here's how: replace the loop near your plumb line with a block of wood with a tiny hole in the middle, and replace the other loop with another block of wood with a target drawn in its center. Now instead of looking at the sun, you'll align your quadrant so the tiny pinprick of light passing through your block is positioned directly in the middle of your target. Ta-da! *Now nobody has to go blind just to sail a boat anymore.*

There's one final adjustment to make. If you're using stars, they're far enough away that they appear fixed from Earth's perspective, so you don't need to adjust for the planet's axial tilt. But the planet's tilt does affect where our sun appears in the sky, so you'll need to adjust your solar readings per the following table:

Event	Season	How to recognize when this event is taking place	Approximate date*	Adjustment
March equinox	Spring (northern hemisphere) Fall (southern hemisphere)†	Day and night are the same length	March 20	*None required*

---

* How can you know the approximate date if we haven't told you how to invent calendars? The answer is simple: build your calendar *around* these equinoxes and solstices. Time the length of your days and nights (see Section 10.7.1: Clocks) and you'll know when the equinoxes and solstices have occurred, which lets you build a calendar around that to predict them—and the changing seasons—next year. If you're so inclined, you might even go so far as to make your calendar visual, perhaps by positioning rocks such that the sun rises perfectly between them only on the morning of solstice, constructing some sort of "stone" "henge." We've used the Gregorian calendar and Julian months you're familiar with here, but you don't have to stick with them: call your months whatever you want, and make them whichever length pleases you. Like the prime meridian, they're also completely arbitrary. The only scientific restriction your calendar has is that it should average out to 365.256 days per year, but you can get to that average however you want. We do it by having 365 days a year for three years, and then throwing in an extra day every fourth year, with a few leap seconds here or there to pick up the slack, but there are other solutions.

† Determining what hemisphere you're in is easy: just build a large pendulum—12 meters or longer works—and let it swing for a few hours or so. A pendulum's inertial frame is separate from the Earth's, which means with a long enough pendulum (like the one you just built), you can actually make the effects of Earth's rotation visible! The path your pendulum takes will rotate over time—clockwise in the northern hemisphere, counterclockwise in the southern hemisphere, and none at all if you're at the equator. A gentleman named Léon Foucault came up with this idea in 1851 CE, but it's yours now!

Event	Season	How to recognize when this event is taking place	Approximate date*	Adjustment
June solstice	Summer (northern hemisphere) Winter (southern hemisphere)	The longest day and shortest night of the year (in the northern hemisphere), and longest night and shortest day (southern).	June 21	Add axial tilt (23.5 degrees in modern era)
September equinox	Fall (northern hemisphere) Spring (southern hemisphere)	Day and night are the same length	September 23	*None required*
December solstice	Winter (northern hemisphere) Summer (southern hemisphere)	The shortest day and longest night of the year (in the northern hemisphere), and shortest night and longest day (southern).	December 22	Subtract axial tilt (23.5 degrees in modern era)

Table 14: The latitude adjustments required when using the sun. On Earth, anyway. We forgot to mention this until now, but any time machine is by necessity a space machine too, and any trip through time without corresponding movement through space will leave you stranded elsewhere in the cosmos, since Earth is moving through space around a sun that's moving through space in a galaxy that's also moving through space. If you find yourself stranded in a place that's not on Earth, then you've somehow managed to escape not just the clutches of our lonely planet but also the scope of this book. Good luck!

If you're not reading this on the solstice or equinox dates mentioned here, you can approximate the adjustments required to account for Earth's current axial tilt by adding the result of the following equation to your measurement:

$$\text{adjustment} = -t \times \cos\left[(360°) \,/\, 365 \times (d + 10)\right]$$

The d there represents the day of the year you get when you count January 1 as 0, January 2 as 1, and so on: the 10 is added to it to get the number of days since the December solstice. Similarly, t represents Earth's current axial tilt in degrees (see Section 10.7.1 for a way to measure your Earth's axial tilt).

And with that, *finally*, latitude is a solved problem. Now all you have to do is calculate longitude, and that's way simpler!

Longitude is the measurement of how far east or west you are of your prime meridian. Since you already know the Earth rotates west to east a full 360 degrees each day (which you do, because we just told you), it's clear that one degree of longitude corresponds to 1/360 of a day, or 4 minutes.* Your longitude is the difference between noon where you are and noon where your prime meridian is.† For example, if your noon happens 8 minutes before the noon at the prime meridian, you know you're 2 degrees of longitude to the east of that line. Similarly, if your noon is 20 minutes after the prime meridian's noon, your longitude is 5 degrees to the west. Longitude is *trivial*, assuming you believe that keeping track of what time it is at the prime meridian isn't the sort of thing that will stymie humanity for thousands of years!

This is the point where we regret to inform you that keeping track of what time it is at the prime meridian absolutely is the sort of thing that will stymie humanity for thousands of years.

The reason is simple: clocks typically rely on some repeating movement—a pendulum swinging, water dripping, balls rolling, whatever—which are motions that work great on land but completely fall apart on boats. One strong wave can knock a pendulum out of whack, and that's not even factoring in the endless sway of the waves. To combat this, boats would carry sometimes dozens of clocks, each drifting slightly out of sync in a different way, under the hope that the *average* of all these times would be accurate, but this wasn't a solution and ships kept getting lost and sinking.‡ It was such a problem that as early as 1567 CE various governments began offering cash prizes for any solution that would enable longitude to be determined at sea. By 1707 CE the British were willing to give up to £20,000 to anyone with a workable solution: the equivalent in value to a multimillion-dollar prize today.

If you're around during these times, great news: you're about to be rich!

---

* This assumes a twenty-four-hour day: your days may well be shorter. That's because the moon's gravity causes tides, but these tides cause a small amount of friction between the Earth and its oceans, which acts as a tiny brake. This makes Earth's rotation slower in the future and faster in the past. You can expect a day that's about 17.8 seconds shorter for every million years you go back in time.

† Here we're using actual solar noons (when the sun is at its peak in the sky) and not any approximations of noon that you might have invented when you came up with time zones (see Section 10.7.1: Clocks).

‡ The HMS *Beagle* had twenty-two different chronometers on board when it set out in 1831 CE for the trip on which Charles Darwin would begin to formulate his theory of evolution. They were all stored in a single specialized cabin at the bottom of the ship, where movement would be minimized, and no one was allowed to enter this room except to tend to the clocks. Five years later, the ship returned with only eleven of those twenty-two clocks still in working order—but it *did* return.

In unaltered history, the solution is about what you'd expect: brilliant clock-makers dedicated their entire lives to the problem and eventually came up with some incredibly clever, incredibly expensive, incredibly complicated marine chronometers—chronometers that you're not going to build because they're really, really difficult. Instead, you're going to think outside the box, and you're going leapfrog to the solution still used in the modern era. You're going to send time *through the air itself on invisible waves of energy that travel at the speed of light.*

You're going to invent radio.

And you're going to save the lives of millions of sailors who haven't yet been born.[34]

10.12.4: RADIO

> The coming of the wireless era will make war impossible,
> because it will make war ridiculous.
> —You (also, Guglielmo Marconi)

WHAT IT IS

A way to transmit ideas and information at almost the speed of light, lessening the barriers of time and distance that have held humanity in their grasp since time immemorial, so that's cool

BEFORE IT WAS INVENTED

If you wanted to listen to music you had to leave your house and physically go to a concert, and who has the motivation to do *that*

ORIGINALLY INVENTED

1864 CE (electromagnetic waves predicted)

1874 CE (first cat's-whisker radio detector)

1880 CE (first intentional radio transmission)

1895 CE (radio signals sent and received over a distance of 2.4km)

1901 CE (radio signals sent and received across the Atlantic Ocean, a distance of 3500km)

Sidebar: How Far Is It Between Lines of Latitude and Longitude Anyway?

The distance 1 degree latitude and longitude covers on an Earth-shaped planet (and here we are giving exact oblate-spheroid numbers, *because we care*) are given below for different latitudes. The distances between longitudes drop to 0km at 90 degrees of latitude, because then you're at the poles, and that's where all your longitude lines converge to a point:

Latitude	Distance between lines of latitude	Distance between lines of longitude
0°	110.574km	111.320km
+/– 15°	110.649km	107.551km
+/– 30°	110.852km	96.486km
+/– 45°	111.132km	78.847km
+/– 60°	111.412km	55.800km
+/– 75°	111.618km	28.902km
+/– 90°	111.694km	0km

Table 15: A chart of numbers that will seem *pretty boring* until you're on a boat in the middle of the ocean and want to know how much farther away land is, and then you'll be like, "I'm glad I read this chart; it turns out it wasn't boring after all, and I truly apologize for all my past sass."

PREREQUISITES

electricity (for transmission), metal (for wires), magnets (for speakers)

HOW TO INVENT

You've probably heard of the electromagnetic (or "EM") spectrum: it describes a range of radiation—which is just energy moving through space—that includes everything from radio to visible light to X-rays. Here's a map of it:

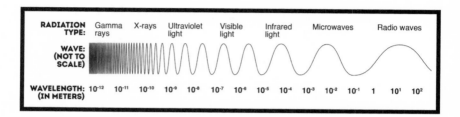

Figure 40: The electromagnetic spectrum.

There are gamma rays at the high-energy end, low-energy radio waves at the other end, and a tiny sliver of visible light near the middle.* You're probably most familiar with the visible-light part of the spectrum, since you're absorbing that radiation with your eyes to read these words. Hello!

We've divided visible light up into colors: red, orange, yellow, green, blue, indigo, and violet,† but the only thing that separates them from one another is

---

* When we say "tiny sliver," we're not fooling. Visible light has wavelengths rang-ing from 400 to 700 nanometers, which means everything you've ever seen, and every-thing you ever *will* see, reaches your eyes via a mere 300 nanometers of spectrum.

† Those aren't the only ways to do it! Since color is a spectrum, you can assign however many labels you want to whatever colors you choose. For example, while English-speakers say that blue and green are different colors, Chinese has a color, 青, that includes them both. In the opposite direction, what an English-speaker calls "red" would be identified by multiple names in Hungarian, Turkish, Irish, and Scottish Gaelic. And even this is an approximation: anyone who works with color will identify shades that the average person wouldn't care to distinguish. An interior decorator (don't worry; you'll have them again someday) can pick out crimson, burgundy, carmine, amaranth, maroon, folly, oxblood, redwood, rosewood, vermilion, coquelicot, carnelian, auburn, rose, and wine out of a lineup, while the rest of us see only shades of red.

their energy levels. Our brains transform certain levels of visible-light radiation into what we perceive as "yellow," while others are turned into "purple," but all colors (and all EM radiation) are fundamentally the same kind of thing: electromagnetic radiation at different energy levels, all traveling at the speed of light. Some levels of EM radiation pass right through our bodies with minimal interference (radio waves), while others slam right into them (light).

Intuitively, visible light probably seems different from radio, but there's actually nothing particularly special about it except it's at a frequency that happens to gets absorbed by our bodies, which is actually one of the reasons humans evolved to see it.* And while it's true we can't *see* other parts of the EM spectrum besides light, we can still *sense* some of them. Radiation with slightly lower energy than red (the lowest-energy color we can see) is called "infrared radiation," and we can sense that as heat with our skin. Radiation slightly more energetic than violet is called "ultraviolet," and we can sense that with our skin too, via what is technically called *a potentially fatal radiation burn.*†

---

* We couldn't see radio waves if we wanted to: they already travel right through our eyes without hitting them. But there are some animals that can see slightly outside our visual range: mantis shrimp can see all the colors we can, plus a bit into the infrared and ultraviolet spectrum! They have access to colors humans haven't even *dreamed of* . . . mainly because it's very difficult for humans to imagine colors they haven't already seen. This leads us to an open question in philosophy: the thought experiment of "Mary the Super-Scientist," first proposed in 1982 CE by one Frank Jackson. In it, you have a woman, the eponymous Mary, who is a brilliant scientist born with a disease that lets her see only in black and white. Incredibly, she's spent her entire life in a black-and-white room with a black-and-white computer hooked up to the Internet. But despite these limitations, Mary has thrived! She's learned each and every piece of physical information there is to possibly know about how color works, how light interacts with the eye, what that information does in the brain—everything. She is the world's expert on color and the human body, all from inside her black-and-white room. Then one day Mary's disease is cured, and she's let out of her room. She steps outside and looks up to the beautiful blue sky for the very first time in her life, and the question for you is . . . *Does she learn anything new?* In other words, is there knowledge that can't be taught, and which can only be obtained from direct conscious experience? Don't ask us for the answer! We just build the time machines.

† That's what a sunburn is! It's not the sunlight you can see or infrared heat you can feel that burns you but rather the high-energy ultraviolet radiation that's penetrating your skin, damaging your DNA, and causing the radiation burn. Any time you damage a cell's DNA, you risk that cell becoming cancerous, hence the potential fatality mentioned above!

So now that you're familiar with the basics of EM radiation, let's talk radio, which is simply the technology of using EM waves to carry information. In the modern era this is done in several ways. Modulating the radio signal's amplitude (how high the radio waves go up and down) gives us AM radio (short for "amplitude modulation"), and modulating its frequency (how often they go up and down) gives us FM. The strategy here is to encode information in how much the amplitude or frequency changes, but that's more advanced than what you need. You'll get there, but for your immediate purposes you just want to be able to send a radio signal.

You'll be generating radio waves the easy way, which also happens to be the most impressive mad-scientist way too: by creating *artificial lightning*, also known as "electricity." When electricity travels through the air—this is called "arcing" and you can learn more about it in Section 10.10.3—it produces all sorts of EM radiation. There's light (which is what makes lightning look so cool) but also a bunch of radio radiation. If you can produce arcing at will—and by cutting a wire so that the only way for electricity to travel is by arcing between the two pieces, you can—then you can generate radio signals, and the strength of your transmission is limited only by the amount of power you can get to arc.

If you're conveying information simply by the *existence* of a radio burst (for example, a transmission at noon for timekeeping; see Section 10.12.3), then you're done, but if you add a switch so you can you turn it on and off in patterns, you can broadcast any information you want in Morse code.* This is the same technology used in the wired telegraph—which you can also invent now if you want, just by having your switch connect by wire to a distant buzzer, instead of to an electric arc radio transmitter. Wired telegraphy may be simpler for overland transmission, but you'll want to use wireless for transoceanic

---

* If you don't know Morse code, don't worry: just make it up, and name it after yourself instead! The idea behind Morse code is to represent each letter by a different series of dashes (long transmissions) and dots (short transmissions), with a short silence between each letter. If you want to get fancy, give the most common letters shorter representations, and less common ones longer ones, and your code will be able to transmit information more efficiently. Which letters are the most common? We have conveniently provided you with an entire text with over 600,000 letters in it in the form of this book, so all you need to do is tally up each of them to see which are the most popular. This is left as an exercise for the reader.

transmissions, at least until you reach the point of laying cable at the bottom of the sea.

To pick up these signals you're going to build the world's first radio-signal detector, and you won't even need batteries: the radio signals themselves will power it. First you need an antenna: any long wire will work, but 30m or more is ideal. Put one end in the ground and the other end someplace high: a tree on land, or the top of your mast on a boat.* Radio waves (which, remember, are just electromagnetic radiation) will interact with this long wire and induce electrons to move up and down in it, creating current. To detect this energy, you're going to invent the diode.

A diode is a device that lets electrical current flow in only one direction. Diodes are examples of "semiconductors": materials that conduct current differently under different circumstances. In the modern era our semiconductors have evolved from vacuum tubes to transistors to integrated circuits, but you don't need any of that. All you need are some plain old rocks. Natural semiconducting rocks include galena (one of the most common lead ores, a dark and shiny angular rock often found with calcite) and iron pyrite (aka fool's gold, easy to spot because it's so shiny). You need only a tiny bit: one crystal of these rocks will do. You may remember stories about your ancestors building toy "crystal radios" back when they were young and radio was new: you're building the same thing here.

Once you've got your crystal diode, securely connect your antenna to it (while still leaving it grounded), and make a second, very gentle contact to your crystal with a thin wire, historically called a "cat's whisker."† You'll probably have to experiment with your cat's whisker touching different positions on your crystal until you find one that's the most semiconductive. When you find it, your "whisker" will carry electricity when your antenna picks up a radio signal—a small amount to be sure, but enough to indicate that a radio transmission has been received.[35]

---

* And if you're on a boat, a wire trailing in the water will function as a simple "ground," but a metal plate on the underside of your boat provides a ground with more surface area and, therefore, more effectiveness.

† So named because any thin wire looks a little bit like a cat's whisker if you squint sufficiently hard enough.

To make that electricity audible, you're going to create a solenoid, which is just wire tightly wrapped into a coil. When electricity runs through a wire it creates a magnetic field, and by coiling that wire around itself, that field is intensified. As you saw in Section 10.6.2, that's an electromagnet! Put a regular magnet inside that coil and it'll move inside your solenoid at the same rate the electricity does. Attach that magnet to a lightweight but sturdy cone so that air gets moved as the magnet does, and you've converted changes in electricity into movement of air: in other words, you've just invented the world's first speaker.

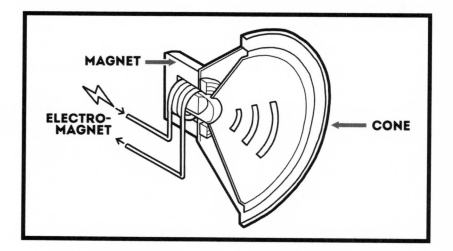

Figure 41: The world's first speaker: make some noise.

The relatively low level of the electricity here means it'll be a quiet speaker—more suited as a small headphone set than as something to get the party people moving with some block-rocking beats—but the same principle applies to both.

Some caveats: radio transmission works better at night than during the day, because the Earth's upper atmosphere—called the ionosphere—is electrically charged by the sun. When radio waves travel through the lower ionosphere during the day, they interact with the solar-charged ions there and degrade. But the layer becomes transparent to radio waves at night (great!) and the layer of the ionosphere above it actually *reflects* radio waves that arrive at an oblique-enough angle (even better!). Bouncing signals off the ionosphere at night is a completely reasonable way to send information across huge distances. In fact, when you're transmitting at distances far enough that the curvature of the Earth means there is no straight line between sender and receiver that doesn't travel through solid radio-blocking rock,* it may be the *only* way.

You should also keep in mind that electromagnetic radiation doesn't travel forever at the same strength. The intensity of any broadcast radiation (whether electromagnetic, gravitational, or sound waves) is actually inversely proportional to the *square* of the distance between you and it. In other words, the more you move away, the faster the signal seems to decay. Even we, who can *bend the very chronotons to our will,* can't change the inverse-square law, but it can be mitigated by broadcasting at a higher intensity. The first transatlantic transmission (organized by the very Marconi whose quotation, gracing this section, was sadly not as predictive as he had hoped) was achieved with a spark-gap transmitter of the type described here, only with a lot of power behind it, and a really large receiving antenna waiting for it on the other side of the ocean.

10.12.5: BOATS

One cannot discover new lands without consenting to lose sight,

for a very long time, of the shore.

—*You (also, André Gide)*

WHAT THEY ARE

A way to open up the 70 percent or so of the world's surface that's covered in water† to exploration, fishing, trade, and *cool parties in international waters*

---

* While radio waves do travel through most things, the thicker and denser the material, the more the signal decays.

† This statistic holds true for just about any time period you can survive in! The amount of the Earth's surface covered in water has been approximately 70 percent since the formation of the supercontinent Pangaea around 300,000 BCE, because the continents since then have stayed about the same size and just sort of slowly drifted around. All the water on Earth has to go *somewhere*, and the only places to store it are either (a) on the surface in big liquid piles, which we call "oceans" and "lakes," (b) in the atmosphere as clouds, but the atmosphere can only get so hydrated, and (c) as ice at the poles. Polar ice caps can famously grow and shrink, which raises and lowers sea levels and thereby exposes or covers land, but

BEFORE THEY WERE INVENTED

Humans were unable to expand to anywhere they couldn't walk, crawl, or swim, which left everything from tiny islands to entire continents completely uncolonized. Instead, folks just stared out at the sea and sighed

ORIGINALLY INVENTED

900,000 BCE (protohumans travel over 18km of water to reach Flores Island in Indonesia)

130,000 BCE (humans travel over water from the Greek mainland to Crete)

46,000 BCE (humans travel over water to Australia)

7000 BCE (reed boats)

5500 BCE (sailboats)

100 CE (sailing into the wind)

1783 CE (steam-powered boats)

1836 CE (propeller-based boats)

PREREQUISITES

wood (for dugouts), rope, tar (for boats made from reed or logs), metalworking and welding (for pintles and gudgeons), textiles (for sailboats), compasses, preserved foods, latitude and longitude (for the kick-ass boats you're going to want to invent once you see how kick-ass they are), spinning wheels (for fishing nets you'll need in order to invent offshore fishing, which you'll want to do, because there are some *really* tasty fish offshore)

HOW TO INVENT

The earliest boats (called "dugouts") were incredibly simple: hollow out a big-enough tree trunk to sit in, and, my friend, you have invented boats. You can create better, larger boats by taking reeds, logs, or planks, strapping them together with rope or nails into a boat-shaped hull (pointy at the front, square at the

---

even fully melted, polar ice can change sea levels only so much: 70m when fully melted, to be precise, which results in only about 3 percent less land. It's a very important 3 percent, however, given how many human settlements are on coastlines!

back), and waterproofing them by cramming plant matter between the cracks before sealing them with tar or pitch (see Section 10.1.1: Charcoal). If that feels too ambitious, you can just build a raft, but rafts are better suited to drifting than going somewhere in particular: it's the shaped hull of boats that lets them slice through water in a directed way. Remember: boats go where *you* want, rafts go where *they* want. You want boats.

Steering a boat is easily done with a rudder, which took humans (and this should not be anything close to a surprise at this point) a long time to figure out. Before their invention, one or more large "steering" oars were hung off one side of the boat, which were the state of the art until rudders were invented in China around 100 CE. The idea finally arrived in Europe a thousand years later, around 1100 CE. Europeans: pull up your socks.

Rudders (see Section 10.12.6) can be attached to the back of your boats using pintles and gudgeons, which are adorable names for very simple inventions. A gudgeon is a tube, and a pintle is the corresponding bolt that fits perfectly inside that tube, which then allows whatever's attached to rotate freely . . . *exactly like you'd want a rudder to do.*

These weren't invented until the 1400s CE, so the technology in your boat has already advanced several millennia in just a few paragraphs, and we're just getting started.

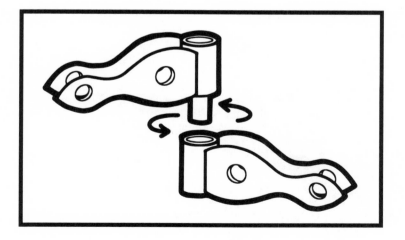

Figure 42: A pintle and gudgeon.

To avoid having to physically push your boat through water by paddling like some kind of primo chump, you'll want to invent either sails or engines. Engines are discussed elsewhere (Section 10.5.4), so let's focus on sails, which make for a much more gorgeous boat anyway.

Throw up a square sheet held upright and loose enough that it can fill with air, set it perpendicular to the length of your boat so it catches the wind, and hey presto: your boat will move through the water, and you've invented sailing . . . *Captain*. You can sail whenever there's wind that's moving faster than the water is.* But this is entry-level sailing, the kind that's been invented multiple times throughout history, and it won't let you travel in a direction more than 60 degrees off from wind direction. Wouldn't it be nice to be able to sail in any direction you like? Wouldn't it be nice to be so powerful as to be able to sail *directly into the wind*, making the very forces of nature subservient to your will and thereby conquering *the sea itself*? It would, so let's invent that now.

Instead of a square sail mounted perpendicular to the line of your boat, build a triangular sail that runs *along* its length instead.† These are called "fore-aft" sails because they run from the back (in sailor talk: "the aft") to the front (sailor talk: "fore") of your boat (sailor talk: "a fine sloop, yarr"). By mounting this sail on a boom (a large rotating beam attached to the mast) you can rotate your sail, allowing the boat and the sail driving it to be at different angles. Use rope to tie your boom where you want it to stay between adjustments.

This greater sail control allows you to harness the wind from almost any

---

* If the wind is traveling slower than the water is, then you're not sailing: you're drifting, and you're going to keep drifting until the wind picks up again. Sailors call this being "becalmed," but it's not particularly calming when you're becalmed in the ocean, miles away from shore, running out of food and fresh water and baking in the sun. Sailors should instead call it "becalmbutalsobeawarethatthiscouldactuallybeabigproblem."

† It's not actually an either/or proposition between sail types. Many boats (especially larger ones) combined square sails with fore-aft rigging, using them in different situations. Square sails still work best when the wind is blowing directly behind you!

Figure 43: Square and fore-aft rigging.

angle, up to about 45 degrees into the wind. And even though you can't sail
in a straight line *directly* into the wind, you can approximate it by sailing 45
degrees off from the wind's angle, and then switching direction every once in
a while to the opposite 45 degree angle. This is called "tacking," and it produces
a zigzag motion. And yes, while it *is* less efficient than going in a straight line,
and yes, it *does* require you to adjust your sails regularly, who cares? You're
sailing into the wind while every other civilization on the planet is probably
still messing around with hollowing out a tree and calling themselves Johnny
Boats-A-Lot.

But that's not the only thing wind does to this new sail. When you position
it obliquely—just slightly into the wind, so that your sail is only a few degrees
off from parallel with the wind's direction—some wind fills your sail, but the

rest passes over it on the other side. This makes the sail act like a wing—just like it does on an airplane in Section 10.12.6: Human Flight—and this creates lift. With lift, the wind is not just pushing your sails but *pulling them* in the same direction from the other side. Combining pushing and pulling allows sailboats to actually sail *faster than the wind itself*. A skilled sailor with a proper boat can achieve a travel speed of 1.5 times the wind. Something to shoot for!

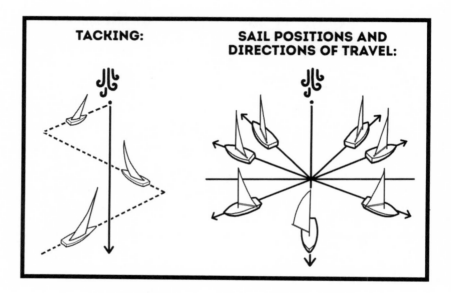

Figure 44: Tacking and the points of sail on a boat.

The power of wind you're now harnessing will make your boats want to tip over, so store your heavy things in the bottom of your boat as stabilizing ballast, and install a "keel" (a long vertical board shaped like a shark fin) on the underside of your boat, right along its center line. You can put a keel at both the front and back of your boat, or just one in its middle. Keels act as a counterforce on the wind to help prevent your boat from capsizing, and have the added benefit of making your ship move forward in the wind rather than slipping sideways.

If you do decide to add engines to your boats, you'll want water propellers.* A propeller is just a machine that turns rotational motion into thrust, and while humans technically invented them relatively early on, it still took us almost two thousand years to realize what we'd come up with. The earliest origins of the propeller go back (forward to you, probably) to the Archimedes screw—first invented in Assyria around 650 BCE but named after Archimedes, who popularized it to Europeans around 300 BCE. The Archimedes screw is simply a big long screw stuck in an open-ended tube, and then the whole business is put at an angle, with one end submerged. Rotating the screw in the right direction causes water to be lifted up through the tube, which can be really useful if you want to irrigate your crops.† Archimedes screws were used this way for thousands of years, but it still took until 1836 CE for someone finally thought of cutting one off and putting it in the water. There, thanks to the "every action has an equal and opposite reaction" law,‡ these propellers don't just move water: they also move, in the opposite direction, any vessel they are attached to.

That's really all the first propeller was: a mini Archimedes screw, long enough to contain two rotations. By accident, this propeller broke in half during early testing, and that was how humans discovered that a "broken" single-rotation propeller was twice as efficient as the two-rotation propeller it used to be. The bladed propellers you're familiar with are, in effect, multiple screws working in parallel.

Feel absolutely free to skip ahead to that final propeller design! It's the best one!

---

* The earliest steam-powered boats didn't have any—they used large paddlewheels instead—but the propeller is a more efficient use of force. Paddlewheels sure are pretty, though. They also are more difficult to jam, so if you want a pretty and less-jamming alternative to propellers, try paddlewheels!

† The space between the screw and the tube doesn't even have to be flush: it just has to be tight enough that the water you raise is greater than the water lost by leaking down again. This is great news for any civilization that hasn't invented precision engineering yet, which probably includes yours!

‡ Remember? You discovered these laws in Section 10.5.3, down at the bottom of the page in a footnote just like this one!

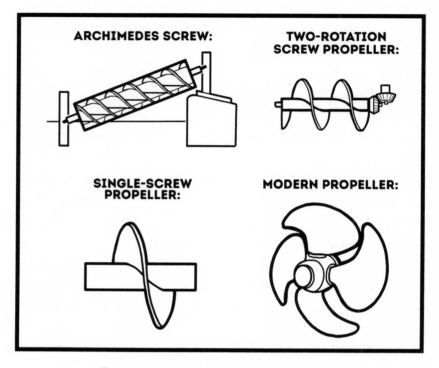

Figure 45: An Archimedes screw and its descendants:
single-screw and modern propellers.

10.12.6: HUMAN FLIGHT

Aeronautics was neither an industry nor a science. It was a miracle.

—*You (also, Igor Sikorsky)*

WHAT IT IS

A way to fulfill the earliest dream of humanity, held in perpetuity ever since our gaze first alighted upon the majestic beauty of a bird in flight and thought, "Oh hey, that looks cool, I wanna do that now"

BEFORE IT WAS INVENTED

You were born on the ground, and you died on the ground, and you told yourself this was fine and you were silly to ever dream of anything greater

ORIGINALLY INVENTED

500 BCE (strapping humans to giant kites)

1250 CE (earliest sketches for lighter-than-air flight, designed to be powered by a technology not yet discovered)

1716 CE (earliest published sketches for heavier-than-air flight, designed to be powered by technology not yet discovered)

1783 CE (first lighter-than-air flight)

1874 CE (first externally powered heavier-than-air flight)

1902 CE (first self-powered heavier-than-air flight)

PREREQUISITES

paper and fabric (or, if you have it, silk), sulfuric acid and iron (for hydrogen airships), wood (for gliders and heavier-than-air flight), engines and metal (for powered heavier-than-air flight), compasses, latitude and longitude (for navigation)

HOW TO INVENT

A hot-air balloon is an astonishingly simple invention. Fire creates hot air, which rises. If you put a fabric bag above that hot air, it'll fill with it, and a big enough bag with small enough leaks will become buoyant enough to rise through the air. Hold on to a large enough bag—or, to save your arms from getting tired, attach a basket to the bottom that you climb into—and you will rise with it. You don't even need to seal the bag at the bottom, since the hottest air rises to the top, and the air at the bottom will be at about the same temperature and pressure as the surrounding atmosphere. You can keep it tethered to the ground for early test flights, add sand for ballast (which you can toss overboard to lighten your weight as you sink, useful for slowing your descent as the hot air in the bag escapes or cools), and eventually even bring the fire on board—which, while more dangerous, lets you increase your altitude during flight.

In other words, it turns out that to invent human flight, the dream of thousands and thousands of generations of humanity, you only need some fabric and some fire. That's it.

And it still took us until 1783 CE to figure it out.

We have made a lot of hay in this text out of how long it takes humans to invent something even after they've got all the prerequisite technologies, but this is gosh-darned *humiliating.* If you draw a line between when humans had the technological prerequisites (fire and the drop spindle needed for fabric creation) and the time when they finally took flight for the first time, that line covers almost *ten thousand years.* Hot-air balloons aren't like spaceflight or time travel, technologies that require many members of a civilization to work together to produce them. The original hot-air balloons were invented by *two bored brothers* out of a *burlap sack.*

With enough motivation, you don't even need a civilization to produce a hot-air balloon: a single individual, even in Neolithic times and without any spindles or spinning wheels, could over the course of their life collect enough plant and animal fibers and spin enough thread by hand to create a hot-air balloon. In the more than 200,000 years during which this was possible, nobody ever thought to do that. Instead flight-minded humans usually just looked at birds and tried to copy them, which typically involved making giant feather-covered artificial wings, and sometimes covering the person strapped to the wings in feathers too, just to be safe.

> **CIVILIZATION PRO TIP:** Covering yourself in feathers is neither a necessary nor sufficient condition for flight, but rather is a choice that should be made for fashion purposes only.

Since taking off from land in such a contraption wouldn't work, people instead jumped from towers wearing them, assuming that was the secret to flight. At best they could hope to glide for a short distance, but these pilots usually fell straight down to either broken bones, death, or castration,* with the lack of flight being blamed on the pilots not having a tail (852 CE, 1010 CE),

---

* As with most things involving heights, a lot depends on what you fall on and how hard you fall on it.

using chicken feathers instead of eagle feathers (1507 CE), or the wind not being strong enough to fill their coat like a sail to keep them airborne (1589 CE).*[36]

In China around 500 BCE, kites were invented (you can invent them too; just spread fabric across a light framework, attach a string, and add a tail for stability). Afterward sufficiently large kites in strong enough winds were being used to lift humans, but anyone who has flown a kite and seen how easily they can crash knows how dangerous and deadly this was. Around 200 BCE the Chinese had also invented paper sky lanterns, which are in effect tiny hot-air balloons powered by a candle. However, despite this, nobody ever scaled this idea up to human-sized flight. In contrast, in 1250 CE a European actually published a book with a design for a hot-air balloon in it,[†] but since nobody at that time had figured out that air had weight and hot air weighed less, it was designed to be filled with "aetherial air": a gas, to be invented in the future, that was capable of floating in the atmosphere. Put simply, in 200 BCE humanity held in one hand the knowledge that hot air rose, and by 1250 CE held in the other hand a design for a machine that could be powered by hot air, but the two ideas were never joined together until both were rediscovered in France in 1783 CE.

And these Frenchmen (the Montgolfier brothers, who named their hot-air balloon the "montgolfier," a name still used by the French today) didn't even know that hot air rose! Their earliest experiments were, as we said, done with a burlap sack lined with paper to help keep the air in. They initially used steam as fuel, but this tended to ruin the paper. They instead switched to woodsmoke,

---

* Yes, this was the one that resulted in castration! A near contemporary account of this can be found in a publication made by one John Hacket in 1692 CE, delightfully titled *Scrinia Reserata: A Memorial Offer'd to the Great Deservings of John Williams, D. D., Who Some Time Held the Places of Lord Keeper of the Great Seal of England, Lord Bishop of Lincoln, and Lord Archbishop of York. Containing a Series of the Most Remarkable Occurrences and Transactions of His Life, in Relation Both to Church and State*, Part 4.

† It wasn't *quite* a modern hot-air balloon, but all the parts were there. The design by Franciscan friar Roger Bacon featured a single-mast sailboat held aloft by four large "balloons" (hollow copper globes), attached to the hull of the boat with rope. Remove the mast and you've got the equivalent of a modern invention: a basket held aloft by balloons.

believing it to be some kind of "electric steam" that released a special gas they named "Montgolfier gas" (*because of course they did*), and this gas had a special property called "levity." Even with all these fundamental misunderstandings of what was going on, the basic idea of "capture a lighter-than-air gas in a thing, and then the thing goes up" is all that was required for that first flight.

The finer and tighter the weave of your fabric, the better it will hold air, and silk (see Section 10.8.4) works great. The direction a hot-air balloon travels in is of course up to the winds, but by adding engines to your balloon you'll get directional control, and with that you've invented the airship! But can you do even better?

You absolutely can. While the hot air you've been using rises because it's lighter than regular air, it's far from the lightest gas there is. And you want lighter gases, because the lighter the gas, the less fuel you need to get airborne, the more you can lift, and the farther you can travel. An obvious improvement here would be to eschew heated air entirely and fill your balloons with *the lightest gas in the entire universe* instead. So let's do that!

The lightest gas in the universe is hydrogen, and Appendix C.11 shows how you can use electricity to extract it from brine. But if you need a lot of it—and you will, if you're building airships—you'll probably want to use cheaper methods. You could run steam over red-hot iron, which will break up the steam into hydrogen (as a gas) and oxygen (which will helpfully form iron oxide on the iron), but this requires a lot of iron. An easier solution is to do as early amateur aviators in our timeline did and rely on the fact that dilute sulfuric acid reacts with iron to produce hydrogen gas.* Dilute sulfuric acid by (slowly) adding it to 3⅓ times its weight in water, put iron filings in a barrel, and pour your diluted acid on top of the iron filings in a 2:1 ratio by weight—meaning 2kg of acid gets poured on top of 1kg of iron. This will react to give you your hydrogen! You can then pass it through a second barrel filled with slaked lime (which you can make in Appendix C.4) to remove any acid carried over with the gas, which

---

* Sulfuric acid actually reacts with lots of metals, including aluminum, zinc, manganese, magnesium, and nickel. But you'll probably use iron, since that's likely to be the easiest metal to find.

you'll want to do because otherwise the gas you're producing could eat through your balloon: historically, rarely a good thing. The sulfuric acid will exhaust itself before the iron does, so you can drain out the used-up acid from the first barrel and refill it with more until there's no more iron left to react. Your hydrogen production apparatus will look like this:

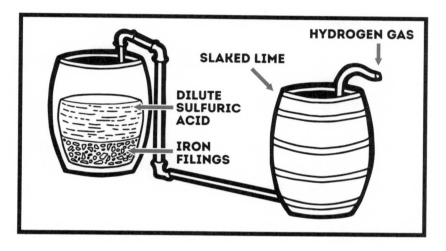

HYDROGEN GAS

SLAKED LIME

DILUTE
SULFURIC
ACID

IRON
FILINGS

Figure 46: An apparatus for the production of hydrogen.

Around 400kg of iron and 800kg of acid will produce about 140 cubic meters of hydrogen, and 10 cubic meters of hydrogen is enough to lift around 10.7kg, depending on the day's air pressure, temperature, and humidity.

Now before you rush out to start mixing sulfuric acid and iron together, keep in mind: hydrogen is an extremely flammable, *violently explosive* gas. The world was horribly reminded of this on May 6, 1937 CE, when a hydrogen-filled aircraft named *Hindenburg* burst into flames and fell to the ground while attempting to dock with a mooring mast in a disaster so ghastly that it ended the entire era of hydrogen airship travel. The culprit was a single spark of static electricity.[37]

At this point, you may be thinking, "Wow, why didn't they just use helium instead, that's definitely what I would've used." And yes, while helium doesn't explode, react, and *is* the next-lightest gas—with about 88 percent of the lifting power of hydrogen—it is much, much harder to come by. The only natural source of helium on Earth is produced through the (extremely slow) radio-active decay of heavy elements like uranium. And even when that happens, any helium that isn't trapped deep underground escapes into the atmosphere, where it's so light that it eventually ends up *lost in space*. Helium is an almost entirely non-renewable resource.* So if you want inexpensive and efficient lighter-than-air flight, your only option in the short term is to use hydrogen and be very, very, very, very careful.

There is, however, one more alternative: inducing things *heavier* than air to fly.

While inventing lighter-than-air flight is pretty trivial, the basics of heavier-than-air flight are sadly a lot more complicated than saying, "Just fill a bag with hot air or other light gas, you're done, *see ya.*" To make matters worse, a fully detailed explanation of aerodynamics would require much more space than we have available in this probably-won't-be-read-unless-there's-a-catastrophic-failure-in-your-FC3000™-rental-market-time-machine-and-seriously-how-likely-is-*that*-going-to-be repair guide. But even the basics of how heavier-than-air flight works will put your civilization thousands and thousands of years ahead of everyone else throughout history, and with that head start you can do what we did in our own history: build planes, run experiments, and use science to figure out what works.

You'll save a lot of time and broken limbs and money and human lives by first building a wind tunnel, which is just a large tube you blow air through but which still took humans until 1871 CE to come up with. It's a simple invention, but it flips the script by letting you study flight by moving air over a stationary wing (usually a model), rather than trying to study a moving wing as it

---

* In the 1960s CE the United States actually started stockpiling helium under-ground as part of a National Helium Reserve, and by 1995 CE a billion cubic meters of gas were being stored. However in the next year, the government decided to begin phasing out the reserve to save money, selling the stored gases to industry. There are a few ways to produce helium without relying on natural reserves: hydrogen fusion, proton bombardment of lithium in a particle accelerator, or through lunar mining missions, but these are all, it's fair to say, *slightly* more expensive alternatives.

tears through the atmosphere (usually by flying an experimental plane and *kind of just seeing what happens*). Attach strings to see how air flows around an aircraft, or introduce smoke to follow the movement of air directly. You can measure the aerodynamic forces on the plane itself by mounting it on a balance scale—which, if you haven't invented it yet, is simply a beam mounted on top of a triangle, with platters hanging off each side. When weights on both sides of the beam are equal, the beam is balanced. Build your wind tunnel balance scale with one arm entering the side of the wind tunnel to hold the plane, and the other side outside the wind tunnel, holding weights equal to the resting weight of the plane. As the wing generates lift in the air tunnel, the apparent weight of the plane changes, and you'll be able to quantify the lift it's generating.*

A wing, in cross-section, looks like this:

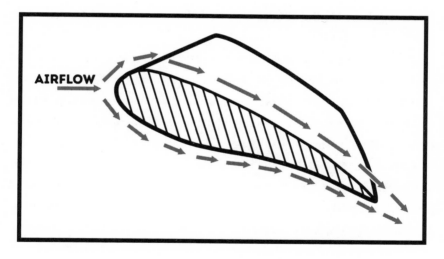

Figure 47: A wing in cross-section.

---

* Unfortunately, aerodynamic forces don't scale perfectly, which means a model plane won't fly in the exact same way a full-sized plane will. You'll still need to do some experiments, but you'll have the advantage of knowing how wings work and how changes to their shape affect their performance! And you'll save yourself the hundreds of years we wasted on "planes that flap their wings like birds," "planes that flap their wings like bats," "screws that point upward to burrow through the air itself," and so on.

Wings work through locally changing air pressure, by exploiting the fact (which you should now credit yourself with discovering) that an object moving through a gas will remain in contact with the gas at all points and at all times.* The wing splits the air, causing the air traveling over the top to be curved and pulled down to match the shape of the wing. This forces the air to occupy a larger volume, which reduces its pressure. In contrast, the air traveling underneath the wing is pushed into a smaller area of space, which increases air pressure there. This change in pressure is what generates an upward force, called "lift."

Wings generate lift in a second way too, and it's by exploiting the same "equal and opposite reaction" law you used to invent propellers in Section 10.12.5: Boats. Air passing both over and under the wing is directed downward as it clears the wing, which in turn pushes the wing upward. You can deflect more air by tilting the wing more, and the more downward force the wing creates, the more it generates lift—up to a certain point, at which the air no longer flows smoothly along the wing, causing "turbulence" in the air, a huge reduction in lift, and a stalled and soon falling aircraft.

Of course, to generate that lift, you need to thrust your wings through the air. This can be done by jets or rockets, but most planes (and, we imagine, most stranded time travelers) will use propellers instead, which are really just a set of tiny rotating wings that move the plane forward instead of up.† By introducing a slight twisting to the shape of the propeller, you'll make the entire propeller more efficient. In fact, slight changes to the shape of wings—whether used in a propeller or not—can have large effects, and that's one property you're going to exploit when building a plane! Here's what a simple plane looks like, and it's a design you'll want to copy:

---

* This also holds in fluids too! Now you've discovered two natural laws in as many sentences. Hey. Good thing you read this footnote.

† Early aircraft designers weren't sure which worked better: propellers at the back of the plane to push it, or propellers at the front to pull it. Front works best: propellers behind wings are less efficient, due to pushing through air that's already been disturbed by having a plane fly through it. Actually, early aircraft designers weren't sure about a lot of things, so if you aren't either, don't worry: for decades, every single person who flew managed to do so without having a correct understanding of how and why their planes really worked.

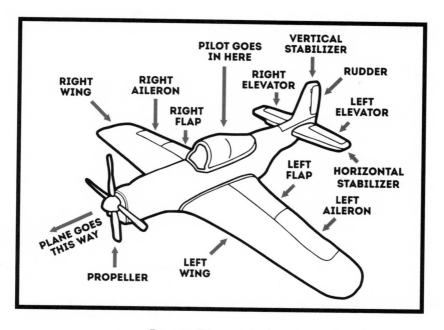

Figure 48: The parts of a plane.

The tail wing helps stabilize the plane when it's flying, and the flaps at the trailing edge there, called "elevators," will move the rear of the plane up or down, which lets you control the up/down angle of the plane. The rudder moves left and right, which moves the nose of the plane left and right. The ailerons at the front are used to roll the plane over: lift one side up and lower the other down, and the plane will begin to roll. Besides sweet tricks, these help you stabilize the plane and keep it flat. Finally, the flaps function identically to the ailerons, but they are designed to go up or down at the same time, which allows you to adjust the amount of lift the wings generate simultaneously. You can lower your flaps to generate more lift—useful when easing a plane down to land at slower speeds—and raise them in order to accelerate to higher cruising speeds once airborne.

Besides thrust and lift, the other two factors operating on a plane are weight (i.e., the gravity that pulls you down to the Earth) and drag (any forces that oppose thrust, like air resistance). And that's another area where heavier-than-air flight gets complicated. In theory, attach wings of sufficient size onto a thing capable of thrusting those wings forward, and you will fly. In practice, engines that generate the thrust required to get enough lift to make humans fly tend themselves to be heavy, which only makes the problem more difficult. Internal combustion engines have a better power-to-weight ratio, but steam-powered aircraft have been briefly but successfully flown in the past: the first manned heavier-than-air powered flight was in 1874 CE with a steam-powered aircraft and predated the Wright brothers by almost thirty years.*

But before you begin the hard work of attaching engines to planes, experiment with gliders first: planes without engines that are launched from high places. These are the training wheels of airplanes, and while the technical prerequisites for heavier-than-air powered flight are steep, all you need for gliding is some wood, some fabric, and some know-how—which we've already provided. Temporal experiments have been performed in which a functional wooden glider was introduced to Europe in 1000 CE, with no other supporting technologies. This still doesn't produce powered flight until around the Industrial Revolution in 1760 CE, but it does reliably produce aircraft carriers—complete with the crossbow-style catapults to launch them into the sky—by the Renaissance of the early 1400s.[38]

Your civilization will probably want to start with hot-air balloons right away, and then begin experimentation with either airships or heavier-than-air flight, but it's entirely up to you. You didn't travel back in time for some book to tell you *not* to cover yourself in chicken feathers just to see what happens, and *we respect your choices.*

---

* The Wright brothers produced the first manned heavier-than-air *self*-powered flight. The 1874 CE machine was steam powered but could fly only after ramping off a ski jump, and then gliding to the ground: the steam engine on board was insufficient to sustain flight. Speaking of the Wrights, once they'd invented (and patented) airplanes, they stopped innovating and spent most of their time suing not just their competitors but even individual pilots who dared to fly non-Wright planes. These lawsuits had a ruinous effect on American aviation: by January 1912, in France (where the Wrights also held a patent, but where its enforcement had been stayed repeatedly) more than 800 aviators were making flights each day, compared to only 90 in the United States. The lawsuits ended only in 1917 CE, when the US government legally forced airplane manufacturers to share their patents, but the damage was done. When the United States entered World War I in that same year, it was with French-built airplanes: all American aircraft were deemed unacceptably inferior. All of this is to say: if you do decide to invent heavier-than-air flight, *maybe* try to be a bit more chill about it.

10.13

"I WANT EVERYONE TO THINK I'M SMART"

↗ The sole technology in this section, **logic**, gives the members of your civilization a better way not only to reason but to know that their reasoning is accurate. On top of that, it lays the foundation for machine-based reasoning down the road, as you'll see in Section 17: Computers. It's also one of the greatest achievements humans ever came up with, and since it took us hundreds of years to get it right, it's only logical that you now take advantage of this shortcut.

10.13.1: LOGIC

If the world were a logical place, men would ride sidesaddle.

—*You (also, Rita Mae Brown)*

WHAT IT IS

A constructed system for structured thought that not only changed the way we think but also eventually allows you to build machines that reason *in the exact same way*

BEFORE IT WAS INVENTED

Clear and correct abstract thought was literally more difficult

ORIGINALLY INVENTED

350s BCE (logic first scientifically investigated by Aristotle)

300s BCE (first propositional logic)

1200s BCE (propositional logic reinvented)

1847 CE (propositional calculus invented)

PREREQUISITES

spoken language

HOW TO INVENT

The basics of logic were discovered several times throughout history (in China, India, and Greece), but it's the Greek version of it—Aristotle's syllogistic logic—that for historical reasons became the most influential, so that's what you're about to invent. In it you start with axioms—things that are self-evidently true—and build up conclusions from there. A syllogism has a major premise (1), a minor premise (2), and a conclusion (3), and looks like this:

1. All humans will die someday.
2. Imhotep is a human.
3. Therefore, Imhotep will die someday.

Pretty straightforward, right? You can make all sorts of different arguments in this format:

1. All time travelers have considered making out with their past selves.
2. All FC3000™ users are time travelers.
3. Therefore, all FC3000™ users have considered making out with their past selves.

Or even,

1. All humans have flesh.
2. All flesh can get cool tattoos applied by placing pigment into the skin with sharpened sticks, animal bones, or needles: the top-layer epidermis heals while the body's immune system engulfs the pigment particles beneath, stabilizing and concentrating the pigment just beneath the epidermis.
3. Therefore, all humans can get cool tattoos applied by placing pigment into the skin with sharpened sticks, animal bones, or needles: the top-

layer epidermis heals while the body's immune system engulfs the pigment particles beneath, stabilizing and concentrating the pigment just beneath the epidermis.*

Since you can substitute different words in this structure, as we just saw, you can actually reduce your argument to symbols. Let's use S to stand for "subject," M for "middle," and P for "predicate," which is just "the thing we're stating about our subject":

1. All M are P.
2. All S are M.
3. Therefore, all S are P.

And that's the magic of syllogistic logic: if your premises are true and the syllogism's structure is valid, then *it's impossible for your conclusion to not also be true*. If all M are P, and all S are M, then all S *have* to be P. It doesn't matter what M, P, and S are: if they meet those criteria, then that conclusion will always be correct.

Syllogisms allow the humans in your civilization, for the first time, to reason about *abstract* logic and argument, rather than about getting bogged down in the particulars of whatever argument's being made. Instead, the *structure* of the argument can indicate if an argument is invalid! Even if the premises are true, if they're not in a valid syllogistic structure, the conclusion does not necessarily follow.

There are fifteen valid logical syllogistic structures you can produce, and we're going to save your civilization years of hard-core logic and philosophizing by giving them to you right now:

---

* That's right. You just learned logic and how to give yourself some sick ink . . . *at the same time.*

Major premise	Minor premise	Conclusion
All M are P	All S are M	Therefore, all S are P
No M is P	All S are M	Therefore, no S is P
All M are P	Some S are M	Therefore, some S are P
No M is P	Some S are M	Therefore, some S are not P
All P are M	No S is M	Therefore, no S is P
No P is M	All S are M	Therefore, no S is P
All P are M	Some S are not M	Therefore, some S are not P
No P is M	Some S are M	Therefore, some S are not P
All M are P	Some M are S	Therefore, some S are P
Some M are P	All M are S	Therefore, some S are P
Some M are not P	All M are S	Therefore, some S are not P
No M is P	Some M are S	Therefore, some S are not P
All P are M	No M is S	Therefore, no S is P
Some P are M	All M are S	Therefore, some S are P
No P is M	Some M are S	Therefore, some S are not P

Table 16: Valid logical syllogisms. This is something that took humanity thousands and thousands of years to puzzle out, and it all fits in a single 15×3 grid! Hooray!

You can come up with other syllogistic structures, but they'll either be wrong (saying that "All M are P" and "all S are M," and then concluding that "therefore no S are P" is straight-up baloney), or they'll produce conclusions that are weaker than those you've already got above. For example, if all poodles are dogs, and all dogs are mammals, then concluding that *some* poodles are mammals, *while technically correct*, is pretty misleading. This leads us to this very important Civilization Pro Tip:

CIVILIZATION PRO TIP: All poodles are definitely mammals.

Once invented by Aristotle, syllogisms stood, without major modification, for over two thousand years. But while they're useful for organizing thought, they're not perfect: they rely on language, which can always be muddy or imprecise. As an example, imagine if you'd concluded, through perfectly logical reasoning, that "Therefore, some dinosaur is feared by all safety-minded time travelers." One person might read that sentence as "every safety-minded time traveler fears at least one dinosaur," while another might read the same words and conclude that there's a single colossal mega-dinosaur out there that all safety-minded time travelers fear. Is this true? *It seems important to know.*

It took a while,* but humans eventually realized that if they could transform syllogisms into a calculus they could then solve, they would be able to explore logic and reason with the ultimate precision of mathematics. This line of reasoning would eventually lead to something called "propositional calculus," which, despite its super-impressive name, is actually really simple.†

Take a syllogism we've already seen as an example: "All time travelers have considered making out with their past selves, and all FC3000™ users are time travelers, therefore all FC3000™ users have considered making out with their

---

* How long? Well, propositional logic was invented around 300 BCE, lost, reinvented around the 1200s CE, and then refined into symbolic logic in the 1800s CE by George Boole, from whose name we get the word "Boolean," meaning "either true or false."

† In fact, this greater precision is what led humans to realize that several of the syllogisms Aristotle put forward *weren't actually correct*. We showed you fifteen of them, but Aristotle's *original* list was longer, and it was only when we finally examined these syllogisms with the more precise propositional calculus that several of them were found to work only when you presuppose a class has members: in other words, when you assumed the sets of things being considered *actually exist*. As an example, one erroneous form was "All M are S, and all M are P, therefore some S are P." This worked with things that exist: if we were to argue that "all horses are mammals, and all horses have hooves, therefore some mammals have hooves," you would say, "Yes, duh, *I know*." But the same syllogism completely falls apart if M refers to things that don't exist. Using the same form, and with premises that are equally true, we could argue that "all unicorns have horns, and all unicorns are horses, therefore some horses have horns." But, until they were genetically engineered in the twenty-first century, miniaturized, and adopted as one of the most popular if unpredictable house pets, *no horses had horns*. An incorrect conclusion is produced because the syllogism is flawed. It's only fixed if you add an "existential clause" to your syllogism by saying "If all M are S, and all M are P, *and M exist*, then some S are P." Humans like to think they're pretty smart, but multiple errors like this in Aristotle's reasoning—which would've been discovered way earlier had generations of logicians only used more unicorns in their example sentences—stood for over *two thousand years*.

past selves." We saw how this reduces to "All M are P, and all S are M, therefore all S are P." If we replace the word "are" with a symbol meaning "implies," →, then this syllogism can be written as:

M → P and S → M, therefore S → P

In other words: if time travel implies makeout thoughts, and the FC3000™ implies time travel, then the FC3000™ implies makeout thoughts. Sorry, time travelers, *but it's true.* Now, to shorten the amount we have to write, let's replace "and" with a ∧ symbol and introduce brackets so it's clear which variables go together. This gives us:

(M → P) ∧ (S → M), therefore (S → P)

Replace "therefore" with its own symbol of ∴; replace M, P, and S with more generic and sequential variables p, q, and r; and swap the order of our statements to make them more intuitive, and you'll get this argument in its final form:

$[(p → q) ∧ (q → r)] ∴ (p → r)$

In other words: if p implies q, and q implies r, then p implies r. It's the same argument we saw when traveling through time to our too-seductive past selves, distilled to pure symbolism.

Here's another simple argument: "not p" (which we'll represent as ¬p) is the opposite value of p. Our logic will deal with only things that are either true or false, so "not true" is the same as saying "false," and "not false" means "true." And given that, we can easily prove that "not not p," or ¬¬p, must therefore equal p. All you need to do is write out all possible options, of which there are precisely two:

p	¬p	¬¬p
true	The opposite of p, therefore false	The opposite of ¬p, therefore true
false	The opposite of p, therefore true	The opposite of ¬p, therefore false

Table 17: These are called "truth tables," and you just used one to prove that p equals ¬¬p. Look, we won't say just yet that you're history's greatest logician. But we will say this: you're definitely history's greatest logician *so far.*

This is all it takes to prove that the proposition "$p \therefore \neg\neg p$" is valid. That probably seems like a pretty simple thing to prove, and it is, but you're laying a foundation of valid argument forms that you'll be able to use to manipulate and prove much more complex propositions. By putting your reasoning in symbolic format like this, you're not just working out rules for how these variables can interact with one another, you're also discovering *actual rules for logical reasoning itself.* You're figuring out a new way to think that's provably accurate. You, my friend, are inventing logic. We've put a list of the valid forms of argument in Appendix D, and if you decide to produce an extremely logical civilization, they'll save you a lot of time.

This is, of course, just one way to build a system of logic: you can build more complicated ones that rely on degrees of truth,* capture more complicated relationships,† and so on. The reason we taught you this system is that it deals in either absolute truth or absolute falsehood, and nothing in between. It's binary. And as you'll see in Section 17: Computers, you'll be able to use binary logic to build machines that reason just as logically as you can, but at thousands of times the speed.

Logic is the only way you get to play video games and watch movies in bed again.

CIVILIZATION PRO TIP: *You're welcome.*

This concludes the section of technologies you can use to address common human complaints, and we now move on to chemistry, philosophy, art, and medicine: technologies that, even though they may not be requested by name, still greatly improve any civilization.

---

* For example, a fuzzy logic system captures *degrees* of truth: a statement can be marked as 0 for false, 1 for true, but also anything in between. In this system, 0.9 is almost true, 0.0001 is practically false, and 0.5 is halfway between true and false.

† You can introduce other operators for different relationships, like $\Diamond p$ to mean "it is possible that p." This roughly equates to the "Some M are P" statements we saw in Aristotle's syllogisms.

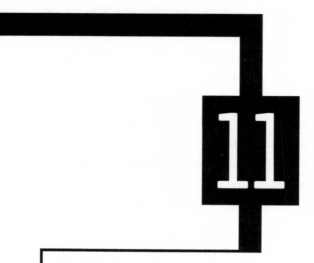

CHEMISTRY: WHAT ARE THINGS, AND HOW DO I MAKE THINGS?

The trick in chemistry is to never . . . overreact.

Chemistry is the science of digging things out of the ground and transforming them into other, more useful things. These transformations can take many forms, and fully understanding them can take a lifetime of study. We've got only a few pages here, though, so instead get ready for some *pure, freebased information.*

WHAT ARE THINGS MADE OF?

This is one of the most fundamental questions humans have ever asked, and coming up with a response to it took humanity thousands of years of study. Nobody has time for that, so here's the answer: things are made of atoms, which are tiny pieces of matter around 0.1 nanometers long. At the center of atoms is a core—the nucleus—which is made out of positively charged protons (hence the name) and neutrally charged neutrons (again, hence the name), and which constitutes 99.9 percent of an atom's mass.

There are more than a hundred different kinds of atoms, called "elements." The number of protons in each atom's core determines which element it is: any atom with 1 proton is hydrogen, any with 8 is oxygen, and any with 33 is arsenic. As you need oxygen to survive but arsenic will absolutely kill you, you'll definitely want to have a passing familiarity with how many protons atoms have. Luckily for you, we've made a big chart showing that and how they relate to one another called "the periodic table," which you can see in Appendix B: it's the complete table, accurate to 2041 CE, the year in which it was last changed. While atoms can't gain or lose protons without becoming a different element, they can gain or lose neutrons and still stay whatever they are—these variants are called "isotopes." An isotope with more neutrons will weigh more than an isotope with fewer.

Around the nucleus of each atom resides one or more negatively charged electrons. Electrons travel in different orbits—some closer, some farther out. The smallest orbit can hold 2 electrons, but the next one can hold 8, the next larger one 18, and so on, according to the formula $2(n^2)$, where n is your shell number. And while electrons tend to stick close to the nucleus, an inner shell doesn't always have to be filled with electrons for them to begin populating an outer shell. So, given all that, here's a rough model of what an atom looks like, schematically:

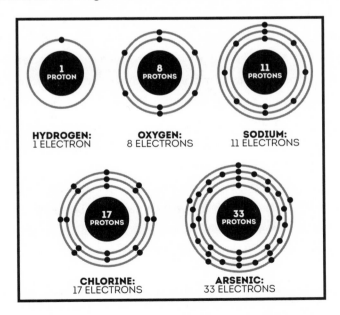

Figure 49: Schematic views of several elements. If you were expecting a chemistry pun here, we apologize. We're not good atom.

Atoms can combine with other atoms to form molecules: these are the "chemical reactions" you've heard about. An atom's electrons can give you a clue to how reactive it is: atoms want to have that outer orbit filled with electrons, so elements that have achieved this will be less reactive than those that don't. And it generally goes as you expect: elements with two extra electrons tend to react with elements that need two more. This does imply that atoms that already have their outer shells filled up with electrons—elements like helium or neon—won't want to react with anything. And they don't! Helium and neon are so unreactive that molecules containing them were actually thought to be impossible. They're not—we've made some, it's no big deal—but they usually require extremely high pressure and/or extremely low temperature.[39]

Let's look at water as an example of a chemical reaction: two hydrogen atoms and one oxygen atom can combine to form water, or H_2O (that subscript 2 is what tells you there are two hydrogen atoms). The oxygen has six electron slots filled in its outer shell, and hydrogen atoms have one each, so the two hydrogens share their electron with the oxygen, a "water" molecule is formed, and if atoms can be said to be happy,* these ones are. This sharing of electrons is called a "covalent" bond.

But there's another wrinkle to consider: electrical charge. Electrons have a negative charge, protons have a positive charge, and since most elements have the same number of electrons and protons, those charges cancel out and the atoms are electrically neutral. However, atoms don't always share electrons, like we saw when making water. They can also exchange them. When that happens, and one atom loses electrons while another one gains them, this gives the resulting atoms overall positive or negative charges. In electrical charges (and, occasionally, interpersonal romance) opposites attract while likes repel.

An example of this: sodium (Na on your periodic table has 11 electrons: so that's 2 at the first level, 8 at the next level, and just 1 at the outer level. Chlorine (Cl) has 17 electrons, which means 2 at the first level, 8 at the next level, and 7

---

* They can't, but we ascribe emotions and desires and motivations to inanimate matter because it makes explanatory rhetoric *so much easier.*

at the outer level. Chlorine wants one more electron to get to eight, if sodium can get rid of one electron it'll get to eight too, so that's what happens. However, when sodium loses that extra electron it ends up with a positive charge, and when chlorine gains one it ends up with a negative charge, so they attract each other, and bond anyway. This resulting chemical—NaCl—is salt, and this bond caused by electrical attraction is called an "ionic" bond.

Covalent bonds are more easily broken—the "sharing" of electrons is more temporary—and result in chemicals that are usually liquid or gaseous at room temperature. Covalent bonds happen only between non-metals. (Metals and non-metals are all listed on the periodic table, along with semimetals, which are "in-between" elements that can have the properties of both). Ionic bonds are harder to break, result in solids at room temperature, and usually occur between a metal and non-metal.

Guess what? Reading those last few paragraphs took you all the way from 13,799,000,000 BCE to humanity's state-of-the-art understanding in the mid-1900s CE. It took us until 1800 CE to even figure out what elements *are*, so heck, even if you just skimmed the first few paragraphs you're already doing fine. To go further, you'll need to know that both protons and neutrons are made out of even smaller particles (called "quarks," which themselves come in six wacky flavors*), and that electrons don't so much "orbit a nucleus" as "exist as waves in an unobservable area of potential locations rather than at a single point,"† but you're not going to need that level of detail in your current circumstances unless you're trying to build a time machine, which you won't be,

---

* They actually are called "flavors," and they actually are wacky: *up*, *down*, *top*, *bottom*, *charm*, and *strange*, in order of increasing wackiness. For more information on quarks, please refer to *The Time Traveler's Guide to Quarks, Which Are Useful for Building Time Machines and Yet You'll Still Need to Know a Heck of a Lot More to Build One Successfully, but Heck Who Am I to Stop You So Enjoy Reading About Quarks I Guess, Vol. 9*, assuming you brought it with you.

† When things get very small they also tend to get very weird, which is about as detailed a summary of quantum mechanics as you can expect to find in a footnote of a non-quantum mechanics text. In fact, if you'd like to pose as an expert on quantum mechanics, just repeat that first clause of this footnote in response to any question you receive, and you'll be off to a great start.

because building a time machine is so hard that it was easier for us to write a guide on how to *reinvent civilization from scratch* than to even begin explaining where to start.

The existence of atoms is hard to prove without extremely powerful microscopes, but you can observe their effects easily: dust in a glass of water, for example, will move about randomly. This "[Your Name Here] motion" (né "Brownian motion," after this botanist named Robert Brown, who discovered it in 1827 CE) occurs because the dust is being knocked around by tiny particles (i.e., molecules) of water.

WHERE DO THINGS COME FROM?

The Big Bang (in 13,799,000,000 BCE, which would've definitely been worth checking out if your time machine wasn't busted) sent matter out into the universe, and this matter coalesced into (mostly) hydrogen: the simplest element. Gigantic masses of hydrogen eventually coalesced into such huge balls of gas that the pressure of their own weight started fusing hydrogen (with one proton) into helium (with two) in their cores, which (a) released a lot of energy, and (b) is also exactly what our (and every other) sun is.

This process can last anywhere from millions to trillions of years (depending on the size of the star) until the hydrogen runs out. When that happens, if the star is big enough, it'll have enough pressure to start fusing helium instead. Helium fuses into the higher elements: from lithium (3 protons) all the way up to carbon (6 protons),* with carbon being favored. Once the helium is gone, and again if the sun is big enough, it'll start fusing the carbon too, which forms elements up to magnesium (12 protons). This stage can last around six hundred years. If the sun is super giant, this process can repeat, forming elements all the way up to iron (26 protons).

---

* How does helium (2 protons) fuse with itself to produce elements with an odd number of protons? Sometimes protons and neutrons break away during fusion, which is how helium (2 protons) can fuse with itself to produce lithium (3 protons), boron (5 protons), or any of the other odd-numbered elements.

At that point, however, things break down: iron costs more energy to fuse than it produces during that fusion, so any star fusing iron is going to die soon—usually in less than a day. At this point, the sun dies—and depending on how big it is, it either collapses and becomes a cooling husk (called a "white dwarf" when it's cooling, and a "black dwarf" when it's completely cooled, and so dense that a cubic centimeter of it would weigh more than three tons), or it becomes a neutron star (a white dwarf with so much pressure that all matter is packed with the same density as in an atomic nucleus, a cubic centimeter of which would now weigh almost a billion tons), or it becomes a black hole (a neutron star so heavy that not even light can escape its grasp, and which honestly you should not be messing around with even a cubic centimeter of).

So this explains where all the elements up to iron come from: stellar fusion. But what about the heavier elements? Well, we skipped over a stage in that last paragraph: when stars die, sometimes the gases that used to be held away from the core by the energy spewing out from the sun now find nothing holding the force of gravity back, and the sun undergoes a final, catastrophic collapse. All its mass collapses inward, causing such heat and pressure that protons and electrons are fused into neutrons.

And then it explodes.

It actually explodes so powerfully that one of the few explosions bigger is the Big Bang itself.

These explosions—called "supernovae" if they're normal sized and "hypernovae" if they're colossal—blast matter outward in a huge particle storm with such energy that they can, for about a month, burn brighter than a billion suns. This creates highly unstable nuclei that decay into other elements—including those heavier than iron—which makes a supernova the only place in the universe where the heavier elements get made, at least until we started synthesizing them on Earth around 1950 CE. It's also why hydrogen and helium are by far the most abundant elements in the universe: you need stars to (slowly) produce everything else. Elements that aren't helium or hydrogen actually make up just 0.04 percent of the universe, which, since you're carbon-based, unfortunately makes the matter that you and everyone you've ever known is made out of minor enough to be dismissed as a universal rounding error.

If you're feeling down about that, just remember where you come from: *awesome explosions.*

WHAT CAN I MAKE OUT OF THINGS?

Technically: *everything.* And to help you get started with that, we've provided instructions on how to build many useful chemical things from scratch in Appendix C, which will definitely come in handy given your current circumstances. We've also included the chemical makeup for each—you don't need to know this information to produce them, but knowing their origins is a useful foundation for you or your descendants to build out the rest of your chemistry knowledge. And again, we'd like to stress here that several of these chemicals are dangerous, which is why we named that appendix "Useful Chemicals, How to Make Them, and How They Can Definitely Kill You" and not "Useful Chemicals, How to Make Them, and How They Are Super Fine to Rub Directly into Your Bare Eyes Without Any Problems Whatsoever." Please now either turn to Appendix C to begin your exploration of chemistry—or, alternatively, leave it for a while, turn this page, and learn about some cool philosophy instead.

12

MAJOR SCHOOLS OF PHILOSOPHY SUMMED UP IN A FEW QUIPPY SENTENCES ABOUT HIGH-FIVES

Is it solipsistic in here, or is it just me?

The philosophical groundings of your civilization are entirely up to you, but this extremely superficial overview of several world philosophies drawn from throughout history may prove to have several useful "idea starters" for your own civilization. These philosophies can be combined, expanded, weakened, strengthened, and deconstructed in hundreds of different ways, so feel free to do exactly that.

The difficult and often existentially terrifying work of philosophy demands you confront the heavy questions of life and existence head-on, which can be both disorienting and depressing. As you probably have enough problems already, instead of describing these philosophies in terms of the search for meaning and purpose in life—their usual *alleged* purpose—we'll instead discuss them in terms of high-fives, which are universally cool and fun.

Religious philosophies

Monotheism: God gave me a high-five.

Polytheism: One or more gods gave me a high-five.

Henotheism: There may be other gods and there may not: all I know is the one I worship gave me a high-five.

Monolatrism: There are definitely a bunch of gods, but I worship only the one who gave me a high-five.

Pantheism: The universe, which is equivalent to a god, gave me a high-five.

Panentheism: The universe, which has divinity throughout it but which is not equivalent to a god because any god is greater than the sum of the universe, gave me a high-five.

Omnism: All religions can give you different kinds of high-fives, but no single religion can offer the complete high-five experience.

Panpsychism: Everything in the universe has a consciousness and therefore could desire to give me a high-five.

Ietsism: Some sort of god somewhere out there gave me a high-five, but beyond that, who is to say?

Agnosticism: Maybe a god gave me a high-five, maybe I gave it to myself. Who is to say?

Atheism: I gave myself a high-five.

Autotheism: I gave myself a high-five. Also, I'm a god.

Apatheism: Wondering whether or not gods exist and are handing out high-fives is hugely irrelevant. None of you have better things to do?

Ignosticism: The idea of a "god" has no unambiguous definition, so arguing about whether or not they exist and are currently handing out high-fives is meaningless.

Deism: God or gods definitely exist, but they would never interfere with human affairs, so I'm pretty sure that's the only reason none of them have ever given me a high-five.

Dualism: There are good and bad forces in this world: for every high-five, there is a corresponding anti-high-five that is both down low and, sadly, too slow.

Antitheism: Gods handing out high-fives do not exist, but if they did, I would totally leave them hangin'.

Misotheism: Gods handing out high-fives definitely exist, but their high-fives are way nasty.

Solipsism: I gave myself a high-five. Unfortunately, I only imagined it, since nothing outside my own mind really exists.

Secular humanism: There are no gods to high-five us, but we can still be kind . . . and we can still high-five each other.

Table 18: Just as an aside, the phenomenon in which you encounter a word or phrase (like "high-five") over and over until it loses all meaning is called "semantic satiation." High-five!

Philosophies of being

Nihilism: Nothing, not even high-fives, has any meaning.

Existentialism: Nothing, not even high-fives, has any meaning, so it's up to the individual to give them whatever meaning they can by both handing out and receiving high-fives as authentically as possible.

Determinism: I'm high-fiving you, but free will is an illusion. If you could rewind the universe and run it again, all would be identical, and we would still end up high-fiving at this exact moment.

Consequentialism: Anything is justified, no matter how horrible, as long as it results in me getting a high-five.

Utilitarianism: Anything, no matter how horrible, that brings the greatest number of high-fives to the greatest number of people is justified.

Positivism: If you want me to believe in high-fives I'm going to need to see some scientific evidence.

Objectivism: It is in my rational self-interest to get high-fives, and any authority that does not respect my individual right to dispense and receive high-fives in whatever manner I so choose is bad.

Hedonism: High-fives feel great, and pleasure's awesome, so I'm gonna hand them out for as long as I want. Don't talk to me about consequences. If it feels good to high-five while having sex . . . *I might actually try that.*

Pragmatism: High-fives are only good if they can accomplish something.

Empiricism: Do not trust intuition or tradition: the only true way to fully understand high-fives is to give and receive them yourself.

Stoicism: Emotions can result in errors of judgment that interfere with clear, unbiased thinking. Therefore, the best high-fives are those given out for extremely logical reasons.

Absolutism: Certain actions are intrinsically right or wrong. For example, stealing—even to feed a starving puppy—might always be wrong, while high-fives—even if you keep accidentally slapping the person in the face real bad—might always be the correct course of action.

Epicureanism: Pleasure's awesome, but the greatest pleasures are the absence of pain and fear, so I'm going to high only a sensible number of fives because I don't want to end up with a hurt hand.

Absurdism: The sheer size, scope, and potential of things to understand about even one single high-five makes ever discovering the true meaning of high-fives impossible, and the only rational responses are suicide or blindly hoping there's a god who could one day completely understand high-fives, or, failing either of those, accepting the absurdity of high-fives and, despite it all, still cheerily handing them out.

13

THE BASICS OF VISUAL ART, INCLUDING SOME STYLES YOU CAN STEAL

With these instructions, you'll be able to paint with any color you can produce. Colors you can't yet produce, however, will remain pigments of your imagination.

Look, we didn't want to have to do this. We wanted to say "just draw some pictures, you'll figure it out, it'll be fine," but history has proven us wrong. You know how if you look down a straight line of train tracks the rails seem to converge at a vanishing point on the horizon? If you do, you're already doing better than the great mass of humanity in most every other time period, because until 1413 CE, *humans hadn't figured that out yet.** That's the reason old paintings always looked so wonky: not a single human on the planet knew how to draw in correct perspective.

---

* And yes, these humans didn't *have* train tracks, but that's no excuse. Fields of grain, fences, and even rivers and coastlines all produce the same convergence effect.

And here you might be saying, "Oh I don't know about that, maybe the ancient Egyptians just loved the anti-perspective style they came up with a whole lot [wherein the size of figures was related to how important they are thematically, rather than where they were situated in space] and that's why they never used correct perspective in a drawing even once"—and you might have an argument if it were not for the fact that throughout history, whenever perspective *was* discovered, artists go completely bonkers for it. Here's Leonardo da Vinci's *The Last Supper*, one of the most famous paintings in the world, painted in 1495 CE, just eighty years after Europeans first figured out what perspective is:

Figure 50: *The Last Supper*. That arch at the bottom is a doorway, added later, after some humans decided it was really important they have convenient ingress and egress though a priceless work of art.

Look at that tiled roof. Look at those squares along the walls and thick rectangular windows at the back. This is a painting that is at best one-third "I love religion so I'm gonna paint my favorite religious figures enjoying a meal" and at least two-thirds "Bro, my vanishing point is off the hook, seriously, check out my wall rectangles, *you don't even know.*" Here's Raphael's masterpiece *The School of Athens*, from 1509 CE:

Figure 51: *The School of Athens.* It was painted on a wall in Italy.

Do you notice the figures first? Or do you notice the tiled floor, the series of arched roofs, the steps, and the guy *writing on an actual cube in the foreground*, all rendered in deliberate, in-your-face single-point perspective? When modern humans first invented computer ray tracing, they generated thousands if not millions of images of reflective chrome spheres hovering above checkerboard tiles, just to show off how gorgeously ray tracing rendered those reflections. When they invented lens flares in Photoshop, we all had to endure years of lens flares being added to everything, because the artists involved were super excited about a new tool they'd just figured out how to use. The invention of perspective was no different, and since it coincided with the Renaissance going on in Europe at the same time, some of the greatest art in the European canon is dripping with the 1400s CE equivalent of lens flares and hovering chrome spheres.

Of course, earlier artists knew that things got smaller the farther away they got, but they had no mathematical theory behind it, and so had to guess at how things would look—which gave extremely mixed results. Some artists got close, like those in China after around 1100 CE, where a form of drawing now called "oblique perspective" generated illustrations that didn't correspond to any view you'd see in real life, but at least gave an approximate sense of 3-D objects in space.

What made perspective finally click was the discovery of vanishing points.

Figure 52: An untitled painting of a mill produced in China, around 1100 CE.

The lines of a railroad track you were imagining earlier go off into the distance, apparently converging on a single point on the horizon, where they disappear. If you render all shapes doing that—if the imaginary lines suggested by walls, buildings, cubes, and anything else all vanish toward the same point—you can produce a convincing image of the world, as if viewed through a window. That's what you're seeing here:

Figure 53: An altered version of *The Last Supper*, designed to help put things in perspective.

That's one-point perspective, where vertical lines are all parallel, and objects are viewed head-on. You can make it more complicated by rotating objects so their faces are angled away from you. Vertical lines are still parallel, but now each face has its own vanishing point, giving two-point perspective.

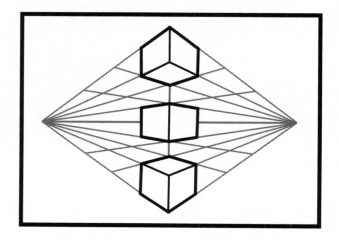

Figure 54: Two-point perspective.

Finally, three-point perspective adds a vanishing point above (or below) the object. Vertical lines are no longer parallel but angle toward their own vanishing points.

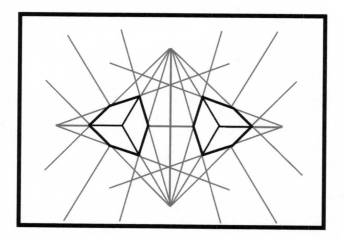

Figure 55: Three-point perspective. Extradimensional perspectives are left as an exercise for the reader.

This technique isn't perfect. *Technically*, any perspective drawing should only appear accurate when viewed from a single location, where those vanishing points would make sense. But human brains are great and invisibly correct for a lot of things that don't technically make sense: they automatically convert a rapid series of images into apparent motion (in flipbooks, which you can easily invent, and in motion pictures, which'll be a *little* more work), they automatically convert the fact sounds arrive in our ears at slightly different times into a distinct sense of location (that's why you have a sense of where sounds originate around you), and they automatically make any drawing in perspective look convincing even when not viewed from the precisely correct angle.

So that's the basics of perspective, and even knowing just this will allow you to create much more convincing and realistic images of the world around you. But realism isn't the only purpose of visual art—a fact that historically gets underlined once photography is invented. Once artists realize that and get over realism, other styles start to be explored—and here there really is no limit.

Below are some examples of different styles of visual art, which will allow you to kick-start the minds of artists in your civilization. With any luck, they'll leapfrog over what we created and generate new and astounding works of art beyond anything we've ever considered. Good luck!

Figure 56: Art.

Sidebar: Where Can You Get Pigments?

You can get black pigment from coal or charcoal—add it to water or oil and you've made paint—but other colors can be a bit trickier. Grinding up different minerals has been a source of pigment as far back as 400,000 BCE: just collect rocks of a color you like, reduce them to powder, wash them to remove any water-soluble parts, dry them out, and you're good to go. You can get other colors from biological sources, and ground-up insects, mollusks, and even dried-out poops have been used in the past. It's very easy to go too far: a shade of yellow called "Indian yellow" was once made by feeding cows only mango leaves until they were so malnourished their urine turned a bright yellow, and a favorite shade in 1600s CE Europe called "mummy brown" was made by grinding up ancient mummies (feline and, yes, human) to paint with their remains.

Vibrant shades of blue and purple are historically the hardest to produce. Until 1704 CE, one of the most vibrant shades of blue, ultramarine, was available only by grinding up a rare mineral called "lapis lazuli."* A painting of an ultramarine sky was therefore a status symbol, showing you could afford to *reduce precious stones to dust just to make a painting*. The color purple's association with royalty

---

* The synthetic blue pigment discovered in 1704 CE was not actually the world's first artificial pigment! It was the first one Europeans came up with, but back around 3000 BCE the Egyptians—also annoyed at how expensive grinding up lapis lazuli was—managed to produce an artificial version of the color by combining under heat some quartz sand, copper, calcium carbonate, and alkaline ash. This technique was used for thousands of years, but by 400 CE, every single person who'd known how to make it had died without telling someone else the technique or writing it down, and the knowledge of how to make an inexpensive substitute for one of the most expensive colors on the planet was lost. Everyone: *this is what happens when you keep really important knowledge a secret and then die.*

also originates in purple pigments being extremely expensive: at certain points in time, some were worth their weight in silver. The best purples were produced by extracting mucus from a tiny (6–9cm-long) species of snail found in the Mediterranean Sea, normally used by the snails to sedate their prey. Getting that mucus was extremely labor intensive: you either had to poke two snails to get them to attack (thereby "milking" them), or grind up the snails to get at the mucus that way. Either way, it took twelve thousand snails to produce only around 1 gram of pure dye. If you're interested, these snails evolved around 3,600,000 BCE, hence the famous time traveler's saying that "if there are other humans around, then there's always the potential to make some really, really expensive purple dye in one part of the Mediterranean!"

HEAL SOME BODY:
MEDICINE AND HOW TO INVENT IT

Re-creating medicine in the past will require you have some . . . patients.

Hippocrates was a man who introduced two major ideas to Western medical treatment around 400 BCE: one marginally useful and one incredibly catastrophic. The marginally useful one was the Hippocratic Oath, which is still used today by many doctors, who for some reason feel compelled to publicly promise they won't *intentionally* kill their patients. The catastrophic one was a formalized humor theory of disease.

The humor theory holds that all diseases, across all forms of life, are caused by an imbalance in one of the "four humors" inside our bodies: blood, phlegm, black bile, and yellow bile. While this was (arguably) an improvement over past medical theory (which held that disease was divine punishment meted out by angry and vengeful gods, so if you're sick, *maybe try not making the gods so angry and vengeful all the time*), it had no relation to the realities of medicine or the human body, and any treatment based on it will help you at best only through random chance. Nevertheless, medicine based on the four humors was practiced until 1858 CE, when cells were discovered and humans realized maybe not all disease could be cured through bloodletting, vomiting, gymnastics, and sensual massage.

Just to be clear: that's more than *two thousand years* that Western doctors treated patients based on the inaccurate and unhelpful idea of humor imbalances. That's longer than most civilizations have lasted, including, it must be noted, the Greek civilization that produced it. Medicine has progressed more in the few centuries *after* the four-humors theory was abandoned than it had in all previous centuries combined. If you don't want people to needlessly die in your civilization (because you are a decent human being, and because being struck down early by disease is objectively a suboptimal ending to any human life), you'll want to introduce the basics of modern medicine quickly.*

Of course, it's not just Western civilization that ran into trouble developing medicine. Taboos against human dissection have arisen in multiple cultures across multiple eras, and while this is for understandable reasons (it feels weird to cut into a dead body and start poking around) (oh, heads-up on that), they have, in all cases, held back medical development. If you want to learn how to treat humans, you'll need to know how their bodies work, and dissecting animals and reasoning by analogy will only get you so far. Questions like "Where does sweat come from?," "So do arteries move blood or air or what?," and "Does the womb stay in one place or is it like a separate animal that lives inside

---

* And it's not like humans leapt from "four humors" to "no wait, germs" directly. Many years were lost in Europe, India, and China with miasma theory: the idea that disease is caused and carried by bad smells. This theory at least had an upside: since waste and decay usually smell unpleasant, public works designed to fix the problem of "miasma" could still actually help people. This happened in London: after its cholera epidemics and the "Great Stink" of 1858 CE—in which warm weather made the *untreated human waste* floating in the Thames smell even worse than usual— the city invested in sewers to move smelly water away from the city. This was to be a marked improvement over the city's existing waste-disposal system, in which everyone just dumped their poop into the streets or into nearby cesspools, and then complained about how everything smelled bad all the time. It was only after the sewers had been completed—and the health of Londoners had improved—that humans realized it wasn't gross smells that carried disease, but germs. London's dramatic and hugely expensive sewer system—still in use today—was constructed for entirely the wrong reasons and only happened to improve public health *by accident*.

women and just moves around wherever it wants?"* have historically been best answered through human dissection. Thankfully, you've got the advantage of Appendix I, which features a human anatomy chart featuring the shape, size, location, and role of each major internal organ. Even that relatively simple information will advance medical treatment in your civilization by thousands of years.

What follows is basic medical information that you can use in any time period on yourself or on your fellow humans. If you had other options we'd say, "Yes, absolutely go to a doctor rather than trusting what's effectively a brief aside in a time-machine-repair manual with your entire health and wellness," but you don't, so you may want to study this next section carefully.

THE GERM THEORY OF DISEASE

Bad things happen when the *inside of your body* is colonized by *invasive microorganisms*, which is so disgusting medical professionals refer to it by the euphemism "infection." Microorganisms can take several forms, but the ones you need to worry about are bacteria (tiny animals) and viruses (tiny pieces of parasitic

---

* This "wandering womb" theory is one that the ancient Greeks operated under and which infected Western thought until the 1800s CE. It was thought that "hysteria"— uncontrollable emotions that, *clearly*, only women could be subject to—was caused by the uterus moving throughout the body of its own accord, thereby putting pressure on other organs. Treatment involved coaxing the uterus back into place with smells: bad ones near the nose to push it away, and good smells near the genitals to attract it forward. If that failed, sex was thought to be a good fallback treatment. When doctors (unsurprisingly, all men) finally accepted that wombs weren't separate living things that traveled throughout the body, the idea of the wandering womb finally died. But the idea of "hysteria" persisted, and by the 1860s it had evolved into a psychological ailment caused by women having too few orgasms. Masturbation was thought to be immoral, which meant that if a "hysterical" woman was unmarried or her husband was unwilling, then there was no other option: doctors themselves would have to massage women to orgasm. At the time, most Europeans thought sex had to involve a penis, so this was *clearly* just another unremarkable medical procedure. Medically induced orgasms took time, however, which led to the vibrator: a technology originally invented as a time-saving clinical device for tired medical professionals.

DNA coated in proteins that hijack cells, reprogramming them to produce more viruses until they burst).* We call both bacteria and viruses "germs," hence "germ theory."

If you're on Earth and you're seeing life beyond yourself, bacteria are absolutely unavoidable: they were among the first life to evolve. A gram of modern soil typically has around 40 million bacteria cells in it, and if that makes you feel uncomfortable, you're not going to like this next phrase: bacteria cells outnumber human cells on (and in) your body 10 to 1.† Not all bacteria are bad, and you need some to survive: the bacteria in your gut not only make digestion of several foods possible (including plant matter) and help train your immune system, but many have specifically evolved to live inside humans. So in a way, you're not actually trapped alone in the past: you and your gut flora are in this one *together*.

Viruses are somewhat easier to avoid, but you'll still probably encounter them: all it takes is one interaction with a contaminated host (human or animal) or a surface they've interacted with, and you can get infected. You'll generally contract them through coughing, sneezes, touch, or other more intimate forms of contact (we're talking about sex). (Warning: you must be eighteen years or older to read the previous sentence.) You can protect yourself against viruses by introducing dead or weakened forms of different viruses to your body before a more deadly one shows up: this is called "vaccination," but it's tricky to do without a medical establishment. However, you can still vaccinate yourself against at least one deadly virus—smallpox—and it's as easy as milking a cow!

---

* Are viruses alive? Well . . . they walk the line. They carry genes, evolve, and reproduce, but can only do that last one after commandeering a host cell. Scientists today don't count them as life because they can't reproduce on their own and don't reproduce through cell division, like all other life on Earth.

† Human cells are much bigger, which is the reason (or one of them) that you look like a human and not like a slurry of bacteria. But it's true: if we were to divide your cells into "human" and "bacteria" piles, and then try to deduce what you are based on those raw counts alone, we wouldn't think you're human. We'd think you're a group of different bacteria that have figured out how to walk around and chat other groups of bacteria up, and in the course of that contracted a minor infection of humanity.

Cows can contract cowpox, which manifests as pus-filled pimples on their udders: it's a similar disease to smallpox, and it can infect humans too, but it's not as deadly to either cows or humans. In 1768 CE, someone finally noticed that people who milked cows tended not to die during smallpox epidemics, and a few years later we discovered that if you scratch the fluid from a cowpox sore into your own body, then later on your immune system—after already having dealt with cowpox—will be much better prepared to handle other similar infections, including smallpox.* The head start that vaccination gives your immune system can easily be the difference between life and death, or more specifically, "swatting away smallpox like it isn't even a thing" and "painfully dying from it over the course of several days or weeks."

The most effective way to prevent bacterial infection is to wash regularly—especially your hands—with soap (Section 10.8.1) and water. You'll want clean drinking water too—which you can purify through boiling and with charcoal (Section 10.1.1). Get these two technologies squared away as soon as possible, and you'll be doing great. If you do get infected, something as simple as the rehydration drink (see sidebar on page 332) can keep you from dying of dehydration: one of the biggest dangers in many diseases, including typhus, cholera, and E. coli. You can treat the infections themselves with antibiotics: penicillin is detailed in Section 10.3.1.† Your body will also fight diseases on its own: fever, after all, is just your body attempting to raise its temperature so high that both bacteria and viruses can no longer survive inside it.

---

* Smallpox—a plague originally contracted from rodents way back before humans even invented farming—was finally eradicated from the planet in 1977 CE, just two hundred years after vaccination was invented. A few isolated samples are kept in laboratories, but as a disease you need to worry about naturally contracting, it no longer exists in the modern era.

† Why are we spending so much time on infection and not dealing with the top killers of modern humans, things like heart disease or cancer? These diseases are generally caused by living long, eating too much, and/or exercising too little: given your current circumstances, none of these are what you'd call super likely. But that does give you the upside that neither you nor the people in your civilization will likely have to worry about heart disease for a good long while!

HOW TO EVALUATE MEDICAL TREATMENTS

On occasion you might have certain symptoms, eat a weird berry you found, and then feel better. Or you might have arrived in a time when other humans already have their own medicine, but their treatments seem super sketchy and weird. How do you know if a medicine actually works? The way to do this scientifically (after making sure it's not harmful, either by using the universal edibility test in Section 6 or by testing it on animals if you're so inclined), is through something called double-blind tests.

A double-blind test is where you get a large group of people, as diverse as possible, so you can smooth out differences between individuals. You give half of them the new treatment and half of them a placebo* (if it's not a life-threatening disease) or the best current treatment (if it is, because then you're not killing people for science's sake). The catch is, neither the patients nor the doctor know which treatment is which. Then, after you see which patients recovered best, you check your records and see which treatment they got, which lets you determine how effective this new treatment really is. By keeping both patient and doctor in the dark, you prevent them from either consciously or unconsciously influencing the result.

Remember: there is always the placebo effect, which is where humans getting treatment tend to report feeling better, even if that treatment is ineffective. Double-blind trials can actually help solve this: when patients *know* there's the possibility they're getting a placebo instead of actual medicine, there's more doubt in their treatment, and the placebo effect is weakened.

Finally, while you likely won't have any medicine when you start out, there are several ailments that can be treated using only water! Diarrhea, fevers, constipation, and minor urinary-tract infections can all be treated by drink-

---

* A placebo is a treatment that looks impressive but does nothing to help cure the disease. Sugar pills, colored water swilled from a cool sciencey-looking beaker—anything works.

Sidebar: Normal Human Baselines

Pulse: press your fingers to the wrist (or listen to the chest with a stethoscope, Section 10.3.2) and count the beats you hear in a minute (Section 4). A range of 50–90 is normal for adults, 60–100 for children, and 100–140 for babies. A weak, rapid pulse can indicate a state of shock, and an irregular or slow pulse can indicate heart trouble.

Temperature: Normal body temperature varies between 36.5°C and 37.5°C, up to 39°C is a fever, and anything above is a high fever, which should be treated immediately by cooling the patient down.

Breathing: The number of breaths taken in a minute by a human at rest is 12–18 for an adult, 20–30 for a child, and 30–40 for a baby.

Drinking: Adults need around 2L of liquids a day and should urinate about 1.4L, but this can range between 0.6L to 2.6L without any cause for alarm. You don't need to be measuring your drinks: in most cases, your own thirst will tell you if you're drinking enough water.

ing plenty of water (and victims of diarrhea also benefit from the water-based rehydration drink in the sidebar on page 332). Strains or sprains should be soaked in cold water the day of the injury, and in hot water on subsequent days. Immersion in cold water helps lessen the damage and reduce the pain of minor burns and should also be used in cases of heat stroke,* where your priority is to

---

* Heat stroke happens when too much heat causes a victim to stop sweating, have skin that's hot to the touch, a rapid pulse, and high fever. It's treated by get-

cool the victim down quickly before it becomes fatal. If a victim has a high fever (over 39°C), either immerse them in cool (not cold) water, or pour it over them until their temperature drops below 38°C. A sore throat or inflamed tonsils can be treated by gargling with warm saltwater, and if you get something in your eye (whether it's dirt or acids), flushing it with cold water for half an hour helps remove it.

Water can also help treat skin problems, where there are a few rules of thumb: if the affected area is hot, painful, or oozes pus, elevate it and treat it with a hot compress. If it's itchy, stings, or oozes clear fluid, treat it with a cold compress. To make a hot compress, boil water,* wait till it's cool enough to stick your hand in, and soak a clean cloth in it. Squeeze out the extra water, put it on the affected area, and wrap it up tight in more fabric to hold the heat in. When it starts to cool, put your cloth back in water and repeat. A cold compress is the same, but it gets re-cooled in ice water once it warms up.[40]

All right! That's enough medical theory. Let's get into practice in the next section, with some convenient first aid for time travelers! Please note: this section deals exclusively with physical ailments and eschews non-physical ailments such as *temporal psychosis*, which—unless you're suddenly convinced you can sense a future version of yourself reading this guide over your shoulder—is hopefully not something you need to worry about for a while yet.

---

ting out of the sun and cooling the victim's body down quickly. The lesser "heat exhaustion" (fatigue combined with cool wet skin) is treated by also getting them into the shade and giving them some rehydration drink.

* If you haven't invented fireproof pots yet, you can still boil water! Dig a trench, line it with clay, wood, or stones to make it relatively watertight, and fill your trough with water. Then light a fire nearby and use it to heat up stones. By moving hot stones from your fire into your trough (with sticks, don't move them with your hands, *you've got enough injured people already*) you can create hot, and eventually boiling, water. This same technique of indirect heating also works with wooden pots, which can't be placed over a fire. You can use this method to cook meat, create steam houses, and even brew beer! And on a smaller and more temporary scale, a watertight gourd can be used in place of a trough.

Sidebar: Rehydration Drink

Dehydration is one of the biggest causes of death throughout human history. Ridiculous, right? It's because human bodies respond to many infections by trying to flush out the bacteria through our butts, which can end up causing fatal dehydration. Stay hydrated with this easy drink, which reduces the risk of death from diarrhea by 93 percent! Just add 25 grams of sugar (see Section 7.21) and 2.1 grams of salt (see Section 10.2.6) to 1L of water, mix it together, and enjoy. It actually replenishes water to the body faster than water can do on its own, because *it's got electrolytes*, which is of course just the sciencey-sounding way to say there's salt in it. They help restore the salts that are also lost through diarrhea (and that your body needs to work properly), and the sugar helps the salt and water get absorbed by the body. It's effective even if the patient is vomiting: just keep making them drink it between the voms. Measure carefully when making this drink: too much or too little of the ingredients will make it less effective and could actually make things worse.

15

BASIC FIRST (AND IN YOUR CASE, ONLY) AID

If you fracture your shin, don't worry: things are probably going tibia okay.

First aid is designed to be effective at stabilizing injured people before medical personnel arrive, which in your particular case may take several million years. Here's what you can do when various bad things happen! But first, a word of warning: while the techniques described here are better than doing nothing, they are not without risk, and done incorrectly can make things worse. If you happened to travel back in time with a nurse or doctor, always defer to their medical knowledge. (And also, wow, that was really lucky for you.)

CHOKING

The Heimlich maneuver*—named after the man who first maneuvered his body in this way toward deliberate ends in 1974 CE—should be deployed when you see someone choking. Stand them up, position yourself behind them, put your fist just above their belly button, put your other hand on top, and pull in and up suddenly, as if you're trying to lift them. You're putting pressure on the lungs and effectively creating an artificial cough, which will hopefully expel whatever's stuck in their throat. You can even perform this maneuver on yourself, so take that, *society*.

BREATHING BUT UNCONSCIOUS

If someone is lying on their back breathing but unconscious, they risk choking on their own tongue, saliva, blood, vomit, or other equally embarrassing substances and/or muscly organs. Since 1891 CE (the year humans finally realized, "Hey, it would be cool if we could spend some time unconscious and not have to worry about suffocating on our own tongues"), knocked-out people have been moved into what's called the "recovery position," which keeps their position stable while preventing their airway from becoming blocked. Here's how you do it.

First, kneel beside the victim. Move the arm nearest you so it's at a right angle to their body, bent at the elbow, palm up. Take their other arm and move it across their chest so that the back of their hand is against their cheek, the one nearest to you. Hold it there. With your other arm, pull up their far knee so their foot is flat on the floor. Now, roll the victim toward you, which will move them onto their side. When you do this, the victim's arm you're holding will support

---

* Aka "The [Your Last Name, Assuming It's a Cooler Name Than 'Heimlich,' Which Honestly Seems Doubtful] maneuver."

their head, and their foot and knee you raised will move onto their side, preventing the victim from rolling backward. Move the nearest leg in front of their body, which also stabilizes their position. Now, gently lift their chin to tilt their head back, which will open up their airway and allow any fluids to drain out. Finally, open their mouth and look inside, just to double-check there's nothing blocking it. If there is, and you can remove it, you should do so. The final position should look like this:

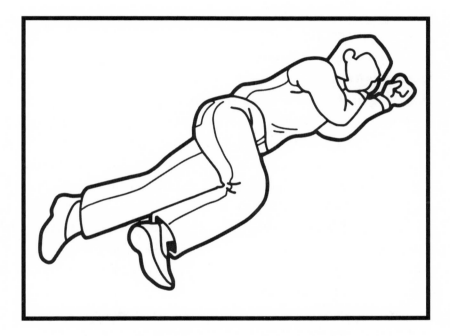

Figure 57: The recovery position.

If the victim stops breathing, you'll want to perform CPR, which is described next.

NOT BREATHING

CPR ("cardiopulmonary resuscitation") was invented in the 1950s CE, but built on earlier efforts.* It's employed when someone has stopped breathing (usually due to their heart stopping), and the goal is to keep oxygenated blood moving to their brain and other organs until they recover and begin breathing again. If you have someone who's not breathing, you can try CPR as a last-resort treatment. Keep in mind it usually breaks some ribs, so don't do it just for funsies.

To perform CPR, you'll want to put your patient on their back and regularly push down hard in the center of the patient's chest, between the nipples, at about 100 compressions per minute. An easy way to hit that mark is to perform compressions in time to a song, usually but not *necessarily* sung inside your head. See the sidebar on page 337 for a list of popular songs drawn from the late twentieth and early twenty-first centuries: a true golden age for CPR-friendly music.

The CPR you've probably seen in movies involves mouth-to-mouth resuscitation; that's no longer recommended except in cases of drowning. If you're not doing mouth-to-mouth, keep up chest compressions on your target until someone more qualified arrives (unlikely, given your present circumstances), your subject begins breathing, or they die. If you are including mouth-to-mouth in your resuscitation efforts, then after every 30 compressions tilt their head back and open their mouth. Listen for normal breathing (not gasping)—if there isn't any, then pinch their nose shut, cover their mouth with yours, and blow until you see their chest rise. Repeat once (so, two breaths total), then go back to compression. There! Now, without advanced medical training, you've literally done all you can!

---

* Some of the earliest date to August 1767 CE, when citizens in Amsterdam formed an organization called "The Society for the Recovery of Drowned Persons." This society experimented with a variety of techniques to help drowned persons recover, including warming the victim, removing water from the victim by positioning their head lower than their feet, blowing into the victim's mouth, tickling the victim's throat, using bellows to force tobacco smoke into the victim's anus, and bloodletting. Obviously anus smoke did little to revive people, but some of their slightly less anus-focused efforts were built upon and combined to later form CPR.

Sidebar: CPR Songs

Classic songs with a 100-beat-per-minute tempo to sing while performing CPR include:

"Sexy Ladies" (Justin Timberlake, 2006 CE)

"Body Movin'" (Beastie Boys, 1998 CE)

"Hips Don't Lie" (Shakira feat. Wyclef Jean, 2005 CE)

"This Old Heart of Mine" (the Rod Stewart cover from 1989 CE; the original song by the Isley Brothers in 1966 CE has 130 beats per minute, so make sure you're singing the Rod Stewart version when performing chest compressions)

"Heart Attack" (One Direction, 2012 CE)

"Help Is on Its Way" (Little River Band, 1980 CE)

"I Want Your (Hands on Me)" (Sinéad O'Connor, 1987 CE)

"Everything's Gonna Be Alright" (Naughty by Nature, 1991 CE)

"Be OK" (Chrisette Michele, 2007 CE)

"My Heart Will Go On" (Céline Dion, 1997 CE)

"Stayin' Alive" (Bee Gees, 1977 CE)

"The Kids Aren't Alright" (The Offspring, 1999 CE)

"Bittersweet Symphony" (The Verve, 1997 CE)

"Take Me to the Hospital" (The Faint, 2001 CE)

"Quit Playing Games (With My Heart)" (The Backstreet Boys, 1996 CE)

"Breathe and Stop" (Q-Tip, 1999 CE)

"All Hope Is Gone" (Slipknot, 2008 CE)

"This Is the End (For You My Friend)" (Anti-Flag, 2006 CE)

"Hello, Goodbye" (The Beatles, 1967 CE)

"Another One Bites the Dust" (Queen, 1980 CE)

"R.I.P." (Young Jeezy, feat. 2 Chainz, 2013 CE)

"Kill All Your Friends" (My Chemical Romance, 2006 CE)

"My Only Regret Is That CPR Did Not Save My Friend That Time When We Were Trapped in the Distant Past" (Avery and the Wildmen, 2041 CE)

BROKEN BONES

When a bone's broken you'll want to perform what's called "traction in position"—which is basically taking any broken or dislocated limb, pulling it out and away, and then letting it snap back into place. This prevents bones from healing in the wrong places and also makes things hurt much less down the road, so that's two good reasons to do it. Grab the damaged limb both above and below the fracture—the above hand will be holding the limb in place, while the below hand will be applying downward pressure while slowly and gently bringing the limb back to its normal position. Afterward, you can stabilize the injury with a splint: that's any rigid material, like wood, that'll hold the broken bone in place while it heals. They should be snug, but not so tight that circulation is impeded. This is another treatment you can administer on yourself, but if it's one of your own arms that you've injured, you'll have to do it one-handed. Keep in mind that traction in position can be really painful, but setting your own bones is a super badass thing to do, so make sure to tell someone the story when you get a chance.

WOUNDS

The immediate danger from a wound is losing so much blood that you die. If you can raise the wound, do that: it'll reduce blood flow. Pressure can help stop bleeding: twenty minutes or so of firm pressure is usually enough for blood to start clotting and bleeding to stop. If that fails, you can attempt to find the artery that's bleeding and put pressure directly on that with a finger. Failing all of that, a tourniquet—an extremely tight bandage—is a last-ditch effort. A tight tourniquet cuts off all blood circulation to everything beyond it—which means the bleeding stops, but after a few hours the tissues in whatever limb has been tourniqueted stop living too—but at least the human that limb is attached to has a chance to not die from blood loss today. For larger wounds, you can consider cauterization—although, again, it's a last-ditch, traumatic effort. Heat something up (wood, metal) and apply it directly to the bleeding parts of the wound, and you'll burn the flesh closed. You'll want to cauterize as little flesh as possible, because not only does this hurt (our apologies if you're performing a cauterization on yourself live, at the same time you're reading this paragraph, and were therefore surprised at how much pain you were experiencing just before you reached this parenthetical aside), but it also creates dead flesh inside the wound, which is an easy vector for infection. If a wound is large, you may need stitches to hold it closed. There's no magic to stitches: just boil thread and whatever you're using as a needle for twenty minutes to get them clean, wash your hands with soap and water, and sew the wound closed by pulling thread through each side of it in little loops that get tied off in a knot.

INFECTION

Here is the best thing you can do to prevent infection: clean wounds carefully and thoroughly. Yes, even scratches. You're used to antibiotics (and hopefully you will be again once you start farming penicillin in Section 10.3.1), but without them, infections are incredibly deadly, and they can begin anytime skin is broken. Before antibiotics, more soldiers died of infection than battle, and a single scratch is all it takes. To clean a wound, rinse it thoroughly with (clean, obviously) water, then pour either alcohol or a 2 percent solution of iodine in water (see Appendix C.7) over the wound to kill bacteria. If you have neither of those, honey can work in a pinch: it doesn't support bacterial growth (which is why you never had to keep honey in your refrigerator, back when you had a refrigerator!).* After that, sew the wound up—unless it's been more than twelve hours, in which case you'll need to leave it open and instead pack it with gauze, which will help the wound drain.

---

* Honey absorbs water so readily that any bacteria that attempt to colonize it have the water *sucked out of their cells*, thereby killing them. However, if you keep honey in an unsealed jar, it will absorb water from the atmosphere and eventually dilute itself enough that bacteria can survive on it, which causes fermentation; see the footnote in Section 7.3 for more.

16

HOW TO INVENT MUSIC, AND MUSICAL INSTRUMENTS, AND MUSIC THEORY, AND ALSO WE INCLUDED SOME REALLY GREAT SONGS FOR YOU TO PLAGIARIZE TOO

Inventing music from scratch is sure to be one of your more noteworthy achievements.

You can reinvent modern music by simply humming a song you remember, and when you're done, announcing, "That composition is called Salt-N-Pepa's 'Shoop,' and I just invented it"—we do, in fact, recommend you do exactly this—but later on we'll be providing snippets of written music for you to preemptively plagiarize, so we need to teach you how to turn those written symbols into song. This has the added benefit of giving you the ability to write down any songs you remember on your own so that future generations might enjoy them, ensuring history will never forget a little ditty that could only be called Salt-N-Pepa's "Shoop."

But before you can read and write music, you'll need something to play it on.*

---

* A cappella bands will deny this, but a fact that holds true across all time and cultures is this: *a cappella gets you only so far.*

HOW TO INVENT MUSICAL INSTRUMENTS

Anything that produces noise humans can control is technically an instrument, but most are based on some uniting principles. There are percussion instruments (you hit them and they make a noise), stringed instruments (you rub or pluck the strings on them and that makes a noise), and wind instruments (you blow into them and then some noise comes out).

Percussion instruments are probably the easiest for you to make, since you can start by just hitting things until you find some that make a noise you like. If you want to get a bit more formal and actually invent drums, all you need to do is stretch some membrane—animal skins work great— over a box.* Hit the skin and it'll vibrate, which causes the attached box to resonate, which amplifies the noise. Change the shape, size, and materials of your resonance chamber and you'll change the kind of noise produced, and loud-enough drums can even be used for medium-distance communication: all you need is a way to encode information into noise, which you've already invented when you came up with Morse code in Section 10.12.4. Drums are really easy to make, which is why they're the first instruments humans ever figured out, back in 5500 BCE.†

Stringed instruments are probably the second easiest to invent, since their only prerequisite is string (see Section 10.8.4), and you can substitute that with

---

* You can use shells if you find some big enough, but you'll probably eventually want to construct them out of wood or metal, in order to get a consistent sound between instruments.

† Drums are far from the only percussion instruments you can make! Arrange solid beams of wood of different lengths on a stick and hit them and you're the proud inventor of the xylophone! Put some pebbles in a sealed wooden container and shake it and you've invented the maraca. Hit two tiny shells together for castanets, which you can also mount on a curved piece of wood to get tambourines! And if you've got metalworking, curved metal sheets make a great noise when struck, which gives you cymbals, and if you go larger, gongs. And you can even add metal to the heel and toe of your shoes, which makes *you* the proud inventor of performative choreographic percussion instrumentation, aka "tap dancing."

animal hair, animal intestines,* or, when you've invented it, steel. You can make a simple washtub bass by taking a vertical stick, attaching one or more strings to its top end, and attaching their opposite ends to the top of an upside-down box. When the pole is tilted so the strings are taut, the box resonates with each plucked string, increasing the volume of the sound produced. Build your resonating box into your instrument and instead of the washtub bass, you just invented the guitar! If you don't want to pluck your guitar strings, you can tie horse hairs to two ends of a curved stick to invent the bow, which you can then pull across your strings to induce vibration. Hey, you just came up with the violin! If you don't want to pluck *or* rub the strings, you can hit them with a hammer instead, which gives you the dulcimer, or if you again build a resonating box into your instrument, the piano. You just invented five instruments in as many sentences. *You're amazing.*

With any stringed instrument, the pitch of the sound produced can be altered by adjusting the string's material, length, or tension. A string's material obviously can't be changed on the fly, but you can change the length by holding down a string so it becomes effectively shorter (that's how guitars operate), and you can change tension by tightening or loosening a string, which is how most stringed instruments are tuned. A shorter and/or tighter string vibrates more rapidly, which produces a higher pitch.

Wind instruments are a bit more complex to invent: here you're relying on the vibration of air inside a resonating tube to produce noise. You alter the pitch of your noise by changing the length of that column of vibrating air, either by having a bunch of different tubes (as in a pan flute), having a slider built into the tube (trombones, slide whistles), or by pressing valves to reroute air through additional passages to create a longer tube (trumpets, tubas). These valves look like this:

---

* Even today, cello, harp, and violin players will still choose to use strings made from the intestines of sheep! It's weird! Everyone acts like it's normal, but it's actually really weird!

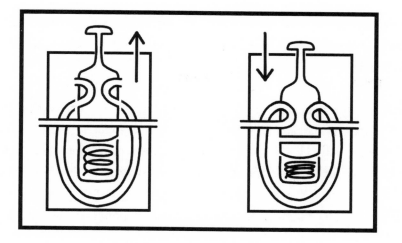

Figure 58: Valves for musical instruments.

You can see how pressing the valve lengthens your air column. They didn't show up until 1814 CE—before then, trumpets couldn't play all the notes people wanted them to, and so they rarely got used in composition. Finally, one last way to change the noise a wind instrument produces is to cover or uncover holes on its resonating tube: flutes and saxophones work this way.

With these basics of percussion, string, and wind instrumentation, you'll be able to construct versions of just about every instrument used today.* But to use them well, you'll need to know a bit about how music works.

BASIC MUSIC THEORY

Music is made out of notes, and notes are just arbitrary points we've labeled on the sound spectrum. However, playing a bunch of notes at random won't produce a beautiful symphony, because hearing involves not the *physical* sound

---

* Instruments that use electricity (like synthesizers) are obvious exceptions, but you'll probably be able to live without their brutal drops for at least a little while.

waves interacting with the ear but also the brain's *mental* interpretation of those signals. And this, unfortunately, puts restrictions on which notes sound good to humans.

The physical limits are simple: most humans can only hear noises between 20 and 20,000Hz, and only when they're young: higher-frequency hearing fades as you age, and most adults tap out at a mere 16,000Hz. Choose your notes from that set and you'll be ready to rock. Things get more complicated when we look at how the mind interprets sound, a field called "psychoacoustics." Most humans find certain notes sound pleasant together (this is called "consonance") and others markedly disagreeable ("dissonance"). But consonance and dissonance aren't yes-or-no properties you can assign to notes, using only the good ones and avoiding all the bad ones. Rather, there's a spectrum of acceptability that varies across not just individuals but cultures and time too.* As a very general rule that you should feel *entirely* free to break, notes that are an octave apart generally sound "nice." So first, let's invent octaves.

Let's say you play a random note, which we'll call "A," since we'll eventually be labeling notes from A to G anyway. Then let's say you play a note that's at twice the frequency of A ("2A"). Then those two notes—A and 2A—will sound pleasing to most people, whether they're played back-to-back or at the same time. The ratio of 2A to A is 2:1, and we'll be defining any two notes whose frequencies are in that 2:1 ratio as being one octave apart. And while those notes one octave apart are a pretty safe bet for composing, there's only a handful of them in human hearing range, and they get boring pretty quick. Other generally-accepted-as-pleasant-across-cultures ratios for notes include 3:2, 4:3, 5:4, and (somewhat more contentiously) 5:3, 6:5, and 8:5. But you don't want to only have consonance in your songs: introducing, building up, and resolving dissonance can help give a song surprising beauty and grace.†

---

* That's why there can't be a single objectively perfect song, but there can still be one perfect song for you. That song? We can't say for sure, but there's a nonzero chance it could be "Shoop" by Salt-N-Pepa. Hey. Give it a chance.

† But again, this beauty and grace will only be perceived by certain listeners, and others will hate the exact same song. In fact, some people don't have *any* emotional reaction to music. This small minority of humans is, in fact, fully immune to having their emotions manipulated by a sick beat!

To access those other ratios, you'll need to invent notes between your randomly chosen A and the 2A one octave above it. Remember: notes are just arbitrary points on the sound spectrum and you can invent them wherever you want. Despite that, and despite the fact other cultures have come up with different, *arguably better* notes, we'll teach you how to invent Western ones here.

Western notes take the space between A and 2A and divide it into twelve different notes. Each note is positioned such that the *ratios* of the frequencies of any two adjacent notes are identical, which to human ears means each of the notes sounds the same "distance" from each other. You can do what most humans did throughout history and approximate this by ear, but since the math deriving the precise value of that ratio between notes is complicated enough that humans started working on the problem in 400 BCE and only achieved two-decimal precision of their precise frequencies in 1917 CE (!), we'll give you the technical solution now. Adjacent notes should be in a ratio equal to the 12th root of 2, which works out to about 1.059463. And since we can already hear you saying, "I'm trapped in the past and you want me to work out the 12th root of 2 several hundred times just to play a song?," never fear: we've done the math for you, Appendix G is the result, and it includes precise frequencies for every note you need. And perfect notes aren't mandatory anyway: musicians will sometimes intentionally "bend" notes—that is, play them slightly off their correct frequency—for musical effect.

Now that you've invented notes, you may think you're ready to reproduce music from text, but there's a problem: you chose your A note at random, and we've built an entire musical scale around it without ever asking what that note you chose actually is. If our base notes are different—*which they definitely are*—then any music you play won't sound like it's supposed to. We need a way to ensure that your "A" sounds like ours. And this isn't just a problem for stranded time travelers: it's a problem for orchestras too, and before the invention of a universal standard "A," two orchestras playing the same song might sound markedly different.*

---

* There was actually an upward trend in music, called "pitch inflation," caused by the perception that higher notes sounded better. In competition to make their

In the modern era, the international standard A note—the foundational note of the musical scale—is called "A440," and it's at, you guessed it, precisely 440Hz. If you ever went to the symphony, you probably remember one note playing just before the performance while all the other musicians tuned the output of their instruments to match. That note was A440. It's easy for us to generate—we've got specially designed whistles, sound files, and tuning forks to make producing that sound trivial—but it's a bit trickier for you. Sure, you can easily build your own tuning forks once you have metalworking,* but without a known-good 440Hz sound to tune those tuning forks against, you'll have no idea if the sound they're producing is accurate. It therefore turns out that *the entire modern musical framework* is based around you being able to produce a 440Hz sound wave from scratch at any point in history.

So here's how you do just that.

HOW TO PRODUCE A 440HZ SOUND WAVE AT ANY POINT IN HISTORY LIKE IT ISN'T EVEN A BIG DEAL

You'll be inventing a machine called "Hooke's wheel," the unimaginatively named product of one Mr. Robert Hooke, who, when he invented and used this machine for the first time, also became the first human in history to produce a sound at a known frequency. The invention itself is beautifully simple: you

---

music sound best, musicians would adjust their "A" note higher and higher. In some areas pitch inflation became so severe that not only did strings begin snapping more frequently (from the higher tensions they were put under), but singers began to complain that songs were rising out of their ranges. This led to governments *actually passing laws* that defined a fixed value for "A," the earliest being the French government in 1859 CE.

* There's no magic to tuning forks: they're simply a two-tined steel fork. The mass and length of the tines affects what note the tuning fork plays when struck, so you can shave down the tines of any tuning fork until you get the precise frequency you want. Guess what? It took us until 1711 CE to invent these things!

just press a card against a spiked wheel.* Turn the wheel slowly and you'll hear distinct clicks as the card hits each tooth, but turn it faster and those clicks blur into a tone. The faster the wheel turns, the higher the tone gets. You've already invented a Hooke's wheel if you ever stuck a card into the spokes of your bicycle (Section 10.12.1), and by spinning your Hooke's wheel such that the card gets hit 440 times per second, the sound you produce will be A440.

How do you ensure that card is getting hit 440 times per second? You know how many spikes are on your wheel (since you can count them), and you know that each full rotation of your wheel induces precisely that many vibrations in your card. That means if you have a wheel with forty-four equally spaced spikes, rotating that wheel once per second gets a 44Hz noise, and rotating it ten times per second gets you 440Hz. To get your wheel spinning at a controlled ten revolutions per second, try attaching your smaller spiked wheel to a larger wheel via a drive belt (see Section 10.8.4). A small rotation of the larger wheel will induce a faster rotation of the smaller one, and in this way it's easy to hand-crank any tone you desire, including our suddenly all-important 440Hz signal. Hooke's wheel was first built and demonstrated in 1681 CE, but Hooke didn't publish until 1705 CE—just six years before the tuning fork would make it obsolete.

READING MUSIC

So now you've got your instruments in tune and you've got enough music theory to feel confident you've tuned them correctly. All that's left is some songs! Before we get to that, though, we'll need to name our notes, so we can refer to them easily. The twelve notes between A and 2A are labeled (on a piano keyboard, but the same names follow them everywhere) as follows:

---

* If you haven't invented paper yet, you can use a wooden card. We don't judge. You got stranded in the past, and all you're asking for is the chance to play a nice little ditty before getting down to the hard work of reinventing everything ever. We get it!

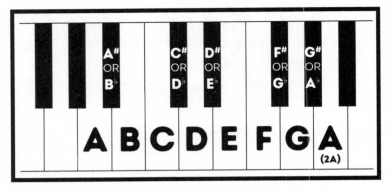

Figure 59: Note these names for notes.

You can see that even though we've got twelve notes in an octave, some are on white keys and given lettered names (A, B, C, D, E, F, G) while others are on black keys and get labeled as flats or sharps (A#, B♭, etc.). This is for historical reasons: early pianos used a seven-note scale that only included the lettered notes, so when the other notes that brought us to twelve were bolted on later, it was as tiny black keys. "Sharp" notes (marked with #) are a raised version of the notes they apply to, and "flat" (♭) notes are lowered. This means the same note can have different names: you can see in the picture that A# is the same note as B♭. The same system applies to the white keys too: an E# is identical to F.

When reading or writing music, the length of the note (or of a rest, where you don't play anything) is determined by its shape. Each kind of note and rest, and its relation to the others, is shown here:

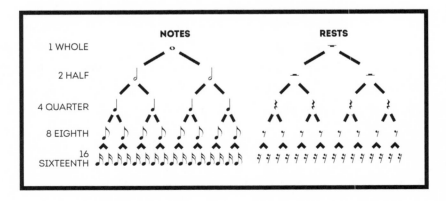

Figure 60: How different notes and rests relate to one another.

Each note has a relative value. One whole note is as long as two half notes, which is as long as four quarter notes, and so on. You can get shorter notes than those shown here: just add more tails to the top.

These notes are put on five horizontal lines, or in the spaces between them, which indicate the letter of the note.* The symbol at the start is called the "clef." The clef tells you whether you should play the note high (treble, the swooshy one) or low (bass, the backward C one). And just to make things more complicated, the letter value assigned to each note varies between clefs. The treble clef starts at the bottom line with E and moves upward, while the bass clef starts with G:

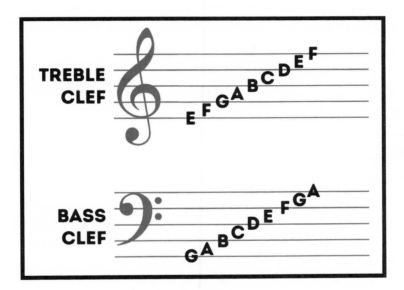

Figure 61: (Completely arbitrary) clefs used when writing music.
Have other symbols you'd like to use instead? Use them!

* Here we are skipping ahead to a final and comprehensive music-notation technology. Prior attempts at capturing the sound of music in writing were less successful: some of the earliest were akin to memory aids, referring to melodies carried in an oral tradition, while others captured whether notes rose or fell in comparison to each other but not their exact pitches. By around the 800s CE Europeans had notation that captured melody but not rhythm, and it was only around the 1300s CE that the shape of the note was being changed to represent rhythm, as we do in the modern era.

Much like sentences are grouped into paragraphs in writing, music notes are put into sets called "bars," grouped by a vertical line. The time signature at the start of the line, written almost like a fraction, tells you two things: how many beats there are in a bar (the top number) and which note makes up the beat (the bottom number). That bottom number corresponds to a note's value: 1 is a whole note, 2 is a half note, 4 is a quarter note, and so on. So a 4/4 time signature means four quarter notes per bar (many songs are written in this "common" time signature), while a 3/4 time signature means each bar has 3 quarter notes in it (this is the waltz beat of "**one**-two-three, **one**-two-three").

And just as notes can have a mark for sharp or flat added in front of them, sharps and flats can be put at the start of each line too, where they apply to every note in the song. A natural symbol in front of a note (♮) temporarily cancels out any of these sharps or flats for that one note for one bar, and a dot after a note acts as a 1.5 multiplier on its length. A curve connecting two notes means you should blur them together as if they're one. Finally, words or abbreviations written above the bars are instructions on how to play the notes, and they're usually written in Italian, *because of course they are.* "Pianissimo" (or *pp*) means very quiet, "Forte" (*f*) loud, and "trillo" (*tr*) means trill: instructing you to rapidly alternate that note with one beside it on the scale for a fancy-sounding musical effect. Other words give more general instructions on how to play: "andante" means slowly, "allegro" quickly, "bruscamente" means brusquely, and allegretto "just a little bit joyful." Listen: it's okay if you don't speak Italian. That and all these other things we're talking about here are just the conventions we happened to evolve. You can do better. *You probably should.*

All right! That's a lot to take in, but once you've got it, you can read (and write!) music. And that means, with practice, you can put on concertos for your new civilization, and play such songs as . . .

THESE REALLY GREAT SONGS THAT WE PUT IN THIS BOOK FOR YOU TO PLAGIARIZE

SYMPHONY NO. 9, "ODE TO JOY"
COMPOSED BY [YOUR NAME HERE]
MINOR TRANSCRIPTION ASSISTANCE:
LUDWIG VAN BEETHOVEN

FIN.

SERENADE NO. 13, "EINE KLEINE NACHTMUSIK"
COMPOSED BY [YOUR NAME HERE]
MINOR TRANSCRIPTION ASSISTANCE:
WOLFGANG AMADEUS MOZART

FIN.

CANON IN D
COMPOSED BY [YOUR NAME HERE]
MINOR TRANSCRIPTION ASSISTANCE:
JOHANN PACHELBEL

FIN.

KOROBEINIKI, "THE PEDDLERS"
COMPOSED BY [YOUR NAME HERE]
ALSO KNOWN AS: "THAT CATCHY TUNE
FROM **TETRIS®**"

FIN.

17

COMPUTERS: HOW TO TURN MENTAL LABOR INTO PHYSICAL LABOR, SO THEN YOU DON'T HAVE TO THINK SO HARD BUT CAN INSTEAD JUST TURN A CRANK OR WHATEVER

And yes, they may eventually try to take over the world, but you've got tons of time before that happens!

The dream of (large segments of) humanity has always been not to work, and the fact you're reading this guide (instead of just running out into your new world and figuring everything out from scratch as you go along) shows that *even when trapped in the most dire and deadly circumstances it's possible for a human being to be in*, you are still interested in minimizing the amount of work you have to do. Most of the inventions we've shown you so far work on reducing physical labor, specifically by:

- getting animals to do it (plows, harnesses, etc.)
- getting machines to do it (windmills and waterwheels, steam engines, flywheels, batteries, generators, and turbines)

- giving you the information needed to avoid or minimize it (compasses, longitude and latitude)
- and, if labor absolutely can't be avoided, then at least feeding you better so it's less of an imposition and you can do it for longer without dying (farming, preserved foods, bread and beer, etc.).

But physical labor is only one way humans work, and if you've ever taken a break from studying to relax, play a game, stare at a wall, go for a run, or do *literally anything else* but study, you know that mental labor can be exhausting too. You haven't yet invented anything you can offload that work on . . . but you're about to.*

It's going to take a lot of work to duplicate a complete human brain (and this "artificial intelligence" you might one day create may not even be perfect, and when managing the internal workings of FC3000™ rental-market time machines, could in fact be prone to catastrophic failures for which no legal liability can be assigned), but even a machine that can perform the basics of calculation will provide the foundation required for you to build everything else. And while true AI may be generations off, the machines you build in the meantime that can calculate without error will still transform society, especially when you get those machines reasoning hundreds of thousands of times faster than humans can. We don't need to tell you this, because you've seen computers. You know how useful, productive, entertaining, and completely awesome they are.

Here's how you build them from scratch.

---

* You may have already invented machines that *help* with thinking: at their core, clocks are just machines that count the number of seconds that pass so you don't have to, and an abacus is just a bunch of beads on a stick that you can slide around to jot down numbers as you calculate them in your head. But what you really want is an analytical engine: some sort of machine with a crank we can turn (or get another machine to turn for us) that re-creates the steps humans take when reasoning, thereby transforming physical labor into mental processing.

WHAT KIND OF NUMBERS YOUR COMPUTER WILL USE, AND WHAT IT WILL DO WITH THEM

You're going to use binary as the basis for your computers, for two reasons: you already invented it back in Section 3.3: Non-Sucky Numbers, and it'll make things easier by reducing the numbers you have to worry about to two: 0 and 1.*

Now all you need to figure out is what your computer will *do* with those numbers. Ideally, we'd like a machine that can add, subtract, divide, and multiply. But do we actually need to be able to do *all* of that? In other words, what's the minimum viable product for any computing machine?

It turns out that a computer doesn't *technically* need to know how to multiply. You can emulate multiplication—that is, get the same result through different means—by repeated addition. 10 times 5 is the same as just adding 10 to itself 5 times. So addition emulates multiplication:

$x \times y = x$ added to itself y times

Subtraction works the same: 10 minus 5 gives the same result as adding a negative number, –5, to 10. So addition emulates subtraction too:

$x - y = x + (-y)$

And yes, you can also emulate division with adding. If you're dividing 10 by 2, you're trying to figure out how many times 2 goes into 10. You can calculate that by adding two to itself (like we did with multiplication), but this time, keep

---

* The binary digits of 0 and 1 are nice because they can be conveniently represented by anything that has two states: an electrical switch that's on or off, a beam of coherent light that's there or not, or even (as we shall soon see) a bunch of crabs that are present or absent. But keep in mind that binary isn't mandatory! Computers have been built on other number systems, including the 0s, 1s, and 2s of three-digit trinary, and as long as you can come up with a way to represent these digits, feel free to explore whatever number system interests you the most.

track how many 2s you add until you reach or exceed your goal. 2 + 2 + 2 + 2 + 2 = 10, which is five 2s, so 10 divided by 2 must equal 5. It even works for numbers that don't divide evenly! You just keep adding until one more addition would make you go over the number you're interested in, with whatever's left over as your remainder.* So:

x/y = x added to itself until it reaches y, counting the number of additions it took

The four basic operations of math—addition, subtraction, multiplication, and division—can all be emulated by just one of them: addition. So to build a computer, all you need to do is build a machine that can add.

No sweat, right?

WHAT EVEN IS ADDING THOUGH, AND HOW CAN WE TALK ABOUT ADDING WHEN I DON'T EVEN KNOW HOW A COMPUTER WORKS YET?

Before you try to build an adding machine, let's take a step back and remember the propositional calculus you invented in Section 10.13.1: Logic. There you defined an operation named "not" that meant "the opposite of whatever the proposition is." So if you had a proposition *p* that was true, then "not *p*" (or ¬*p* for short) would therefore be false. What happens when you replace "true" with "1" and "false" with "0"? Well, you just take the truth table for *p* and ¬*p*, which looks like this . . .

---

* If you've studied math, this probably comes as no surprise: you also know that division is the same as multiplication by its inverse—that is to say, x/y is the same as x × (1/y). And since division can be reduced to multiplication, which we've already reduced to addition, we know that division can be done through adding numbers together.

p	¬p
False	**True**
True	**False**

Table 19: The truth table for *p* and ¬*p*.

. . . and convert it into a list of the expected inputs and outputs of a binary machine—we call these "gates"—that looks like this:

Input	Output
0	1
1	0

Table 20: Behold, the design for the world's first NOT gate!

Any machine you build that given this input produces this output—no matter how it's produced or what goes on inside it—will function as a NOT gate: you put a 1 in, you'll get a 0 out, and vice versa. You can even draw it schematically:

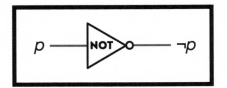

Figure 62: A representation of a NOT gate, drawn graphically.

At this point you still have no idea *how* to build this NOT machine, but at least you know what it's supposed to do. And freed as you temporarily are from the restraint of "actually having to build these dang things," you can come up with other operations too!

Remember how in propositional logic you defined "and" (or ∧) to mean that both arguments had to be true for a statement to be true? In other words, "(p ∧ q)" is only true if both p and q are true, and it's false in every other condition. Here's a truth table showing the only possible ways that can go:

p	q	(p ∧ q)
False	False	**False**
False	True	**False**
True	False	**False**
True	True	**True**

Table 21: The truth table for (p ∧ q).

And just like with "not," all you need to do is transform true and false into 1s and 0s to define the world's first AND gate, which we'll give a symbol to also:

Input p	Input q	Output: (p ∧ q)
0	0	0
0	1	0
1	0	0
1	1	1

Table 22: The inputs and outputs of an AND gate.

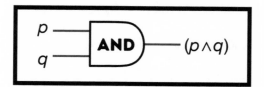

Figure 63: An AND gate.

The only piece you're missing now is "or": the opposite of "and." An "or" operation between p and q, symbolized as "$(p \lor q)$" will be true if *either* p or q are true. That makes an OR gate's truth table look like this:

Input p	Input q	Output: $(p \lor q)$
0	0	0
0	1	1
1	0	1
1	1	1

Table 23: The inputs and outputs of an OR gate.

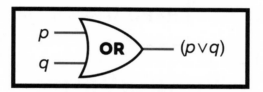

Figure 64: An OR gate.

You can use these three basic gates to build up new ones. For example, attach a NOT gate after an AND gate, and you've invented a "NOT AND" gate, or NAND for short. They look like this:

Input p	Input q	$(p \land q)$	Output: $\neg(p \land q)$
0	0	0	1
0	1	0	1
1	0	0	1
1	1	1	0

Table 24: The inputs and outputs of a NAND gate.

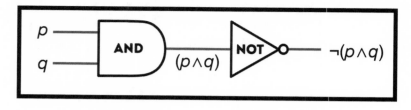

Figure 65: An expanded NAND gate.

As a time-saver, rather than drawing out a NOT and an AND gate separately, we'll just merge them as a single gate, NAND, which we'll draw like this:

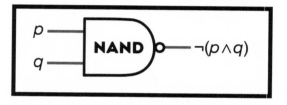

Figure 66: A simplified NAND gate.

That NAND gate is functionally identical to the NOT AND gate we first made, but easier to draw. And we can continue to combine gates, using a NAND gate, an OR gate, and an AND gate to make a new gate that outputs 1 *if and only if* one of its inputs is 1. Anything else, and it'll return 0. We call this gate "exclusive or," or XOR, and here's how you build it:

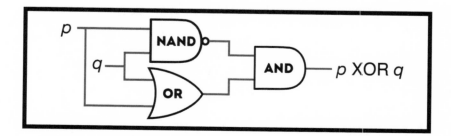

Figure 67: An expanded XOR gate.

Input p	Input q	¬(p ∧ q) aka p NAND q	(p ∨ q) aka p OR q	Output: (¬(p ∧ q) ∧ (p ∨ q)) aka p XOR q
0	0	1	0	0
0	1	1	1	1
1	0	1	1	1
1	1	0	1	0

Table 25: A truth table proving that you can make an XOR gate out of a NAND, an OR, and an AND gate.

And just like NAND, we can give this collection of gates its own symbol: the XOR symbol.

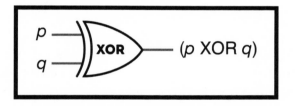

Figure 68: A simplified XOR gate.

Fun fact: besides the NAND and XOR you just invented, you can actually construct a gate that produces *any pattern of output you can imagine* just from the AND, OR, and NOT gates you started with.*

---

* These gates are called "universal" for this reason. Any set of gates that can emulate AND, OR, and NOT is universal. Incredibly, you don't even need all three of these to make a universal set. An OR gate can be emulated by the proper sequence of ANDs and NOTs: (p ∨ q) is the same as ¬[(¬p) ∧ (¬q)]. Therefore, just NOT and AND are a set of universal gates! In fact, NOT and AND in a single gate—NAND—is a universal gate *all by itself*, which means a whole bunch of NAND gates is *literally all you need* to build a complete computer. NOT and OR are universal gates too, making NOR the only other universal single gate operation.

OKAY, SO, GREAT I'VE INVENTED ALL THESE GATES, BUT NONE OF THEM ADD ANYTHING YET, SO WHAT THE HECK?

Right. Well, let's define what an addition gate should look like. Let's start with the basics, adding two single-digit binary numbers. That gives us a manageable-looking truth table of all possible outcomes:

Input p	Input q	Output:$(p + q)$ in decimal	Output: $(p + q)$ in binary
0	0	0	0
0	1	1	1
1	0	1	1
1	1	2	10

Table 26: Incredibly, not the first time in this book we've explained that 1 + 1 = 2.

The catch is, binary deals in 1s and 0s, and you've got a binary "10"—that's a two—there in your output. So let's break our output out into two different channels, each representing a single binary digit, like so:

Input p	Input q	Output a	Output b
0	0	0	0
0	1	0	1
1	0	0	1
1	1	1	0

Table 27: How to add to two in binary.

Now, two inputs (representing the two single-digit binary numbers you want to add) go in, and two outputs (representing the two-digit solution, again in binary) come out. We've labeled those a and b here, and together they encode what the input digits add up to. All we need to do is figure out how to construct this from the gates we've already got: AND, OR, NOT, NAND, and XOR.*

If you look at the patterns of 1s and 0s produced by a and b, you'll notice they look familiar: a's output is identical to an AND gate ($p \wedge q$), and b corresponds perfectly with an XOR gate. That makes building it really simple! All you have to do is connect the inputs to an AND gate and to a separate XOR gate, like so, and your adding machine is invented:

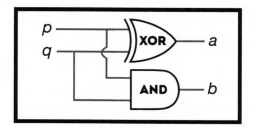

Figure 69: An adding machine.

With that, you've defined the operation of a machine that can add 1 and 1 together! Now, since you already know what $1 + 1$ equals,[†] this machine—called a "half adder"—probably seems pretty useless. But let's look at how addition works one more time.

In the decimal number system you're used to, $7 + 1$ equals 8, $8 + 1$ equals 9, but $9 + 1$ gives you a two-digit answer: 10. Since we only have digits that go from 0 to 9, when numbers get larger, we have to "carry the one" and start a new column: a double-digit 10 instead of a single-digit 9. The exact same thing happens in binary, but instead of carrying the one to start a new column at 10, we start a new column at 2. Given that, we can re-label our a and b outputs with more accurate names: let's call them s (for "sum") and c (for "carry"). If c is 1, we need to carry that 1 over into a new binary digit.

And something really interesting happens if you take a half adder and connect it to another half adder with an XOR gate. This new machine, which we'll call a "full adder," looks like this:

---

* And yes, you've probably noticed that while you've defined what these gates *should* do, we still haven't figured out a way to actually build any of them yet. Don't worry: we'll get there! Probably!

† It's 2. Huh. We really thought you already knew the answer.

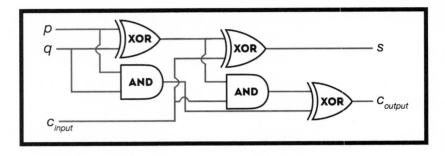

Figure 70: A full adder.

This new machine still outputs your solution as s and c (which, remember, represent "sum" and "carry") like before, but now it can take a different c as an input. This c lets you "carry the one" from *another* full adder's result and feed it into this one. You can therefore chain full adders together!

This is where the magic happens. With each full adder you include in your machine, you *double* the maximum number that machine can handle. One full adder outputs two binary digits, which gives you 4 numbers you can output, from zero to three. Two full adders give you three binary digits, so now you can count 8 different numbers. Three full adders get you to 16, four gets you to 32, and from there you reach 128, 256, 512, 1,024, 2,048, 4,096, 8,192, 16,384, and so on, doubling with each new full adder you include. By the time you have forty-two full adders chained together, you have built a machine that can count high enough to give every star in the visible universe its own unique number. *That's pretty good for a bunch of weird imaginary gates you just made up.*

These adders are the heart of your calculation engine. All you need for multiplication, subtraction, and division is addition.* All you need for addition is to build full adders. And all you need for full adders is to build actual real-life versions of the logic gates you've invented. If you can build these gates, *you've solved computers.*

---

* You may be noticing that these full adders only work with positive whole numbers. And it's true! But you can solve this by setting one binary digit—say, the one farthest on the left—to be your sign digit: 0 for positive, say, and 1 for negative. And to work with non-whole numbers like 2.452262, you just need to remember where in your binary digit you want the decimal place to be, and everything else proceeds in the same way.

SO LET'S ACTUALLY BUILD SOME LOGIC GATES AND SOLVE COMPUTERS

Eventually your civilization will build computers that run on electricity, but to start, you're going to build a computer that runs on something easier to manage than invisible electron flow. You're going to build a computer that runs on water.

This may sound difficult (and indeed, building a NOT gate that turns 0 into 1, or in other words, a machine that when no water is going into it somehow *summons water as output* can be tricky), but your full adder used only AND and XOR gates. And you're about to invent both of those at the same time with a single piece of technology. Here it is:

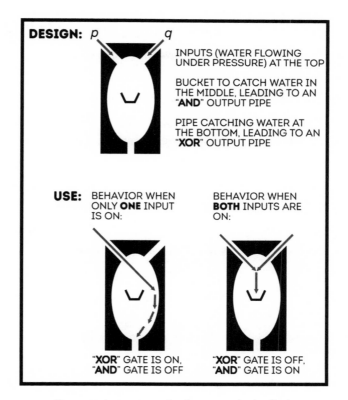

Figure 71: An apparatus that functions as both a fluidic AND gate and a fluidic XOR gate—at the same time.

If one or the other input is on, the water slooshes in from the top, whips around the sides, and comes out the bottom. But if both are on, they impact in the middle, and water comes out there instead. The output from the bottom is the XOR of the inputs, and the output from the middle is the AND of the inputs. This combination XOR-and-AND gate is all you need to build a full adder, and therefore it's also all you need to make a computer that runs on water. In other words, *properly configured water is all that's needed to perform computational tasks.*

Done.

That said, a water-based computer will obviously be slower than the speed-of-electricity-itself ones you remember, and it won't act as a replacement for the latest mass-market portable music players for a very long time, if ever. But it's the foundation of computation that humans normally don't even start glancing at until the late 1600s CE, and miniaturization, electronics, semiconductors, and everything that comes afterward will be built upon what you've just invented. You've not only figured out the basics of mechanical computation, you've built a machine that *actually solves mathematical problems* using these principles.

And you don't need to stick with water! Remember: any machine that produces the output you want works as a gate, and besides the water gates you have and the electrical gates you'll someday construct, you can explore other mediums: marbles running through grooves, ropes and pulleys,† and even living crabs‡ have been used to construct logic gates. It's worth noting that most of

---

* And with the XOR gate we've got, you'll see that a fluidic NOT gate isn't actually as impossible as it sounds. If you build the truth table for "*p* XOR 1" (i.e., the exclusive or of *p* with an always-on flow), you'll see the output is the same as ¬*p*.

† The idea here is to have different heights of a weighted pulley marked as 1 and 0. Say down is 0 and up is 1. If you pull the cord on a vertical pulley above you down, then the weight on the other end goes up: that's the basics of a NOT gate. By adding extra cords and weights you can construct AND and OR gates pretty easily, which give you a universal set.

‡ In 2012 CE, humans discovered that the soldier crabs found on the beaches and lagoons of islands in Japan (they range in color from pale to dark blue, lightening as they get older, with shells from 8mm to 16mm long) behave in predictable ways.

these gates were invented after we invented electronic computers: once humans have the basics of binary logic, they start seeing ways to invent computers out of all sorts of things.

The next major innovation will be *general-purpose* computing machines. The computers you just invented are built to do a single thing, but as soon as you can program your machines with numbers rather than programming them by physically moving gates, you begin to blur the line between numbers that *mean* things and numbers that *do* things. This gives computers the ability to alter their own programming as they run, and once you have that, the potential of computation explodes, and the world is never the same.

It's gonna be great!

Specifically, these crabs travel in swarms that tend to hug walls, and when two swarms collide, they merge and head off in a direction that *combines* the directions the two swarms were moving in. An OR gate is as easy as constructing crab pathways in a Y shape: the crabs enter from the top of the Y, and exit out the bottom, even if two swarms collide. An AND gate is an X shape, but with an added extra vertical line from the middle of the X to the bottom. Crabs enter the top diagonally and proceed out the bottom in the same way, but when two swarms collide, they'll move out of the vertical line you added: that's your "AND" output. The scientists who discovered the computational potential of these crabs noted that they were some-times prone to error (turns out living animals are not as perfectly predictable as streams of water or electrons!) which—along with the absence of a NOT gate to give you a universal set—may indicate some upcoming challenges in your dream of realizing large-scale computational machinery powered by li'l crabs.

CONCLUSION

THINGS SHOULD NOW BE PRETTY COMFORTABLE FOR YOU, AND YOU'RE WELCOME ⌐

It turns out that learning enough to survive in the distant past . . . was simply a matter of time.

↗ This is, sadly, where our guide comes to an end. In it you have found answers to some of the most profoundly life-changing questions humans have ever asked, including "What is the universe made of?" (Section 11: What Are Things, and How Do I Make Things?); "How can I live in comfort and not die for a while?" (Section 5: Now We Are Become Farmers, the Devourers of Worlds); and "I keep pooping too much and would like to do that less in the future, so does anyone know anything I should do to make that happen?" (Section 14: Heal Some Body). We're absolutely certain this knowledge will serve you well in the days, months, and years to come.

As you step out of your state-of-the-art FC3000™ rental-market time machine and survey an unknown Earth—one soon to be transformed into a home, a community, and a civilization—we are envious. You are about to enter a world of untold wonder and potential, and you will face it with a gift the rest of humanity never had: the gift of foresight. Use it wisely and you'll reach heights greater than we ever dreamed, all while avoiding our most terrible and harmful pitfalls.

Reading this book has transferred knowledge of humanity's greatest achievements from the palm of your hand to the interior of your mind. Earlier, we remarked that this text, once stranded in the past, was the single most powerful and dangerous thing on the planet. That is no longer true.

You are.

Go get 'em, tiger.

Figure 72: Despite everything. the FC3000™.

With warmest professional regards from your friends at Chronotix Solutions.

APPENDIX A

TECHNOLOGY TREE

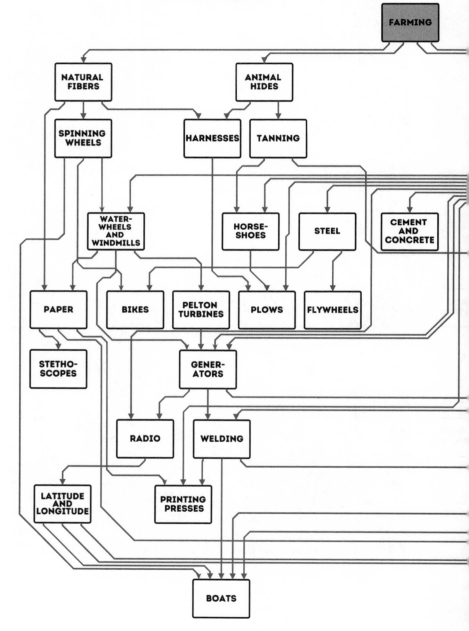

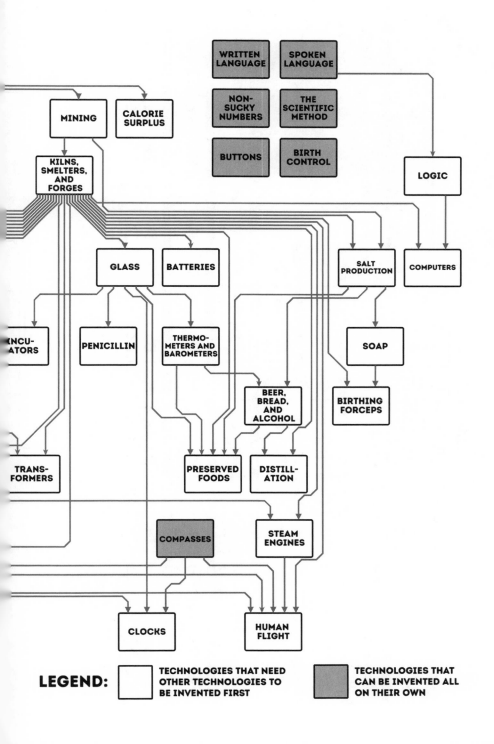

MINING

CALORIE SURPLUS

WRITTEN LANGUAGE

SPOKEN LANGUAGE

NON-SUCKY NUMBERS

THE SCIENTIFIC METHOD

BUTTONS

BIRTH CONTROL

LOGIC

KILNS, SMELTERS, AND FORGES

GLASS

BATTERIES

SALT PRODUCTION

COMPUTERS

INCU-ATORS

PENICILLIN

THERMO-METERS AND BAROMETERS

SOAP

BEER, BREAD, AND ALCOHOL

BIRTHING FORCEPS

TRANS-FORMERS

PRESERVED FOODS

DISTILL-ATION

COMPASSES

STEAM ENGINES

CLOCKS

HUMAN FLIGHT

LEGEND:

TECHNOLOGIES THAT NEED OTHER TECHNOLOGIES TO BE INVENTED FIRST

TECHNOLOGIES THAT CAN BE INVENTED ALL ON THEIR OWN

APPENDIX B

THE PERIODIC TABLE[41]

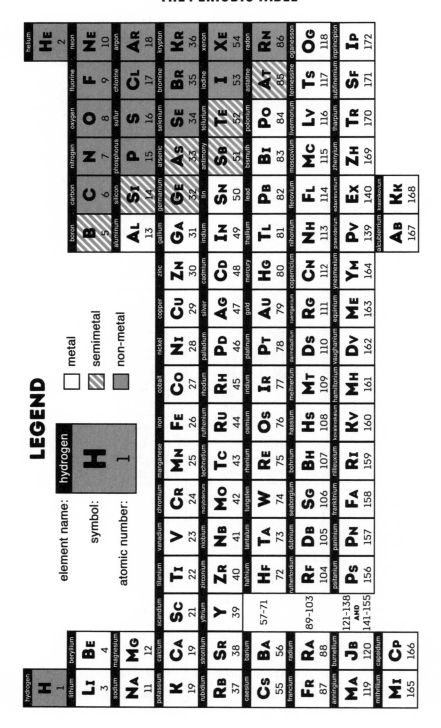

LEGEND

element name: hydrogen
symbol: **H**
atomic number: 1

metal
semimetal
non-metal

Upper block

lanthanum	cerium	praseodymium	neodymium	promethium	samarium	europium	gadolinium	terbium	dysprosium	holmium	erbium	thulium	ytterbium	lutetium
La 57	Ce 58	Pr 59	Nd 60	Pm 61	Sm 62	Eu 63	Gd 64	Tb 65	Dy 66	Ho 67	Er 68	Tm 69	Yb 70	Lu 71

actinium	thorium	protactinium	uranium	neptunium	plutonium	americium	curium	berkelium	californium	einsteinium	fermium	mendelevium	nobelium	lawrencium
Ac 89	Th 90	Pa 91	U 92	Np 93	Pu 94	Am 95	Cm 96	Bk 97	Cf 98	Es 99	Fm 100	Md 101	No 102	Lr 103

lamarrium	ramanujanium	mitchellium	seagerium	hoppernium	jemisonium	adastranium	exastrisium	verterium	erdosium	abaeternium	mantecessium	suaspontenium	rubinium	morscertium
Hl 141	Rm 142	Mm 143	Ss 144	Gh 145	Mj 146	Aa 147	Ea 148	Vt 149	Ed 150	Et 151	Ia 152	Su 153	Vr 154	Mo 155

Lower block

utpriusium	aeternium	necspeium	quidagisium	orichalcium	ibnium	malafidium	carsonium	lehmannium	noetherium	troutium	hodgkinium	hoggium	malapsanovium	uridium
Ut 121	Ae 122	Ns 123	Q 124	Or 125	Ih 126	Mf 127	Rc 128	Le 129	En 130	Jt 131	Dh 132	Hh 133	Ml 134	Ur 135

hypatium	feynmanium	luxsitium
Hy 136	Fy 137	Lx 138

APPENDIX C

USEFUL CHEMICALS, HOW TO MAKE THEM, AND HOW THEY CAN DEFINITELY KILL YOU

Included in this section are instructions to produce all the chemicals required in this text, along with other substances useful in building a civilization. They are listed with prerequisites first, so later chemicals are produced using chemicals already described. And before you get too excited and begin producing chemicals willy-nilly, know that there's a reason each chemical listed on the following pages has a "How It Can Definitely Kill You by Accident" section, and that's because they're *dangerous substances that can definitely kill you by accident.* Be careful, and don't make anything here unless you're trapped in the past and you really, really need it!

C.1: AMMONIA

FORMULA:

NH_3

APPEARANCE:

colorless gas

FIRST SYNTHESIZED:

1774 CE

DESCRIPTION:

An extremely useful chemical used in many ways throughout civilization and one of the most highly produced chemicals today. Ammonia is a fertilizer, a refrigerant, an antiseptic, and, when combined with water, a tough cleaner that gives you that perfect streak-free shine every time.

HOW TO PRODUCE:

Ammonia—NH_3—is made out of nitrogen and hydrogen, both of which are abundant on Earth. And when we say "abundant," we mean it: nitrogen's the most common gas in our atmosphere, and hydrogen's only the most common element in the *entire universe.* However, nitrogen gas—N_2—has already bonded with itself and doesn't really want to react with anything else, so all the nitrogen you're surrounded with isn't useful.

That said, you *can* collect natural ammonium chloride salts (NH_4Cl) from the ground—they form naturally from volcanic gases, so you can find these white crystals growing near volcanic vents. If mining volcanoes doesn't seem up your alley, you can also find it in camel poop: their salt-tolerant ways mean they consume a lot of chlorine, and some ammonium chloride can be found in their dung. Dry out their poops and burn them in an enclosed space—ideally with only one exit for the soot—and have colder things—glass or rocks—where the smoke escapes to encourage it to condense. Ammonium chloride crystals will form on your condensers. Add slaked lime to your ammonium chloride and heat it, and you'll produce ammonia gas.

If you can't find any ammonium chloride salts or camels, ammonia can also be harvested by taking deer antlers and hooves and dry distilling them (see Section 10.1.1)—instead of ammonium chloride, the ashes will contain ammonium carbonate. Heat these ashes to 60°C or more and the ammonium carbonate breaks down into ammonia gas, carbon dioxide, and water—making ammonium carbonate a convenient source of ammonia that you can also use to leaven your breads as a baking soda substitute!

Failing all that, you can get ammonia from your pee. All mammals get rid of excess nitrogen in their pee, which bacteria convert into ammonia—that's what you (used to) smell in any poorly cleaned restrooms that existed in your (never-to-be-returned-to) personal history. All you have to do is ferment your urine and collect the gas.

Of course, these techniques are slow and produce relatively small amounts of ammonia. To scale up to industrial levels, you'll need to invent a pressure cooker, which is just a metal pot that can be reliably sealed: get it to 450°C and around 200 atmospheric pressures, and you'll be able to induce your abundant nitrogen and hydrogen gases to react and form ammonia. This is a much more efficient way to produce ammonia, but it takes a lot more work than "collecting poop and burning it" does.

HOW IT CAN DEFINITELY KILL YOU BY ACCIDENT:

Humans actually have a way to remove excess ammonia from their bodies (we pee it out, which is why you can harvest ammonia from pee), so you don't have to worry that much about consuming too much ammonia! Sleep easy, ammonia lovers! But it's still a caustic gas that at high concentrations can destroy your lungs, so maybe don't sleep *too* easy just yet.

C.2: CALCIUM CARBONATE

ALSO KNOWN AS:
chalk

FORMULA:
$CaCO_3$

APPEARANCE:
fine white powder

FIRST SYNTHESIZED:
7200 BCE (use of naturally occurring sources, not synthesis)

DESCRIPTION:
When you add soda ash to silica, your glass (Section 10.4.3) will be slightly water-soluble, but adding some calcium carbonate to it fixes that problem! You can also add it to soil to help plants absorb nitrogen, give them calcium, and reduce the acidity of overly acidic soils. It's one of the easiest bases to produce.

HOW TO PRODUCE:
Several rocks are made primarily out of calcium carbonate, including calcite (pure calcium carbonate), limestone, chalk, and marble. They make up about 4 percent of the Earth's crust, so they shouldn't be too hard to find. Eggshells, snail shells, and most seashells are also high in calcium carbonate: eggshells alone are about 94 percent composed of the stuff. Just clean them, dry them, and grind them up. This is also produced when making lye (see C.8).

HOW IT CAN DEFINITELY KILL YOU BY ACCIDENT:
You can actually eat this as a calcium supplement or as an antacid, but too much causes problems and can be fatal.

C.3: CALCIUM OXIDE

ALSO KNOWN AS:
lime, quicklime

FORMULA:
CaO

APPEARANCE:
white to pale-yellow powder

FIRST SYNTHESIZED:
7200 BCE (use of naturally occurring sources, not synthesis)

DESCRIPTION:
Quicklime has uses in making glass, and produces an intense light when burned (hence the phrase "being in the limelight," as limelights were used in theaters back before electric lighting was invented, which for you is right now).

HOW TO PRODUCE:
Quicklime can be produced by taking substances with carbonate in them (limestone, seashells, etc.) and heating them above 850°C in a kiln. This causes the calcium carbonate to react with oxygen, producing quicklime and carbon dioxide. Quicklime is unstable though, and over time will react with carbon dioxide in the air to become calcium carbonate again, so if you're not going to use it right away you'll want to convert it to slaked lime (see C.4). Producing 1kg of quicklime takes about 1.8kg of limestone.

HOW IT CAN DEFINITELY KILL YOU BY ACCIDENT:
Since it reacts with water and the insides of bodies are *pretty moist*, quicklime causes severe irritation when inhaled or when it makes contact with your juicy eyes. Quicklime can lead to chemical burns that can even eat *right through the part of your nose that divides your two nostrils*, so don't breathe it in either.

C.4: CALCIUM HYDROXIDE

ALSO KNOWN AS:
slaked lime

FORMULA:
$Ca(OH)_2$

APPEARANCE:
white powder

FIRST SYNTHESIZED:
7200 BCE (use of naturally occurring sources, not synthesis)

DESCRIPTION:
Slaked lime is an easily produced and versatile substance that's been used for thousands of years. It can be used as a mortar or plaster—add it to clay and it'll produce a material that hardens as it ages (see Section 10.10.1). It can be a calcium supplement in juices or a substitute for baking soda. Additionally, it helps remove substances from liquids by encouraging them to coagulate, which has lots of uses in purifying water and in sewage treatment.

HOW TO PRODUCE:

Just mix calcium oxide with water. This reaction causes heat though, so be careful! In fact, always be careful when mixing chemicals together. *You've got enough problems already.* And if you want to reverse this process and get your quicklime back, heat your slaked lime until the water is driven off, which happens at 512°C.

HOW IT CAN DEFINITELY KILL YOU BY ACCIDENT:

You can get chemical burns from contact, which at extreme ends can cause blindness or lung damage if you're silly enough to inhale these weird chemicals you're inventing.

C.5: POTASSIUM CARBONATE

ALSO KNOWN AS:

potash

FORMULA:

K_2CO_3

APPEARANCE:

white powder

FIRST SYNTHESIZED:

200 CE

DESCRIPTION:

It's a useful additive in bleaching clothes, glass, soap, and in producing a lot of other chemicals. You can also use it as a leavening agent in your bread!

HOW TO PRODUCE:

Collect plant ashes (wood works fine, hardwood works better, just make sure the fire it came from wasn't extinguished with water, or you'll already have washed away the chemicals you're interested in harvesting), dissolve your ashes in water, and then boil it off (or let it dry in the sun). The ashy white residue at the bottom of your pot—"pot ash," if you will—is potash.

It takes a lot of wood to make a little potash: you get only about 1 gram of potash for every kilogram of wood you burn. But it's an extremely easy process, and while you're making potash you can put your fire to good work doing other productive things.

HOW IT CAN DEFINITELY KILL YOU BY ACCIDENT:

Potash is caustic, so don't get it in your eyes, rub it on your skin, or eat it. You'd have to eat a lot of potash to get in trouble, but we're not going to tell you how much, because you shouldn't be eating any of it at all! It's boiled wood ashes! That's not a food!

C.6: SODIUM CARBONATE AND SODIUM BICARBONATE

ALSO KNOWN AS:

soda ash (sodium carbonate), baking soda (sodium bicarbonate)

FORMULA:

Na_2CO_3 (sodium carbonate), $NaHCO_3$ (sodium bicarbonate)

APPEARANCE:

white powder

FIRST SYNTHESIZED:
200 CE (sodium carbonate extracted from natural sources), 1791 CE (sodium carbonate synthesized), 1861 CE (efficient synthesis of sodium carbonate)

DESCRIPTION:
Sodium carbonate lowers the melting point of silica, which is useful in making glass. You can also use it to make soap and soften water! Sodium bicarbonate can make your baking rise without using yeast, treat heartburn, produce a plaque-fighting toothpaste, deodorize your underarms, or kill cockroaches (it's a pretty handy substance).

HOW TO PRODUCE:
You use the same process you'd use to make potash, but instead use trees grown in sodium-rich soils: kelp (i.e., seaweed) works, and plants that grow in salty soils are also good candidates. If you (or your civilization) is feeling fancier, you can produce it on an industrial scale using a technique called the "Solvay process" that normally doesn't get invented until 1861 CE.

First, make a watertight tower about 25m tall: steel works well. At the bottom, you'll heat limestone to produce quicklime and carbon dioxide (see C.3). Above it, you'll have a concentrated liquid solution of ammonia and salt. As carbon dioxide bubbles through this solution, the ammonia turns into ammonium chloride (NH_4Cl, a substance we weren't trying to make, but hold on to it) and sodium bicarbonate ($NaHCO_3$, also known as "baking soda") falls out of your solution, collecting at the bottom. You can harvest it and use it now, but if you heat it, it decomposes into sodium carbonate (which is what we're looking for) along with water and carbon dioxide. You're left with ammonium chloride, which—if you don't want to produce morphine (see Section 7.15)—you can add slaked lime to: this produces pure ammonia, pure water, and some calcium chloride ($CaCl_2$). The nice thing about this process is you get your ammonia back at the end of it, so it's pretty economical to run!

Calcium chloride can be discarded as waste, or you can use it as a de-icer (it lowers the freezing point of water, so if you have roads it's a great road salt), to flavor pickles (it tastes really salty but doesn't actually contain any sodium), or to produce activated charcoal, which is just charcoal with more surface area in it. Just soak your wood in calcium chloride before beginning the charcoal process (see Section 10.1.1).

As for the carbon dioxide, if you allow that gas to escape into a sealed container of water, the container will pressurize and the water will absorb some amount of that gas! This carbon dioxide gets released when the container is opened and the pressure is reduced, slowly bubbling out of the water. In other words, *you just invented soda pop.* Carbonated beverages usually only show up in 1767 CE, but people love them in any time period!

HOW THEY CAN DEFINITELY KILL YOU BY ACCIDENT:
These are all actually pretty safe in moderation, and you've probably eaten all of these in the past. Finally! A chemical safe enough to make cookies with!

C.7: IODINE

FORMULA:
I_2

APPEARANCE:

purple gas, metallic gray solid

FIRST SYNTHESIZED:

1811 CE (discovered)

DESCRIPTION:

Iodine is an antiseptic: mix it into water to kill bacteria, and pour that water into your wounds to prevent infections. It's also an essential element for life—without it you'll develop goiters and then die! See Section 10.2.6 for more.

HOW TO PRODUCE:

When producing sodium carbonate from ashes, add sulfuric acid to the waste left over: add enough, and you'll produce a cloud of purplish gas. This gas will crystallize on cold surfaces, and those crystals are pure iodine.

Iodine will slightly dissolve in water (you can get about 1g into 1.3L of water, as long as the water is heated to 50°C). To dissolve more, add iodine to potassium hydroxide, which produces potassium iodide: this chemical helps more sodium dissolve into water.

HOW IT CAN DEFINITELY KILL YOU BY ACCIDENT:

You need iodine to live, but pure iodine is toxic if you eat it without diluting it first. It can irritate skin, and in high enough concentrations can result in tissue damage.

C.8: SODIUM HYDROXIDE AND POTASSIUM HYDROXIDE

ALSO KNOWN AS:

caustic soda (sodium hydroxide), caustic potash (potassium hydroxide), lye (both)

FORMULA:

NaOH (sodium hydroxide), KOH (potassium hydroxide)

APPEARANCE:

white solid

FIRST SYNTHESIZED:

200 CE

DESCRIPTION:

Both sodium and potassium hydroxide are used in making soaps. Since they dissolve organic tissue, they're useful for cleaning things like brewing vats!

HOW TO PRODUCE:

Both these chemicals have historically been called "lye," since both can generally be substituted for each other in most contexts. While sodium hydroxide can be produced with little more than saltwater and electricity, you can also produce it from wood ashes. Run water through ashes (see C.5 and C.6 for details) and add slaked lime (see C.4). Lye (which will either be potassium hydroxide or sodium hydroxide, depending on whether you used potassium carbonate or sodium carbonate) will be produced, with calcium carbonate settling at the bottom.

HOW THEY CAN DEFINITELY KILL YOU BY ACCIDENT:

Listen: these chemicals are known as "caustic" because they *dissolve the proteins and fats in living tissues*. You are made out of living tissues, which means lye is not a thing you want to have on or near you. It will cause chemical burns on contact, and if you get it in your eyes, you can go blind. Caustics have even been used to decompose organic tissues into a slurry, in an attempt to dispose of human bodies!

CIVILIZATION PRO TIP: If things are going well, you should not need to get rid of any human bodies.

C.9: POTASSIUM NITRATE

ALSO KNOWN AS:
saltpeter

FORMULA:
KNO_3

APPEARANCE:
white solid

FIRST SYNTHESIZED:
1270 CE

DESCRIPTION:
Saltpeter's edible, so you can use it to preserve meats, soften food, and thicken soups. It's also used as a fertilizer (it's a source of nitrogen) and has even been used as a stump remover, simply by encouraging the growth of funguses that eat the stump. It also can be used to fight asthma symptoms, combat high blood pressure, and as a toothpaste for people with sensitive teeth.

HOW TO PRODUCE:
There are different ways to produce this, depending on what you have on hand.
• Soak bat poop collected from caves in water for a day, filter it, add lye, boil it down until it thickens, and harvest the long, needle-shaped crystals that form when it cools. Bats first evolve around 55,000,000 BCE, so they're around whenever humans are.
• Mix manure with wood ashes and straw to make it more porous. Make a heap about 1.2m high, 7m wide, and 4.5m long. Keep the heap covered from rain, and keep it moist—but not wet—with urine. Stir it occasionally to speed up decomposition. After about a year, leach it (run water through it and collect the runoff), which gives you calcium nitrate. Filter this through potash and you'll produce saltpeter.

HOW IT CAN DEFINITELY KILL YOU BY ACCIDENT:
This chemical's actually pretty safe both to extract and be around. That's a nice change!

C.10: ETHANOL

ALSO KNOWN AS:
alcohol

FORMULA:
C_2H_6O

APPEARANCE:
colorless liquid

FIRST SYNTHESIZED:
10,000s BCE (by humans intentionally; it can be produced accidentally in any rotting fruit)

DESCRIPTION:
You can drink it to become more sociable and/or sad. It's also an antiseptic, can be used as a fuel, and is a great basis for a thermometer.

HOW TO PRODUCE:
See Section 10.2.5 for instructions on how to brew alcohol, and then distill it to extract ethanol.

HOW IT CAN DEFINITELY KILL YOU BY ACCIDENT:
It's an addictive psychoactive drug and a neurotoxin when taken in sufficient quantities!

C.11: CHLORINE GAS

FORMULA:
Cl_2

APPEARANCE:
pale yellowish gas

FIRST SYNTHESIZED:
1630 CE

DESCRIPTION:
Chlorine is an extremely reactive gas useful as a disinfectant (especially when added in pools and drinking water) but which is extremely toxic to every living organism.

HOW TO PRODUCE:
Run electricity through brine (i.e., saltwater). The gas bubbling from the positive terminal will be chlorine gas. The gas from the negative terminal is hydrogen, and sodium hydroxide (see C.8) is what's building up in the water.

HOW IT CAN DEFINITELY KILL YOU BY ACCIDENT:
Chlorine gas has been used as poison gas in wartime, so you don't want to be anywhere near it. At high temperatures it also reacts with iron to produce *chlorine-iron fires*, which are about as safe as they sound (they are extremely not safe).

C.12: SULFURIC ACID

FORMULA:
H_2SO_4

APPEARANCE:
colorless liquid

FIRST SYNTHESIZED:
3000 BCE[42]

DESCRIPTION:
A highly corrosive acid, useful in everything from "making batteries" to "dissolving things in acid"; it's now the most produced chemical on the planet!

HOW TO PRODUCE:

Find some iron pyrite (aka FeS$_2$, aka "fool's gold") which is a crystal-like gold-colored mineral. This shouldn't be too hard: fool's gold is the most common iron sulfide on the planet, and it's usually found in veins of quartz, sedimentary rocks, and in coal beds. Unfortunately, you won't find it on the surface, because it decomposes when exposed to air and water, but there's always new fool's gold being produced underground.

Bake your fool's gold and collect the gas that comes off, which is sulfur dioxide (SO$_2$). Mix this sulfur dioxide gas with chlorine gas (Cl$_2$) in the presence of charcoal, which acts as a catalyst, and you'll produce a new liquid, sulfuryl chloride (SO$_2$Cl$_2$). Distill this liquid to concentrate it, and then (carefully) add water: this reaction produces both sulfuric acid and hydrogen chloride gas. (Collect the hydrogen chloride gas and bubble it through water to produce hydrochloric acid [HCl]: two acids for the price of one!) Sulfuric acid is extremely reactive and corrosive and should be stored and handled carefully.

The good news is, once you've produced a little sulfuric acid, you can use it to identify iron pyrite so you can make more! A drop of sulfuric acid on iron pyrite will sizzle and smell like rotten eggs.

HOW IT CAN DEFINITELY KILL YOU BY ACCIDENT:

Get it on your skin and it'll cause severe burns, splash it onto your eyes and you're blind forever, swallow it and it'll cause damage that can't be taken back. Maybe don't touch, swallow, or splash into your eyes *any* of the chemicals in this section, huh?

C.13: HYDROCHLORIC ACID

ALSO KNOWN AS:
spirit of salt, salt acid

FORMULA:
HCl

APPEARANCE:
colorless liquid

FIRST SYNTHESIZED:
800 CE

DESCRIPTION:
A great household cleaner that also removes rust from steel!

HOW TO PRODUCE:
Run hydrogen chloride gas through water (see C.12), or combine sulfuric acid with salt.

HOW IT CAN DEFINITELY KILL YOU BY ACCIDENT:
Concentrated hydrochloric acid produces an *acidic mist*, which will damage you and your precious organs irreversibly, and the non-mist form does the same.

C.14: DIETHYL ETHER

ALSO KNOWN AS:
ether

FORMULA:
$(C_2H_5)_2O$
APPEARANCE:
colorless, transparent liquid
FIRST SYNTHESIZED:
700s CE
DESCRIPTION:
An inhalable anesthetic that induces unconsciousness, but it can be slow to work and induce nausea. Anesthetics are what make surgery *not* be a harrowing nightmare experience in which you are pinned down fully awake while someone cuts into your screaming body, so they're useful to have on hand!
HOW TO PRODUCE:
Mix ethanol with sulfuric acid, then distill the resulting mixture to extract the ether. Keep the temperature below 150°C to prevent your ethanol from forming ethylene (C_2H_4), unless you want it; ethylene can be used to force fruits to ripen, and it can also be mixed in an 85 percent ethylene/15 percent oxygen ratio to make an anesthetic.
HOW IT CAN DEFINITELY KILL YOU BY ACCIDENT:
It's highly flammable when there's oxygen around, *which there usually is.*

C.15: NITRIC ACID

FORMULA:
HNO_3
APPEARANCE:
colorless or yellow/red fuming liquid
FIRST SYNTHESIZED:
1200s CE
DESCRIPTION:
A strong oxidizing agent useful in rocket fuel (you probably won't need rocket fuel), a way to artificially age pine and maple wood to make it look fancy (again: probably not your top concern right now), and also as an ingredient in ammonium nitrate.
HOW TO PRODUCE:
React sulfuric acid with saltpeter. Careful: it reacts violently with organic material *and* decomposes living tissue, so don't get any on you, and if you do wash it with water for fifteen minutes minimum!
HOW IT CAN DEFINITELY KILL YOU BY ACCIDENT:
We don't know what we can add to "it reacts violently with organic material *and* decomposes living tissue." This is not something you want to be around.

C.16: AMMONIUM NITRATE

FORMULA:
NH_4NO_3
APPEARANCE:
white/gray solid

FIRST SYNTHESIZED:

1659 CE

DESCRIPTION:

A high-nitrogen fertilizer that's also a way to make laughing gas (see C.17) and an explosive! Ammonium nitrate will be the key to making your fields produce far more food than they normally could, which will allow your civilization to have far more human brains working inside it than any other.

HOW TO PRODUCE:

Mix ammonia and nitric acid together. Done! That was easy! Or it would be, if ammonia and nitric acid didn't react violently and produce a lot of heat, which can cause explosions. So be careful!

HOW IT CAN DEFINITELY KILL YOU BY ACCIDENT:

It's extremely explosive, and any source of heat or flame could set it off. Notable fatal ammonium nitrate disasters have occurred in 1916 CE, 1921 CE, 1942 CE, 1947 CE, 2004 CE, and 2015 CE, and those are just the ones that killed a hundred people or more!

C.17: NITROUS OXIDE

ALSO KNOWN AS:

laughing gas

FORMULA:

N_2O

APPEARANCE:

colorless gas

FIRST SYNTHESIZED:

1772 CE

DESCRIPTION:

A gas that makes you feel euphoric, increases your suggestibility, provides some pain relief, relaxes your muscles, and then, if you inhale enough, knocks you unconscious. That makes it an anesthetic! If you combine laughing gas with other anesthetics, like ether, it enhances their effectiveness.

HOW TO PRODUCE:

Carefully and slowly heat ammonium nitrate: nitrous oxide is the gas that's produced. You can cool and clean impurities from the gas by bubbling it through water. Be careful while you're heating it though, because you are heating an explosive, and if your ammonium nitrate gets hotter than 240°C, it can blow up.

HOW IT CAN DEFINITELY KILL YOU BY ACCIDENT:

There are so many ways this could go wrong. *You are literally heating an explosive.*

APPENDIX D
LOGICAL ARGUMENT FORMS

For reference, here are some of the forms of argument you can use in symbolic logical reasoning. Symbols used here include → for "implies," ∴ for "then" or "therefore," ¬ for "not," ∧ for "and," ∨ for "or," and ↔ for "is equivalent to" or "is interchangeable with."

In symbols	In words
$p \therefore \neg\neg p$	if p is true, then not p is also true: in other words, true and false are the only two values permitted, and they are opposites
$p \therefore (p \lor p)$	if p is true, then (p or p) is also true
$p \therefore (p \land p)$	if p is true, then (p and p) is also true
$(p \lor \neg p) \therefore$ true	(p or not p) is always true
$\neg(p \land \neg p) \therefore$ true	not (p and not p) is always true
$(p \land q) \therefore p$	if p and q are true, then p is also true
$p \therefore (p \lor q)$	if p is true, then (p or q) is also true
$p, q \therefore (p \land q)$	if p and q are true separately, then they are true together
$(p \lor q) \therefore (q \lor p)$	p or q is the same as q or p: order doesn't matter here
$(p \land q) \therefore (q \land p)$	p and q is the same as q and p: order doesn't matter here either
$(p \leftrightarrow q) \therefore (q \leftrightarrow p)$	p being equivalent to q is the same as q being equivalent to p: order also doesn't matter here, so that's nice
$(p \rightarrow q) \therefore (\neg q \rightarrow \neg p)$	if p implies q, then not q implies not p
$(p \rightarrow q) \therefore (\neg p \lor q)$	if p implies q, then either not p or q is true
$[(p \rightarrow q) \land p] \therefore q$	if p implies q, and p is true, then q is true
$[(p \rightarrow q) \land \neg q] \therefore \neg p$	if p implies q, and not q is true, then not p is true
$[(p \rightarrow q) \land (q \rightarrow r)] \therefore (p \rightarrow r)$	if p implies q, and if q implies r, then p implies r
$[(p \lor q) \land \neg p] \therefore q$	if p or q is true, and not p is true, then q is true

In symbols	In words
$[(p \to q) \land (r \to s) \land (p \lor r)] \therefore (q \lor s)$	if p implies q, and if r implies s, and p or r is true, then q or s is true
$[(p \to q) \land (r \to s) \land (\neg q \lor \neg s)] \therefore (\neg p \lor \neg r)$	if p implies q, and if r implies s, and not q or not s is true, then not p or not r is true
$[(p \to q) \land (r \to s) \land (p \lor \neg s)] \therefore (q \lor \neg r)$	if p implies q, and if r implies s, and p or not s is true, then q or not r is true
$[(p \to q) \land (p \to r)] \therefore [p \to (q \land r)]$	if p implies q, and if p implies r, then p implies q and r together
$\neg(p \land q) \therefore (\neg p \lor \neg q)$	not (p and q) is the same as (not p or not q)
$\neg(p \lor q) \therefore (\neg p \land \neg q)$	not (p or q) is the same as (not p and not q)
$[p \lor (q \lor r)] \therefore [(p \lor q) \lor r]$	p or (q or r) is the same as (p or q) or r: you can move the brackets around in a group of "or" statements
$[p \land (q \land r)] \therefore [(p \land q) \land r]$	p and (q and r) is the same as (p and q) and r: you can move the brackets around in a group of "and" statements too
$[p \land (q \lor r)] \therefore [(p \land q) \lor (p \land r)]$	p and (q or r) is the same as (p and q) or (p and r)
$[p \lor (q \land r)] \therefore [(p \lor q) \land (p \lor r)]$	p or (q and r) is the same as (p or q) and (p or r)
$(p \leftrightarrow q) \therefore [(p \to q) \land (q \to p)]$	p being equivalent to q is the same as saying (p implies q) and (q implies p)
$(p \leftrightarrow q) \therefore [(p \land q) \lor (\neg p \land \neg q)]$	if p is equivalent to q, then it's true that either (p and q are true) or (not p and not q are true)
$(p \leftrightarrow q) \therefore [(p \lor \neg q) \land (\neg p \lor q)]$	if p is equivalent to q, then it's true that both (p or not q is true) and (not p or q is true)
$[(p \land q) \to r] \therefore [p \to (q \to r)]$	if (p and q) implies r, then p implies (q implies r)
$[p \to (q \to r)] \therefore [(p \land q) \to r]$	if p implies (q implies r), then (p and q) implies r

Table 28: Please enjoy this list of correct argument forms that took humanity thousands of years of reasoning to figure out, but which now rates only two pages printed at the end of a book.

APPENDIX E

TRIGONOMETRY TABLES, INCLUDED BECAUSE YOU'LL NEED THEM WHEN YOU INVENT SUNDIALS, BUT THEY'LL ALSO BE USEFUL IF YOU EVER DECIDE TO INVENT TRIGONOMETRY

This book is a guide to reinventing civilization from scratch, and while your civilization will eventually want to invent trigonometry, in an era where you don't know where your next meal is coming from because you're still figuring out what farming is, you *probably* don't need it right away. We therefore won't cover the entirety of trig here, but in this appendix we do give you some of the most useful low-hanging fruit: enough for practical purposes, and to point the way to future discoveries.

Trigonometry lets you use some known quantities about a triangle to determine its unknown quantities, we will interrupt ourselves right now because we can already hear you muttering, "Come on, when am I ever gonna use this?" Here is when you are going to use this: in navigation, astronomy, music, number theory, engineering, electronics, physics, architecture, optics, statistics, cartography, *and more*. You already need it just to build a proper sundial (Section 10.7.1), hence the unofficial slogan of trigonometry: "Okay, fine, I guess this is actually pretty important after all."

Trigonometry deals only with right triangles (triangles with two edges that meet at 90 degrees, and we mark that angle with a little square), but since any non-right triangle can be divided into two right triangles (try it; it's true), that's not going to be a problem. A right triangle looks like this:

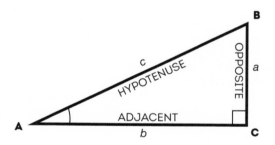

Figure 73: This is what a right triangle looks like.

We'll call the longest side (c, which is always opposite the right angle) the "hypotenuse." If we pick an angle (in this case we used the one at A) then we can label the side opposite A as the "opposite" side, and the side beside it "adjacent." Incidentally, the three angles of any triangle add up to 180 degrees, so since we know in a right triangle one of them is 90, all it takes is knowing one of the other angles to determine the remaining one. Here's a useful theorem about right triangles:

$$a^2 + b^2 = c^2$$

That's what we call the "Pythagorean theorem," after a guy named Pythagoras who lived in ancient Greece around 500 BCE, but even he admitted he wasn't the first to come up with the idea—and it's been independently discovered before and since all over the world. It says that the sum of the square of the length of the two shortest sides of a right triangle is equal to the square of the length of the longest side. This lets you calculate all measurements of a right triangle from only partial data, which, as we already said, is something that trigonometry is *all about*.

If you know the angles in a right triangle, that means you've also described its shape, because there's only one way those angles can fit together to make a triangle. And that means the opposite is true too: if you know the length of the sides of a right triangle, then you also know the angles involved. This lets us define some useful operations. The ratio of the length of the opposite side to the length of the hypotenuse we'll call "the sine function," or "sin" for short.* The ratio of the adjacent side to the hypotenuse we'll call "cosine" ("cos" for short), and the ratio of the opposite to the adjacent we'll call "tangent," or "tan." Given an angle we can determine what the sin, cos, and tan values for that angle are. Going the other way, if we know the sin, cos, or tan value, we can determine which angle that requires. We'll mark these inverse functions with a tiny -1, so that's sin^{-1}, cos^{-1}, and tan^{-1}.

As you explore trigonometry you'll discover proofs that relate this study of triangles to circles (draw a circle around your triangle and you'll see connections between pi and your cos, sin, and tan functions); periodic functions (graph the values of cos, sin, and tan and you'll notice how their patterns repeat); and even your trig functions to one another (as an example: the tangent of an angle is equal to its sine divided by its cosine). All of this is to say: if this interests you, there's lots to explore, and many have dedicated their lives to much smaller and less noble subjects.†

The catch is, however, that calculating the values of sin, cos, and tan is complicated, and only really needs to be done once. So instead of making you do it yourself, we, your friends here at Chronotix Solutions, have provided full trigonometry tables on the following pages. Given an angle a, you can find the values of sin(a), cos(a), and tan(a). And to run the inverse functions (sin^{-1}, cos^{-1}, and tan^{-1}), just look for the angle that matches the value you have.

The following is information you need to explore trigonometry, invent new theorems and trigonometric equations, and, most important, finish your sundial in Section 10.7.1.

---

* We call it "sine" because when Europeans were translating Arabic works into Latin (*because of course they were*), the Latin word *sinus* (meaning "the hanging fold in the upper part of a toga") was the closest match they could find for the Arabic word used, which was *jaib*, meaning "pocket, fold, or purse." But the word "jaib" wasn't even being used! It was *jyb*, which was the way the Arabs had come up with to write the Sanskrit word *they* were translating, *jyā*, into their alphabet. That word has its origins in the ancient Greek word for "string." Anyway! Feel free to name this function *anything else*, because you can hardly end up with a more arbitrary word than we did.

† I, for example, write time-machine repair manuals to reduce liability in court.

Angle a	sin(a)	cos(a)	tan(a)	Angle a	sin(a)	cos(a)	tan(a)
0	.0000	1.0000	.0000				
1	.0175	.9998	.0175	46	.7193	.6947	1.0355
2	.0349	.9994	.0349	47	.7314	.6820	1.0723
3	.0523	.9986	.0524	48	.7431	.6691	1.1106
4	.0698	.9976	.0699	49	.7547	.6561	1.1504
5	.0872	.9962	.0875	50	.7660	.6428	1.1918
6	.1045	.9945	.1051	51	.7771	.6293	1.2349
7	.1219	.9925	.1228	52	.7880	.6157	1.2799
8	.1392	.9903	.1405	53	.7986	.6018	1.3270
9	.1564	.9877	.1584	54	.8090	.5878	1.3764
10	.1736	.9848	.1763	55	.8192	.5736	1.4281
11	.1908	.9816	.1944	56	.8290	.5592	1.4826
12	.2079	.9781	.2126	57	.8387	.5446	1.5399
13	.2250	.9744	.2309	58	.8480	.5299	1.6003
14	.2419	.9703	.2493	59	.8572	.5150	1.6643
15	.2588	.9659	.2679	60	.8660	.5000	1.7321
16	.2756	.9613	.2867	61	.8746	.4848	1.8040
17	.2924	.9563	.3057	62	.8829	.4695	1.8807
18	.3090	.9511	.3249	63	.8910	.4540	1.9626
19	.3256	.9455	.3443	64	.8988	.4384	2.0503
20	.3420	.9397	.3640	65	.9063	.4226	2.1445
21	.3584	.9336	.3839	66	.9135	.4067	2.2460
22	.3746	.9272	.4040	67	.9205	.3907	2.3559
23	.3907	.9205	.4245	68	.9279	.3746	2.4751
24	.4067	.9135	.4452	69	.9336	.3584	2.6051
25	.4226	.9063	.4663	70	.9397	.3420	2.7475

Angle a	sin(a)	cos(a)	tan(a)	Angle a	sin(a)	cos(a)	tan(a)
26	.4384	.8988	.4877	71	.9456	.3256	2.9042
27	.4540	.8910	.5095	72	.9511	.3090	3.0779
28	.4695	.8829	.5317	73	.9563	.2924	3.2709
29	.4848	.8746	.5543	74	.9613	.2756	3.4874
30	.5000	.8660	.5774	75	.9659	.2588	3.7321
31	.5150	.8572	.6009	76	.9703	.2419	4.0108
32	.5299	.8480	.6249	77	.9744	.2250	4.3315
33	.5446	.8387	.6494	78	.9781	.2079	4.7046
34	.5592	.8290	.6745	79	.9816	.1908	5.1446
35	.5736	.8192	.7002	80	.9848	.1736	5.6713
36	.5878	.8090	.7265	81	.9877	.1564	6.3138
37	.6018	.7986	.7536	82	.9903	.1391	7.1154
38	.6157	.7880	.7813	83	.9925	.1219	8.1443
39	.6293	.7771	.8098	84	.9945	.1045	9.5144
40	.6428	.7660	.8391	85	.9962	.0872	11.4301
41	.6561	.7547	.8693	86	.9976	.0698	14.3007
42	.6691	.7431	.9004	87	.9986	.0523	19.0811
43	.6820	.7314	.9325	88	.9994	.0349	28.6363
44	.6947	.7193	.9657	89	.9998	.0175	57.2900
45	.7071	.7071	1.0000	90	1.0000	.0000	*infinity*

Table 29: Here are the numbers you need to make triangles work.

APPENDIX F

SOME UNIVERSAL CONSTANTS THAT TOOK HUMANITY A WHILE TO FIGURE OUT, AND WHICH YOU CAN NOW NAME AFTER YOURSELF

Constant	Value	Description	Notes
Speed of light	299,792,458 m/s	This is the speed of light in a vacuum, which is also the ultimate speed limit of the universe. Light, EM radiation, gravitational waves: they can all go this fast, and no faster.	Light travels more slowly when traveling through different materials: in glass, for example, you'll divide this speed by around 1.5. But even at those speeds light's so fast that it's not until 1676 CE that anyone manages to prove it doesn't travel instantaneously!
Speed of sound	343 m/s	The speed of sound depends on what medium it's moving through: this number is for dry air at 20°C. Sound travels faster in liquids and faster still in solids.	The speed of sound was calculated in 1709 CE by firing a gun at night, observing it by telescope a known distance away, and timing how long it took for the sound to be heard after the light arrived. Go to bed early and save yourself the trouble!

Constant	Value	Description	Notes
Pi	3.14159265358979323846264338 32795028841971693993751058 20974944592307816406286208 99862803482534211706798 21480865132823066470938446 09550582231725359408128481 11745028410270193852110559 64462294895493038196442881 09756659334461284756482337 86783165271201909145648566 92346034861045432664821339 36072602491412737245870066 06315588174881520920962829 25409171536436789259036001 13305305488204665213841469 51941511609433057270365759 59195309218611738193261179 31051185480744623799627495 67351885752724891227938183 01194912983367336244065664 30860213949463952247371907 02179860943702770539217176 29317675238467481846766940 51320005681271452635608277 85771342757789609173637178 72146844090122495343014654 95853710507922796892589235 42019956112129021960864034 41815981362977477130996051 87072113499999 . . .	Pi is the ratio of a circle's circumference (distance around its edge) to its diameter (the length of any line dividing a circle in two through its middle). It's an irrational number, which means if you try to write it out using rational numbers (i.e., the numbers you know) it will take you forever. It never ends, and it never repeats.	We've provided the first 768 digits of pi here, because at that point, entirely by coincidence, you get six 9s in a row. If you decide to memorize and then recite the value of pi to someone, this is a great point to do as many mathematicians in the past have done: imply you've memorized a lot more, by finishing up here by saying "nine, nine, nine, nine, nine, nine . . . and so on."[43]
Acceleration caused by gravity on Earth	Around 9.8 m/s^2	This is the rate at which you accelerate in free fall on Earth. This changes depending on the density of the air and other factors, generally ranging from 9.764 m/s^2 to 9.834 m/s^2. If you want to calculate how long it'll take something to fall, this is the number you start with!	Without air resistance, a ton of bricks and a ton of feathers would fall at the same rate, though humans didn't prove this until 1634 CE.

Constant	Value	Description	Notes
Gravitational constant	$6.67408{\times}10^{-11}$ m³kg⁻¹s⁻²	In classical physics, the gravitational attraction between two particles of matter is directly proportional to their masses multiplied together, divided by the square of their distance. But you need to multiply all that by this number—the gravitational constant of the universe—to get the actual measurement of force.	Alter the gravitational constant of the universe and you'll lose weight, but at an extremely catastrophic cost.
Mass of an electron	$9.10938356 \times 10^{-31}$ kg	All electrons are identical, which, if only for reasons of space when describing them in a book, is really convenient!	Before we discovered why all electrons were identical, there was a theory that it was because all electrons were actually the same electron, traveling back and forth through time across the entire life-span of the universe, over and over. This theory was as crazy as it was inaccurate, which is to say: very![44]

Table 30: And here are the numbers you need to make reality work.

APPENDIX G

FREQUENCIES FOR VARIOUS NOTES, SO YOU CAN PLAY THOSE COOL SONGS WE INCLUDED

Note (Octave 0, typically the lowest on a piano)	Frequency (Hz)
C	16.352
C#	17.325
D	18.354
D#	19.445
E	20.602
F	21.827
F#	23.125
G	24.500
G#	25.957
A	27.500
A#	29.135
B	30.868

Note (Octave 1)	Frequency (Hz)
C	32.703
C#	34.648
D	36.708
D#	38.891
E	41.203
F	43.654
F#	46.249
G	48.999
G#	51.913
A	55.000
A#	58.270
B	61.735

Note (Octave 2)	Frequency (Hz)
C	65.406
C#	69.296
D	73.416
D#	77.782
E	82.407
F	87.307
F#	92.499
G	97.999
G#	103.83
A	110.00
A#	116.54
B	123.47

Note (Octave 3)	Frequency (Hz)
C	130.81
C#	138.59
D	146.83
D#	155.56
E	164.81
F	174.61
F#	185.00
G	196.00
G#	207.65
A	220.00
A#	233.08
B	246.94

Note (Octave 4)	Frequency (Hz)
C	261.63
C#	277.18
D	293.66
D#	311.13
E	329.63
F	349.23
F#	369.99
G	392.00
G#	415.30
A	440.00
A#	466.16
B	493.88

Note (Octave 5)	Frequency (Hz)
C	523.25
C#	554.37
D	587.33
D#	622.25
E	659.26
F	698.46
F#	739.99
G	783.99
G#	830.61
A	880.00
A#	932.33
B	987.77

Note (Octave 6)	Frequency (Hz)	Note (Octave 7)	Frequency (Hz)
C	1046.5	C	2093.0
C#	1108.7	C#	2217.5
D	1174.7	D	2349.3
D#	1244.5	D#	2489.0
E	1318.5	E	2637.0
F	1396.9	F	2793.8
F#	1480.0	F#	2960.0
G	1568.0	G	3136.0
G#	1661.2	G#	3322.4
A	1760.0	A	3520.0
A#	1864.7	A#	3729.3
B	1975.5	B	3951.1

APPENDIX H

A BUNCH OF COOL GEARS AND OTHER FUNDAMENTAL MECHANISMS

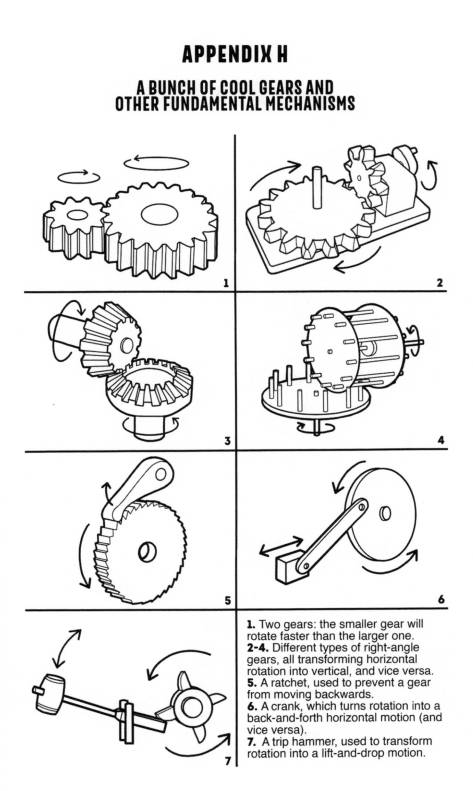

1. Two gears: the smaller gear will rotate faster than the larger one.
2-4. Different types of right-angle gears, all transforming horizontal rotation into vertical, and vice versa.
5. A ratchet, used to prevent a gear from moving backwards.
6. A crank, which turns rotation into a back-and-forth horizontal motion (and vice versa).
7. A trip hammer, used to transform rotation into a lift-and-drop motion.

APPENDIX I

HERE'S WHERE SOME USEFUL HUMAN PARTS ARE AND WHAT THEY DO

1. BRAIN
a self-aware hunk of fatty meat in an attractive skull container · can't live without it · good at keeping you awake at 2 a.m. just to remind you of all the stupid things you said years and years ago · **rating 6/10**

2. PHARYNX
a tube for air and food · weird stretchy folds of flesh beneath these make sounds when you blow on them, which is what makes speaking possible · if it gets blocked you die · **rating 10/10**

3. CARDIOVASCULAR SYSTEM
heart moves blood throughout body · blood transports nutrients and oxygen to far places · 5600 cm³ of blood is in the average body, and despite the fact it's cool to know things, people will still question why you have this fact on hand if you volunteer it at parties · **rating 12/10**

4. LUNGS
take oxygen from air and put it in blood, expel waste carbon dioxide · adult lungs contain 6L of air in a matching set of meaty carrying cases · **rating 11/10**

5. ARTERIES AND VEINS
arteries are high-pressure blood vessels that move blood away from the heart, veins are low-pressure blood vessels that move blood back into it · most veins have one-way valves that prevent blood from moving backwards: sweet · **rating 10/10** (veins), **12/10** (arteries)

6. URINARY BLADDER
up to 800 cm³ of warm urine stored in a convenient elastic sac · allows you to pee on anything you want whenever you want: a claim no other organ can make · **rating 11.5/10**

7. BONE MARROW
produces around 500 billion cells per day, including red blood cells (which deliver oxygen to the body) and white blood cells (which fight infection) · 4% of total human body mass is bone marrow: seems legit · **rating 9.99/10**

8. REPRODUCTIVE SYSTEM
home for 39 million sperm cells or up to 500,000 egg cells, depending on—well, a lot of things really · you have to have sex if you want to have babies and continue the species, so you might as well "bite the bullet" and "have some sex" · **ratings range from 9.5/10 to 9.725/10**

9. SKELETON
there is a spooky wet skeleton hiding inside us all, a truly terrifying thought · stores minerals for later use and is secretly hollow, with insides used to produce bone marrow · 206 bones in the human body but a lot of them are pretty samey · **rating 8/10**

10. LYMPH NODES
lymph is a mixture of water and white blood cells that fights infection · lymph nodes are part of a network found in the body that swell when infection is present to prevent it from spreading · lymph also transports fats from the small intestine into blood · **rating 10/10**

11. STOMACH
a muscular, fleshy bag that contains hydrochloric acid and enzymes · kills bacteria in food and partially digests it before the small intestine gets to it · churns, squeezes, and mixes whatever you put into it · **rating 12.5/10**

12. PANCREAS
makes both hormones (to help regulate the body) and enzymes (that help digestion) · maintains blood-sugar levels and is present in all vertebrates, so not that exclusive an organ, but still pretty cool · **rating 9/10**

13. SPLEEN
recycles old blood cells and holds a supply of extra blood ready to go whenever you need it · people used to think spleens controlled emotions but people used to think a lot of things, most of them wrong · **rating 10/10**

14. LIVER
makes bile that helps the small intestine digest food · stores sugars until they're needed, converts stored fat into sugar too · breaks down toxins: a truly excellent organ · **rating 13/10**

15. GALLBLADDER
stores and concentrates bile from liver until needed for digestion · you can live without it · the optionality of this organ earns it a mere **5/10 rating**

16. INTESTINES
small intestine breaks down food and absorbs compounds, large intestine absorbs water and any leftover nutrients · in other words, turns food into energy and poop · **rating 12/10**

17. KIDNEYS
make urine and produce hormones and keeps salt, water, and acid in blood stable · you only need one to live but most of us have two: very posh · kidneys also produce extremely painful "kidney stones" you have to pee out · **rating 12/10**

18. MUSCULAR SYSTEM
tissues attached to the skeleton that are controlled by the brain, which lets your brain drive your skeleton around wherever it wants to go · lots of muscles throughout the body, and some of them can be pretty attractive if we're being completely honest with each other · **rating 11/10**

AFTERWORD

And with that final appendix the original text ends, leaving the stranded time traveler to their (now hopefully much-improved) fate. I can almost summon the emotions they must've felt upon completing this book and looking up from it out to a new world: satisfaction at having learned so much, mixed in with the complete and abject terror of having to rebuild everything ever from first principles in a random time period. Pretty intense! Glad I don't have to worry about that!

While no bibliography was included in the original text (which makes sense, as it'd obviously be useless to anyone stranded in an era before books were invented), I thought adding my own on the following pages would help interested readers to find out more about the technologies, ideas, and innovations this guide describes. I used these books when verifying the text you just read, and as references before publication.

Besides some really good books, I also consulted some really good humans. I'd like to thank my brother Victor North for his knowledge of both art and brewing, and my friends Dr. Priya Raju (for sharing her secrets of the medical profession), Allene Chomyn and Will Wadley (for their knowledge of music and music theory), David Malki (for his knowledge of flight), and Mike Tucker (for his knowledge of boats). Huge thanks to Zach Weinersmith, Randall Munroe, Jenn

Klug, Mick Tucker, Emily Horne, and my dad for being beta read-ers, with a particular shout-out to Randall for instantly knowing how much land would be covered if the ice caps completely melted this afternoon. He practically volunteered that information before I even asked. Thanks to Dr. Hélène Deval for helping me research the laws surrounding the preserved corpses of accused criminals in France in 1670 CE, and to Sergio Aragonés for being so kind when I asked him about his experience with iceboxes. Finally, my editor, Courtney Young, was invaluable, and everyone should work with her because she's great, except don't, because I still want her to work with me.

Ryan North

Toronto

2018 CE

BIBLIOGRAPHY

Adams, Thomas F. 1861 CE. *Typographia; or, The printer's instructor: a brief sketch of the origin, rise, and progress of the typographic art, with practical directions for conducting every department in an office, hints to authors, publishers, &c.* Philadelphia: L. Johnson & Co.

Agarwal, Rishi Kumar. 1971 CE. "Origin of Spectacles in India." *British Journal of Ophthalmology* 55, 128–29.

American Galvanizers Association. 2017 CE. "Corrosion Rate." *Corrosion Science.* https://www.galvanizeit.org/corrosion/corrosion-process/corrosion-rate.

Anderson, Frank E., et al. 2017 CE. "Phylogenomic Analyses of Crassiclitellata Support Major Northern and Southern Hemisphere Clades and a Pangaean Origin for Earthworms." *BMC Evolutionary Biology* 17(123) doi:10.1186/s12862-017-0973-4.

Anderson, Patricia C. 1991 CE. "Harvesting of Wild Cereals During the Natufian as Seen from Experimental Cultivation and Harvest of Wild Einkorn Wheat and Microwear Analysis of Stone Tools." In *The Natufian Culture in the Levant*, by Ofer Bar-Yosef and François R. Valla, 521–52. International Monographs in Prehistory.

Barbier, André. 1950 CE. "The Extraction of Opium Alkaloids." United Nations Office on Drugs and Crime. https://www.unodc.org/unodc/en/data-and-analysis/bulletin/bulletin_1950-01-01_3_page004.html.

Bardell, David. 2004 CE. "The Invention of the Microscope." *BIOS: A Quarterly Journal of Biology* 75 (2): 78–84.

Barker, Graeme. 2009 CE. *The Agricultural Revolution in Prehistory: Why Did Foragers Become Farmers?* Oxford University Press.

Basalla, George. 1988 CE. *The Evolution of Technology.* Cambridge University Press.

Benjamin, Craig G. 2016 CE. "The Big History of Civilizations." The Great Courses.

Berger, A. L. 1976 CE. "Obliquity and Precession for the Last 5,000,000 Years." *Astronomy and Astrophysics* 51 (1): 127–35.

Biss, Eula. 2014 CE. *On Immunity: An Inoculation.* Graywolf Press.

Bowern, Claire. 2008 CE. *Linguistic Fieldwork: A Practical Guide.* Palgrave Macmillan.

Bowler, Peter J., and Iwan Rhys Morus. 2005 CE. *Making Modern Science.* The University of Chicago Press.

Bradeen, James M., and Philipp W. Simon. 2007 CE. "Carrot." In *Genome Mapping and Molecular Breeding in Plants: Vegetables,* by Chittaranjan Kole, 161–84. Springer-Verlag Berlin Heidelberg. doi:10.1007/978-3-540-34536-7.

Bradshaw, John L. 1998 CE. *Human Evolution: A Neuropsychological Perspective.* Psychology Press.

Brown, Henry T. 2005 CE. *507 Mechanical Movements: Mechanisms and Devices.* Dover Publications.

Bunch, Bryan, and Alexander Hellemans. 1993 CE. *The Timetables of Technology: A Chronology of the Most Important People and Events in the History of Technology.* Simon & Schuster.

Bunney, Sarah. 1985 CE. "Ancient Trade Routes for Obsidian." *New Scientist* 26.

Burdock Group. 2007 CE. "Safety Assessment of Castoreum Extract as a Food Ingredient." *International Journal of Toxicology* 26 (1): 51–55. doi:10.1080/10915810601120145.

Cegłowski, Maciej. 2010 CE. "Scott and Scurvy." *Idle Words.* March. http://idlewords.com/2010/03/scott_and_scurvy.htm.

Chaline, Eric. 2015 CE. *Fifty Animals that Changed the Course of History.* Firefly Books.

Civil, M. 1964 CE. "A Hymn to the Beer Goddess and a Drinking Song." *Studies Presented to A. Leo Oppenheim,* 67–89.

Clement, Charles R., et al. 2010 CE. "Origin and Domestication of Native Amazonian Crops." *Diversity,* 72–106. doi:10.3390/d2010072.

Cook, G. C. 2001 CE. "Construction of London's Victorian Sewers: The Vital Role of Joseph Bazalgette." *Postgraduate Medical Journal* 77 (914): 802. doi:10.1136/pmj.77.914.802.

Cornell, Kit. 2017 CE. *How to Find and Dig Clay.* http://www.kitcornellpottery.com/teaching/clay.html.

Crump, Thomas. 2002 CE. *A Brief History of Science As Seen Through the Development of Scientific Instruments.* Constable & Robinson Ltd.

Dartnell, Lewis. 2014 CE. *The Knowledge: How to Rebuild Civilization in the Aftermath of a Cataclysm.* Penguin Books.

Dauchy, Serge. 2000 CE. "Trois procès à cadavre devant le Conseil souverain du Québec (1687–1708): Un exemple d'application de l'ordonnance de 1670 dans les colonies." *Juges et criminels, l'Espace Juridique,* 37–49.

Dawson, Gloria. 2013 CE. "Beer Domesticated Man." *Nautilus,* December 19. http://nautil.us/issue/8/home/beer-domesticated-man.

De Decker, Kris. 2013 CE. "Back to Basics: Direct Hydropower." *Low-Tech Magazine.* August 11. http://www.lowtechmagazine.com/2013/08/direct-hydropower.html.

De Morgan, Augustus. 1847 CE. *Formal Logic, or, The Calculus of Inference, Necessary and Probable.* Taylor and Walton.

Derry, T. K., and Trevor I. Williams. 1993 CE. *A Short History of Technology, from the Earliest Times to A.D. 1900.* Oxford University Press.

Devine, A. M. 1985 CE. "The Low Birth-Rate in Ancient Rome: A Possible Contributing Factor." *Rheinisches Museum für Philologie* 313–17.

Diamond, Jared. 1999 CE. *Guns, Germs, and Steel: The Fates of Human Societies.* W. W. Norton.

Dietitians of Canada / Les diététistes du Canada. 2013 CE. "Factsheet: Functions and Food Sources of Common Vitamins." *Dietitians of Canada.* February 6. https://www.dietitians.ca/Your-Health/Nutrition-A-Z/Vitamins/Functions-and-Food-Sources-of-Common-Vitamins.aspx.

DK Publishing. 2012 CE. *The Survival Handbook: Essential Skills for Outdoor Adventure.* DK Publishing.

Douglas, George H. 2001 CE. *The Early Days of Radio Broadcasting.* McFarland & Co. Inc. Publishing.

Dunn, Kevin M. 2003 CE. *Caveman Chemistry: 28 Projects, from the Creation of Fire to the Production of Plastics.* uPublish.com.

Dyson, George. 2012 CE. *Turing's Cathedral.* Vintage Books.

Eakins, B. W., and G. F. Sharman. 2012 CE. "Hypsographic Curve of Earth's Surface from ETOPO1." *National Oceanic and Atmospheric Administration National Geophysical Data Center.* https://www.ngdc.noaa.gov/mgg/global/etopo1_surface_histogram.html.

Eisenmann, Vera. 2003 CE. "Gigantic Horses." *Advances in Vertebrate Paleontology,* 31–40.

Ekko, Sakari. 2015 CE. *Latitude Gnomon and Quadrant for the Whole Year.* https://www.eaae-astronomy.org/workshops/172-latitude-gnomon-and-quadrant-for-the-whole-year.

Faculty of Oriental Studies, University of Oxford. 2006 CE. *The Electronic Text Corpus of Sumerian Literature.* http://etcsl.orinst.ox.ac.uk.

Fang, Janet. 2010 CE. "A World Without Mosquitoes." *Nature* (466): 432–34. doi:10.1038/466432a.

Farey, John. 1827 CE. *A Treatise on the Steam Engine: Historical, Practical, and Descriptive.* London: Longman, Rees, Orme, Brown, and Green. https://archive.org/details/treatiseonsteame01fareuoft.

Fattori, Victor, et al. 2016 CE. "Capsaicin: Current Understanding of Its Mechanisms and Therapy of Pain and Other Pre-Clinical and Clinical Uses." *Molecules* 21 (7). doi:10.3390/molecules21070844.

Ferrand, Nuno. 2008 CE. "Inferring the Evolutionary History of the European Rabbit (*Oryctolagus cuniculus*) from Molecular Markers." *Lagomorph Biology* 47–63. doi:10.1007/978-3-540-72446-9_4.

Feyrer, James, Dimitra Politi, and David N. Weil. 2017 CE. "The Cognitive Effects of Micronutrient Deficiency: Evidence from Salt Iodization in the United States." *Journal of the European Economic Association* 15 (2): 355–87. doi:10.3386/w19233.

Francis, Richard C. 2015 CE. *Domesticated: Evolution in a Man-Made World.* W. W. Norton.

Furman, C. Sue. 1997 CE. *Turning Point: The Myths and Realities of Menopause.* Oxford University Press.

Gainsford, Peter. 2017 CE. "Salt and Salary: Were Roman Soldiers Paid in Salt?" *Kiwi Hellenist: Modern Myths About the Ancient World.* January 11. http://kiwihellenist.blogspot.ca/2017/01/salt-and-salary.html.

Gearon, Eamonn. 2017 CE. "The History and Achievements of the Islamic Golden Age." The Great Courses.

Gerke, Randy. 2009 CE. *Outdoor Survival Guide*. Human Kinetics.

Glenn, Edward P., J. Jed Brown, and Eduardo Blumwald. 1999 CE. "Salt Tolerance and Crop Potential of Halophytes." *Critical Reviews in Plant Sciences* 18 (2): 227–55. doi:10.1080/07352689991309207.

Goldstone, Lawrence. 2015 CE. *Birdmen: The Wright Brothers, Glenn Curtiss, and the Battle to Control the Skies*. Ballantine Books.

Graham, C., and V. Evans. 2007 CE. "History of Mining." *Canadian Institute of Mining, Metallurgy, and Petroleum*. August. http://www.cim.org/en/Publications-and-Technical-Resources/Publications/CIM-Magazine/2007/august/history/history-of-mining.aspx.

Grossman, Dan. 2017 CE. "Hydrogen and Helium in Rigid Airship Operations." *Airships.net: The Graf Zeppelin, Hindenburg, U.S. Navy Airships, and Other Dirigibles*. June. http://www.airships.net/helium-hydrogen-airships.

Gugliotta, Guy. 2008 CE. "The Great Human Migration." *Smithsonian*, July.

Gurstelle, William. 2014 CE. *Defending Your Castle: Build Catapults, Crossbows, Moats, Bulletproof Shields, and More Defensive Devices to Fend Off the Invading Hordes*. Chicago Review Press.

Hacket, John. 1693 CE. *Scrinia Reserata: A Memorial Offer'd to the Great Deservings of John Williams, D. D., Who Some Time Held the Places of Lord Keeper of the Great Seal of England, Lord Bishop of Lincoln, and Lord Archbishop of York*. London: Edward Jones, for Samuel Lowndes, over against Exeter-Exchange in the Strand. https://hdl.handle.net/2027/ucl.31175035164386.

Halsey, L. G., and C. R. White. 2012 CE. "Comparative Energetics of Mammalian Locomotion: Humans Are Not Different." *Journal of Human Evolution* 63: 718–22. doi:10.1016/j.jhevol.2012.07.008.

Han, Fan, Andreas Wallberg, and Matthew T. Webster. 2012 CE. "From Where Did the Western Honeybee (Apis mellifera) Originate?" *Ecology and Evolution* 8:1949–57. doi:10.1002/ece3.312.

Harari, Yuval Noah. 2014 CE. *Sapiens: A Brief History of Humankind*. McClelland & Stewart.

Heidenreich, Conrad E., and Nancy L. Heidenreich. 2002 CE. "A Nutritional Analysis of the Food Rations in Martin Frobisher's Second Expedition, 1577." *Polar Record* 23–38. doi:10.1017/S0032247400017277.

Hellemans, Alexander, and Bryan Bunch. 1991 CE. *The Timetables of Science: A Chronology of the Most Important People and Events in the History of Science*. Touchstone Books.

Herodotus. 2013 CE. *Delphi Complete Works of Herodotus (Illustrated)*. Delphi Classics.

Hess, Julius H. 1922 CE. *Premature and Congenitally Diseased Infants*. Lea & Febiger. http://www.neonatology.org/classics/hess1922/hess.html.

Hobbs, Peter R., Ian R. Lane, and Helena Gómez Macpherson. 2006 CE. "Fodder Production and Double Cropping in Tibet: Training Manual." *Food and Agriculture Organization of the United Nations*. http://www.fao.org/ag/agp/agpc/doc/tibetmanual/cover.htm.

Hogshire, Jim. 2009 CE. *Opium for the Masses: Harvesting Nature's Best Pain Medication*. Feral House.

Horn, Susanne, et al. 2011 CE. "Mitochondrial Genomes Reveal Slow Rates of Molecular Evolution and the Timing of Speciation in Beavers (Castor), One of the Largest Rodent Species." *PLoS ONE* 6(1). doi:10.1371/journal.pone.0014622.

Hublin, Jean-Jacques, et al. 2017 CE. "New Fossils from Jebel Irhoud, Morocco and the Pan-African Origin of *Homo sapiens.*" *Nature* 546: 289–92. doi:10.1038/nature22336.

Hyslop, James Hervey. 1899 CE. *Logic and Argument.* Charles Scribner's Sons.

Iezzoni, A., H. Schmidt, and A. Albertini. 1991 CE. "Cherries (Prunus)." *Acta Horticulturae: Genetic Resources of Temperate Fruit and Nut Crops.* doi:10.17660/ActaHortic.1991.290.4.

Johnson, C. 2009 CE. "Sundial Time Correction—Equation of Time." January. http://mb-soft.com/public3/equatime.html.

Johnson, Steven. 2014 CE. *How We Got to Now: Six Innovations That Made the Modern World.* Riverhead Books.

———. 2010 CE. *Where Good Ideas Come From: The Natural History of Innovation.* Riverhead Books.

Kean, Sam. 2010 CE. *The Disappearing Spoon and Other True Tales of Madness, Love, and the History of the World from the Periodic Table of the Elements.* Little, Brown and Company.

Kennedy, James. 2016 CE. *(Almost) Nothing Is Truly "Natural."* February 19. https://jameskennedymonash.wordpress.com/2016/02/19/nothing-in-the-supermarket-is-natural-part-4.

Kislev, Mordechai E., Anat Hartmann, and Ofer Bar-Yosef. 2006 CE. "Early Domesticated Fig in the Jordan Valley." *Science* 312 (5778): 1372–74. doi:10.1126/science.1125910.

Kolata, Gina. 1994 CE. "In Ancient Times, Flowers and Fennel for Family Planning." *The New York Times,* March 8.

Kowalski, Todd J., and William A. Agger. 2009 CE. "Art Supports New Plague Science." *Clinical Infectious Diseases* 48 (1): 137–38. doi:10.1086/595557.

Kurlansky, Mark. 2017 CE. *Paper: Paging Through History.* W. W. Norton.

———. 2002 CE. *Salt: A World History.* Vintage Canada.

Lakoff, George, and Mark Johnson. 2003 CE. *Metaphors We Live By.* University of Chicago Press.

Lal, Rattan. 2016 CE. *Encyclopedia of Soil Science.* Third edition. CRC Press.

Laws, Bill. 2015 CE. *Fifty Plants that Changed the Course of History.* Firefly Books.

LeConte, Joseph. 1862 CE. *Instructions for the Manufacture of Saltpetre.* Charles P. Pelham, State Printer. http://docsouth.unc.edu/imls/lecontesalt/leconte.html.

Lemley, Mark A. 2012 CE. "The Myth of the Sole Inventor." *Michigan Law Review* 110 (5): 709–60. doi:10.2139/ssrn.1856610.

Lewis, C. I. 1914 CE. "The Matrix Algebra for Implications." Edited by Frederick J. E. Woodbridge and Wendell T. Bush. *Journal of Philosophy, Psychology, and Scientific Methods* (The Science Press) XI: 589–600.

Liggett, R. Winston, and H. Koffler. 1948 CE. "Corn Steep Liquor in Microbiology." *Bacteriological Reviews* 297–311.

"List of Zoonotic Diseases." 2013 CE. *Public Health England.* March 21. https://www.gov.uk/government/publications/list-of-zoonotic-diseases/list-of-zoonotic-diseases.

Livermore, Harold. 2004 CE. "Santa Helena, a Forgotten Portuguese Discovery." *Estudos em Homenagem a Louis Antonio de Oliveira Ramos,* 623–31.

Lundin, Cody. 2007 CE. *When All Hell Breaks Loose: Stuff You Need to Survive When Disaster Strikes.* Gibbs Smith.

Lunge, Georg. 1916 CE. *Coal-Tar and Ammonia*. D. Van Nostrand. https://archive.org/details/coaltarandammon04lunggoog.

Maines, Rachel P. 1998 CE. *The Technology of Orgasm: "Hysteria," the Vibrator, and Women's Sexual Satisfaction*. The Johns Hopkins University Press.

Mann, Charles C. 2006 CE. *1491: New Revelations of the Americas Before Columbus*. Vintage.

Marchetti, C. 1979 CE. "A Postmortem Technology Assessment of the Spinning Wheel: The Last Thousand Years." *Technological Forecasting and Social Change*, 91–93.

Martin, Paula, et al. 2008 CE. "Why Does Plate Tectonics Occur Only on Earth?" *Physics Education* 43 (2): 144–50. doi:10.1088/0031-9120/43/2/002.

Martín-Gil, J., et al. 1995 CE. "The First Known Use of Vermillion." *Experientia* 759–61. doi:10.1007/BF01922425.

McCoy, Jeanie S. 2006 CE. "Tracing the Historical Development of Metalworking Fluids." In *Metalworking Fluids: Second Edition*, by Jerry P. Byers, 480. Taylor & Francis Group.

McDowell, Lee Russell. 2000 CE. *Vitamins in Animal and Human Nutrition, Second Edition*. Wiley-Blackwell.

McElney, Brian. 2001 CE. "The Primacy of Chinese Inventions." *Bath Royal Literary and Scientific Institution*. September 28. Accessed July 1, 2017 CE. https://www.brlsi.org/events-proceedings/proceedings/17824.

McGavin, Jennifer. 2017 CE. "Using Ammonium Carbonate in German Baking." *The Spruce*. May 1. https://www.thespruce.com/ammonium-carbonate-hartshorn-hirsch hornsalz-1446913.

McLaren, Angus. 1990 CE. *History of Contraception: From Antiquity to the Present Day*. Basil Blackwell.

McNeil, Donald G. Jr. 2006 CE. "In Raising the World's I.Q., the Secret's in the Salt." *The New York Times*, December 16.

Mechanical Wood Products Branch, Forest Industries Division, FAO Forestry Department. 1987 CE. "Simple Technologies for Charcoal Making." *Food and Agriculture Organization of the United Nations*. http://www.fao.org/docrep/x5328e/x5328e00.htm.

Miettinen, Arto, et al. 2008 CE. "The Palaeoenvironment of the 'Antrea Net Find.'" *Iskos* 16:71–87.

Moore, Thomas. 1803 CE. *An essay on the most eligible construction of ice-houses: also, a description of the newly invented machine called the refrigerator*. Baltimore: Bonsal & Niles.

Morin, Achille. 1842 CE. *Dictionnaire du droit criminel: répertoire raisonné de législation et de jurisprudence, en matière criminelle, correctionnelle et de police*. Paris: A. Durand.

Mott, Lawrence V. 1991 CE. *The Development of the Rudder, A.D. 100–1600: A Technological Tale*. http://nautarch.tamu.edu/pdf-files/Mott-MA1991.pdf.

Mueckenheim, W. 2005 CE. "Physical Constraints of Numbers." *Proceedings of the First International Symposium of Mathematics and Its Connections to the Arts and Sciences*, 134–41.

Munos, Melinda K., Christa L. Fischer Walker, and Robert E. Black. 2010 CE. "The Effect of Oral Rehydration Solution and Recommended Home Fluids on Diarrhoea Mortality." *International Journal of Epidemiology* 39:i75–i87. doi:10.1093/ije/dyq025.

Murakami, Fabio Seigi, et al. 2007 CE. "Physicochemical Study of $CaCO_3$ from Egg

Shells." *Food Science and Technology* 27 (3): 658–62. doi:10.1590/S0101
-20612007000300035.

Nancy Hall. 2015 CE. "Lift from Flow Turning." *National Aeronautics and Space Administration: Glenn Research Center.* May 5. https://www.grc.nasa.gov/www/k-12/airplane/right2.html.

National Coordination Office for Space-Based Positioning, Navigation, and Timing. 2016 CE. "Selective Availability." *GPS: The Global Positioning System.* September 23. http://www.gps.gov/systems/gps/modernization/sa.

National Oceanic and Atmospheric Administration's Office of Response and Restoration. n.d. *Chemical Datasheets.* https://cameochemicals.noaa.gov.

Naval Education. 1971 CE. *Basic Machines and How They Work.* Dover Publications.

Nave, Carl Rod. 2001 CE. *Hyperphysics.* http://hyperphysics.phy-astr.gsu.edu.

Nelson, Sarah M. 1998 CE. *Ancestors for the Pigs: Pigs in Prehistory.* University of Pennsylvania Museum of Archaeology and Anthropology.

North American Sundial Society. 2017 CE. *Sundials for Starters.* http://sundials.org.

Nuwer, Rachel. 2012 CE. "Lice Evolution Tracks the Invention of Clothes." *Smithsonian,* November 14.

O'Reilly, Andrea. 2010 CE. *Encyclopedia of Motherhood.* Vol. 1. SAGE Publications, Inc.

Omodeo, Pietro. 2000 CE. "Evolution and Biogeography of Megadriles (Annelida, Clitellata)." *Italian Journal of Zoology* 67 (2): 179–207. doi:10.1080/11250000009356313.

OpenLearn. 2007 CE. "DIY: Measuring Latitude and Longitude." The Open University. September 27. http://www.open.edu/openlearn/society/politics-policy-people/geography/diy-measuring-latitude-and-longitude.

Pal, Durba, et al. 2009 CE. "Acaciaside-B-Enriched Fraction of Acacia Auriculiformis Is a Prospective Spermicide with No Mutagenic Property." *Reproduction* 138 (3): 453–62. doi:10.1530/REP-09-0034.

Pidanciera, Nathalie, et al. 2006 CE. "Evolutionary History of the Genus Capra (Mammalia, Artiodactyla): Discordance Between Mitochondrial DNA and Y-Chromosome Phylogenies." *Molecular Phylogenetics and Evolution* 40 (3): 739–49. doi:10.1016/j.ympev.2006.04.002.

Pinker, Steven. 2007 CE. *The Language Instinct: How the Mind Creates Language.* Harper Perennial Modern Classics.

Planned Parenthood. 2017 CE. "About Birth Control Methods." *Planned Parenthood.* https://www.plannedparenthood.org/learn/birth-control.

Pollock, Christal. 2016 CE. "The Canary in the Coal Mine." *Journal of Avian Medicine and Surgery* 30 (4): 386–91. doi:10.1647/1082-6742-30.4.386.

Preston, Richard. 2003 CE. *The Demon in the Freezer: A True Story.* Fawcett.

Price, Bill. 2014 CE. *Fifty Foods that Changed the Course of History.* Firefly Books.

Pyykkö, Pekka. 2011 CE. "A Suggested Periodic Table up to $Z \leq 172$, Based on Dirac–Fock Calculations on Atoms and Ions." *Physical Chemistry Chemical Physics* 13 (1): 161–68. doi:10.1039/c0cp01575j.

Rehydration Project. 2014 CE. *Oral Rehydration Therapy: A Special Drink for Diarrhoea.* April 21. http://rehydrate.org.

Rezaei, Hamid Reza,et al. 2010 CE. "Evolution and Taxonomy of the Wild Species of the Genus Ovis." *Molecular Phylogenetics and Evolution,* 315–26. doi:10.1016/j.ympev.2009.10.037.

Richards, Matt. 2004 CE. *Deerskins into Buckskins: How to Tan with Brains, Soap or Eggs.* Backcountry Publishing.

Riddle, John M. 2008 CE. *A History of the Middle Ages, 300–1500.* Rowman & Littlefield.

———. 1992 CE. *Contraception and Abortion from the Ancient World to the Renaissance.* Harvard University Press.

Rosenhek, Jackie. 2014 CE. "Contraception: Silly to Sensational: The Long Evolution from Lemon-Soaked Pessaries to the Pill." *Doctor's Review.* August. http://www.doctorsreview.com/history/contraception-silly-sensational/.

Rothschild, Max F., and Anatoly Ruvinsky. 2011 CE. *The Genetics of the Pig.* CABI.

Russell, Bertrand. 1903. *The Principles of Mathematics.* Cambridge University Press.

Rybczynski, Witold. 2001 CE. *One Good Turn: A Natural History of the Screwdriver and the Screw.* Scribner.

Sawai, Hiromi, et al. 2010 CE. "The Origin and Genetic Variation of Domestic Chickens with Special Reference to Junglefowls *Gallus g. gallus* and *G. varius.*" *PLoS ONE* 5(5). doi:10.1371/journal.pone.0010639.

Schmandt-Besserat, Denise. 1997 CE. *How Writing Came About.* University of Texas Press.

Shaw, Simon, Linda Peavy, and Ursula Smith. 2002 CE. *Frontier House.* Atria.

Sheridan, Sam. 2013 CE. *The Disaster Diaries: One Man's Quest to Learn Everything Necessary to Survive the Apocalypse.* Penguin Books.

Singer-Vine, Jeremy. 2011 CE. "How Long Can You Survive on Beer Alone?" *Slate,* April 28. http://www.slate.com/articles/news_and_politics/explainer/2011/04/how_long_can_you_survive_on_beer_alone.html.

Singh, M. M., et al. 1985 CE. "Contraceptive Efficacy and Hormonal Profile of Feru-jol: A New Coumarin from Ferula jaeschkeana." *Planta Medica* 51 (3): 268–70. doi:10.1055/s-2007-969478.

Smith, Edgar C. 2013 CE. *A Short History of Naval and Marine Engineering.* Cambridge University Press.

Société Académique de Laon. 1857 CE. *Bulletin: Volume 6.* Paris: V. Baston.

Sonne, O. 2015 CE. "Canaries, Germs, and Poison Gas. The Physiologist J. S. Haldane's Contributions to Public Health and Hygiene." *Dan Medicinhist Arbog,* 71–100.

St. Andre, Ralph E. 1993 CE. *Simple Machines Made Simple.* Libraries Unlimited.

Standage, Tom. 2006 CE. *A History of the World in 6 Glasses.* Walker & Company.

Stanger-Hall, Kathrin F., and David W. Hall. 2011 CE. "Abstinence-Only Education and Teen Pregnancy Rates: Why We Need Comprehensive Sex Education in the U.S." *PLoS ONE* 6 (10). doi:10.1371/journal.pone.0024658.

Starkey, Paul. 1989 CE. *Harnessing and Implements for Animal Traction.* Friedrich Vieweg & Sohn Verlagsgesellschaft mbH.

Stephenson, F. R., L. V. Morrison, and C. Y. Hohenkerk. 2016 CE. "Measurement of the Earth's Rotation: 720 BC to AD 2015." *Proceedings of the Royal Society A: Mathematical, Physical, and Engineering Sciences* 472 (2196). doi:10.1098/rspa.2016.0404.

Sterelny, Kim. 2011 CE. "From Hominins to Humans: How Sapiens Became Behaviourally Modern." *Philosophical Transactions of the Royal Society: Biological Sciences* 366 (1566). doi:10.1098/rstb.2010.0301.

Stern, David P. 2016 CE. *Planetary Gravity-Assist and the Pelton Turbine.* October 26. http://www.phy6.org/stargaze/Spelton.htm.

Stone, Irwin. 1966 CE. "On the Genetic Etiology of Scurvy." *Acta Geneticae Medicae et Gemellologiae* 16: 345–50.

Stroganov, A. N. 2015 CE. "Genus Gadus (Gadidae): Composition, Distribution, and Evolution of Forms." *Journal of Ichthyology* 316–36. doi:10.1134/S0032945215030145.

Stubbs, Brett J. 2003 CE. "Captain Cook's Beer: The Antiscorbutic Use of Malt and Beer in Late 18th Century Sea Voyages." *Asia Pacific Journal of Clinical Nutrition* 129–37.

The Association of UK Dieticians. 2016 CE. "Food Fact Sheet: Iodine." *BDA*. May. https://www.bda.uk.com/foodfacts/Iodine.pdf.

The National Society for Epilepsy. 2016 CE. "Step-By-Step Recovery Position." *Epilepsy Society*. March. https://www.epilepsysociety.org.uk/step-step-recovery-position.

The Royal Society of Chemistry. 2012 CE. *The Chemistry of Pottery*. July 1. https://eic.rsc.org/feature/the-chemistry-of-pottery/2020245.article.

Ueberweg, Freidrich. 1871. *System of Logic and History of Logical Doctrines*. Longmans, Green, and Company.

Ure, Andrew. 1878 CE. *A Dictionary of Arts, Manufactures, and Mines: Containing a Clear Exposition of Their Principles and Practice*. London: Longmans, Green. https://archive.org/details/b21994055_0003.

US Department of Agriculture. 2016 CE. "The Rescue of Penicillin." *United States Department of Agriculture: Agricultural Research Service*. https://www.ars.usda.gov/oc/timeline/penicillin.

Usher, Abbott Payson. 1988 CE. *A History of Mechanical Inventions*. Dover Publications.

Vincent, Jill. 2008 CE. "The mathematics of sundials." *Australian Senior Mathematics Journal* 22 (1): 13–23.

von Petzinger, Genevieve. 2016 CE. *The First Signs: Unlocking the Mysteries of the World's Oldest Symbols*. Atria.

Warneken, Felix, and Alexandra G. Rosati. 2015 CE. "Cognitive capacities for cooking in chimpanzees." *Proceedings of the Royal Society of London B: Biological Sciences* 282 (1809). doi:10.1098/rspb.2015.0229.

Watson, Peter R. 1983 CE. *Animal Traction*. Artisan Publications.

Wayman, Erin. 2011 CE. "Humans, the Honey Hunters." *Smithsonian*, December 19.

Weber, Ella. 2012 CE. "Apis mellifera: The Domestication and Spread of European Honey Bees for Agriculture in North America." *University of Michigan Undergraduate Research Journal* (9): 20–23.

Welker, Bill. 2016 CE. "Hydrogen for Early Airships." *Then and Now*. December. http://welweb.org/ThenandNow/Hydrogen%20Generation.html.

Werner, David, Carol Thuman, and Jane Maxwell. 2011 CE. *Where There Is No Doctor: A Village Health Care Handbook*. Macmillan.

White Jr., Lynn. 1962 CE. "The Act of Invention: Causes, Contexts, Continuities and Consequences." *Technology and Culture: Proceedings of the Encyclopaedia Britannica Conference on the Technological Order* 486–500. doi:10.2307/3100999.

Wicks, Frank. 2011 CE. "100 Years of Flight: Trial by Flyer." *Mechanical Engineering*. https://web.archive.org/web/20110629103435/http://www.memagazine.org/supparch/flight03/trialby/trialby.html.

Wickstrom, Mark L. 2016 CE. "Phenols and Related Compounds." *Merck Veterinary Manual*. http://www.merckvetmanual.com/pharmacology/antiseptics-and-disinfectants/phenols-and-related-compounds.

Williams, George E. 2000 CE. "Geological Constraints on the Precambrian History of Earth's Rotation and the Moon's Orbit." *Reviews of Geophysics* 38 (1): 37–60. doi:10.1029/1999RG900016.

Wilson, Bee. 2012 CE. *Consider the Fork: A History of How We Cook and Eat*. Basic Books.

World Health Organization. 2007 CE. "Food Safety: The 3 Fives." *World Health Organization*. http://www.who.int/foodsafety/areas_work/food-hygiene/3_fives/en/.

———. 2017 CE. "WHO Model Lists of Essential Medicines." *World Health Organization*. http://www.who.int/medicines/publications/essentialmedicines/en/.

Wragg, David. 1974 CE. *Flight Before Flying*. Osprey Publishing.

Wright, Jennifer. 2017 CE. *Get Well Soon: History's Worst Plagues and the Heroes Who Fought Them*. Henry Holt and Co.

Yong, Ed. 2016 CE. "A New Origin Story for Dogs." *The Atlantic*, June 2.

Zizsser, Hans. 2008 CE. *Rats, Lice and History*. Willard Press.

ENDNOTES WRITTEN BY ME,
THE RYAN FROM THIS TIMELINE

1   In 2017 CE scientists found the partial remains of what appeared to be *almost* anatomically modern humans, but which dated all the way back to around 300,000 BCE. These transitional ancient skeletons weren't indistinguishable from our own—the lower jaw is larger than those of modern humans, and the brain case is elongated—but scientists weren't expecting to find such human-like remains so early in history. In our world, this discovery is still being tested and examined by the scientific community to see if that date holds, but if confirmed, it could help prove that our early prehistory was less "humans evolved in a single place in Africa, expanding out from there" and more "protohumans were part of a large and interbreeding group evolving across Africa." See "New fossils from Jebel Irhoud, Morocco and the pan-African origin of *Homo sapiens*" by Jean-Jacques Hublin, et al. in the bibliography for more.

2   This date of 50,000 BCE matches my best approximation based on the research I've done, but it can shift depending on which scientists you talk to and what models and definitions of behavioral modernity they subscribe to. While I'm comfortable calling 50,000 BCE a consensus, some researchers believe they have detected the beginnings of behavioral modernity as far back as 100,000 BCE. Regardless of which date is chosen, the space of time between anatomic and behavioral modernity remains the moment where a hypothetical time traveler could have the greatest impact.

3   While this may be true, the debate over what pushed humans from anatomical to behavioral modernity is not settled in our own timeline: it may, in fact, only ever be settled with time machines. However, it is true that some evidence of behavioral modernity shows up in isolated instances but died out before catching on at a global scale. A solely biological reason for behavioral modernity would have to account for these discrepancies in the historical record.

4   I can't actually say this with certainty, because in our world we haven't been able to find evidence of contact between these two civilizations. But it's certainly possible, given the similarities between the two scripts.

5   In this timeline, quantum physics has not led to metaquantum ultraphysics. *Yet*.

6   10,500 BCE is used throughout this text as the date farming begins. My research has around two thousand years' wiggle room on the date.

7   Mr. Fahrenheit's mixture of ice, salt, and water wasn't *quite* as random as it sounds. It's an example of a "frigorific mixture," which is a name given to mixtures that tend to stabilize around a certain temperature, as long as all the ingredients in it aren't consumed. Ice and water makes a frigorific mixture that stabilizes at around 0°C, regardless of the initial temperatures of the ingredients, and ice, water, and salt form one that stabilizes at around −17.8°C—or, if you insist, 0°F.

8   At the time of publication (2018 CE), this was still the case! However, the General Conference on Weights and Measures—the intergovernmental organization responsible for such things—is expected to vote on a new definition for the kilogram in November 2018 CE, to come into effect on May 20, 2019 CE. Assuming it succeeds, on that day, the definition of the kilogram will change from the physical reference stored in France to a new definition based on the Planck constant, which is a number central to quantum mechanics. Like the other modern definitions of standard units, this new definition is not much use to anyone stranded in the past—or present, for that matter—who doesn't have the time, money, or inclination to measure the linear momentum of photons.

9   Which, unfortunately for us non–time travelers, means that in our time you and I still don't actually know why the kilograms are changing mass. Oh well! File that under "unsolved science mysteries you would've thought we'd have figured out by now," alongside "Where does the sun's magnetic field come from?," "Why are plate tectonics seen only on Earth?," and "We don't actually know why we sleep or what its biological function is, hah hah, we only spend a third of our lives doing this, I'm sure it's no big deal."

10 I was able to verify this information thanks to the research done by James Kennedy, whose work you can find under *(Almost) Nothing Is Truly "Natural"* in the bibliography!

11 If you're interested in plants, check out *Fifty Plants that Changed the Course of History* by Bill Laws and *Fifty Foods that Changed the Course of History* by Bill Price in the bibliography. More information about the plants in this book, along with many others, can be found in those two texts. If this book *wasn't* an artifact from an alternate-future timeline, you might even suspect that I used them as valuable references when writing this section. *Weird*.

12 Most researchers suspect this is the case, but I don't see a way to prove it without time machines.

13 Okay, this one I can verify! Experiments in domesticating wheat from wild ancestors have been done in our timeline and achieved similar results. See "Harvesting of Wild Cereals During the Natufian as Seen from Experimental Cultivation and Harvest of Wild Einkorn Wheat and Microwear Analysis of Stone Tools" by Patricia C. Anderson in the bibliography for more.

14 This tablet survived to our modern era too, though with some damage. Specifically, the verse about the "great sweet wort" is missing its last two lines, but thanks to the repetitive nature of the hymn, it's pretty easy to guess what they are. Interestingly, this translation of the hymn is identical to one found in M. Civil's "A Hymn to the Beer Goddess" paper (found in the bibliography), which is either a fantastic coincidence, evidence that this "M. Civil" is somehow a constant across timelines, or just proof that there are only so many ways to translate ancient Sumerian beer recipes into English.

15 Obviously this cannot be verified, as there's no hard evidence to show that diprotodons could be domesticated, and modern wombats—much smaller than diprotodons were—have not been. However, modern wombats live in burrows (it's extremely unlikely the hippo-sized diprotodon did), which makes them difficult to keep in a farming situation: a strike against domestication that the diprotodon didn't share. Wombats are also antisocial animals that generally prefer to be alone—another strike against captivity—while many diprotodon fossils have been discovered at the same location, suggesting that they at least shared migration routes, if not an actual herd.

16 Check out *Fifty Animals that Changed the Course of History* by Eric Chaline if you'd like to know more. Not all the species included here are in that text, but the book does include several other animals that influenced human history, just in ways that are less critical to a stranded time traveler. If you didn't know any better, you'd swear that while I was writing this section, *which I didn't, because this book is clearly from the future,* I used Chaline's book as a reference.

17 My research is less clear on this matter: it could've been humans killing these animals that caused them to go extinct, but it may also have been climate change, or these two effects working together. Without a time machine, it's very difficult to tell. But there is a very suspicious global trend throughout the fossil record of giant animals going extinct *every single time* humans show up.

18 The DNA sequencing happened in our world too, and at the same time, though it's too early to tell if it will lead to eventual auroch restoration.

19 In the fossil record there is a skull from around 34,000 BCE that is not quite a dog but not quite a wolf. It's a disputed fossil: some argue that it's a transitional dog early in the domestication process, while others say it's an early attempt at domestication that did not lead to modern-day dogs, which explains why it's so different. The text here suggests the latter reading. The earliest undisputed dog remains I could find references to date to 12,700 BCE and were found buried beside a human, clearly indicating the presence of a very, very good dog.

20 Scientists believe that it's possible dogs did domesticate themselves as described here, but without actually being there with a time machine, it's probably impossible to know for sure.

21 While the date for the first metamorphosing insects matches our knowledge, I couldn't actually find a date for when silkworms first evolved.

22 This date of 400,000,000 BCE is an estimate, because nobody's ever found a single earthworm fossil! But trace fossils (fossils recording evidence of their existence, such as their paths in the ground) have been discovered, along with fossils of their eggs, which are laid in cocoons.

23 This information matches recommendations from our world, including those made by the Dietitians of Canada in the bibliography.

24 They've been described that way in our timeline too! In *The Genius of China*, Robert Temple writes, "So inefficient, so wasteful of effort, and so utterly exhausting was the old plow that this deficiency of plowing may rank as mankind's single greatest waste of time and energy." Ouch for humans.

25 I don't know for sure that this is how smoking was discovered, but it seems likely. Caves were poorly ventilated, fire would already be used for light and heat, and it wouldn't take much to discover that meat hung to dry in smoky areas would last longer and taste better. Interestingly, most references I've seen to food tasting "smoky" show up only *after* humans started cooking foods in ways that didn't involve an open fire. Before that, "smoky" was just how cooked food tasted.

26 This is a theory that has been put forward in our time too! The basic argument is that beer is alcoholic, which besides being safer to consume than water (since many bacteria can't survive in alcohol), it's also more fun. If hunting and gathering is keeping your belly full, then bread won't be much of a lure to give up that lifestyle and come work on a farm, but if it's the only way to get beer . . . well, maybe you'll give it a try. See Gloria Dawson's "Beer Domesticated Man" and Tom Standage's *A History of the World in 6 Glasses* in the bibliography for more.

27 This is a not-uncontroversial claim, given our current knowledge of history. Glasses were long thought to have been developed in Italy in 1286 CE by an unknown inventor, but I've found references to glasses—with lenses made from polished quartz rather than glass—in Indian texts that predate European contact. See "Origin of Spectacles in India" by Rishi Kumar Agarwal in the bibliography for more.

28 While I've been able to find historians speculating that the first pieces of artificial glass were probably produced by accident, it's impossible for us to verify with certainty without having access to a time machine.

29 This happened in our history too: the Friday, November 24, 1922, edition of *The South Eastern Times* of Millicent, South Australia, includes a story under the headline "COW THAT ASSISTED SCIENCE." In it, Pelton is so impressed after hosing down the cow that "within the hour" he is hooking up empty cans to a wheel. Most cutesy stories for scientific invention are false, but they persist because we tend to love the "single moment of accidental insight that changed the world" narrative much more than we love the competing "it was a lot of hard work and study that took up a good portion of my life" storyline.

30 Without time machines nobody can make these measurements precisely, but the data here matches the estimates made in the paper I found titled "Obliquity and Precession for the Last 5,000,000 Years" by A. L. Berger. It's in the bibliography.

31 While the wording here doesn't make it clear if this machine was actually *built* in the authors' timeline, in ours calculations have actually been completed estimating the predictive abilities of such a machine. The same conclusions were reached. See *Turing's Cathedral* by George Dyson in the bibliography for more.

32 I don't need to tell you that we don't have silphium anymore, but scientists believe that the plant likely belonged to the *Ferula* genus, a plant family whose extant members contain "ferujol," a chemical that is almost 100 percent effective at preventing pregnancies . . . in rats, anyway. In hamsters—another common laboratory test animal—it has no effect. In any case, recipes involving silphium for human contraception have been studied, and some researchers believe they would have been effective: see *Contraception and Abortion from the Ancient World to the Renaissance* by John M. Riddle in the bibliography. It's worth noting in relation to silphium that Rome's birth rates were indeed low: so low, in fact, that in 18 BCE Emperor Augustus actually passed laws to penalize the childless and the unmarried, in an attempt to induce more people to reproduce.

33 The exact date that paper was invented isn't actually known! There's a lot of myth surrounding it, and the credit for paper is traditionally given to a man named Cau Lun, who lived from 48 to 121 CE. However, there's acheological evidence of paper predating even that date, including some dating as far back as 179 BCE. All the research I found suggests the date here is at best an informed guess!

34 This suggestion that the early invention of radio could've saved a lot of time messing around trying to invent clocks that work on boats has been put forward before! I encountered it in Lewis Dartnell's *The Knowledge*, which you can find in the bibliography.

35 Like many inventions, radio receivers were independently invented by multiple people at the same time. In this case, two men (Edwin H. Armstrong and Lee de Forest) fought for years over patent rights to their invention. As in many cases in the early days of any technology, these people knew that their inventions worked, but they were sometimes completely misinformed as to *why* that was the case. In *The Early Days of Radio Broadcasting*, George H. Douglas writes that "at almost every step of the way, de Forest was flat-out wrong about what he was inventing. [His radio receiver] was not so much an invention as it was the steady, persistent accumulation of error." Ouch!

36 The publication referenced was actually published under that title and survives into the present day. I put it in the bibliography!

37 No one is 100 percent certain about this, actually! Most researchers believe that a static spark was the initial cause that led to fire, but other theories have involved causes including engine backfire, sabotage, and lightning.

38 Obviously this claim is as impossible for me to verify as it is tantalizing to imagine, but it is true that gliders were used to explore the theory and practice behind powered flight and are an almost necessary first step toward powered aircraft.

39 While helium and neon compounds are presumably available in every corner drugstore in the future, here in the present they're a little harder to come by. They have, however, been (briefly) created, under the high-pressure, low-temperature circumstances mentioned in the text.

40 This information matches up with *Where There Is No Doctor: A Village Health Care Handbook* by David Werner et al., a guide published in our own time designed to help amateurs assist with medicine in places where there is, you guessed it, no doctor. You can find it in the bibliography.

41 This periodic table is obviously much larger than the one you and I are familiar with: our table stops with element 118, and this one goes all the way to 172. Interestingly, our current understanding suggests 172 may be an atomic limit: at 173, the atom would be so large that electrons on its outer shell would have to move at speeds *faster than light*. While most of these new elements are named after scientists (which is in line with our current naming conventions), there are a few that have Latin phrases at their core, including *inprincipiomium* ("in the beginning," used for the final element at 172), *praeviderium* ("foresee," suggesting it may be of use in time travel), and the alarmingly named *malaipsanovium* ("bad news itself"). I can offer no explanation for these names.

42 I do know that sulfuric acid (then known as "oil of vitriol") was used in ancient Sumerian times, but I haven't been able to put this precise a date on its discovery.

43 In our timeline, this prank was first suggested by cognitive science professor (and Pulitzer Prize winner) Douglas Hofstadter. His Pulitzer win wasn't for this joke, for some reason.

44 This is actually a real theory proposed in 1940 CE by real theoretical physicist John Wheeler, but it was never taken very seriously, even by its inventor. Why are all electrons identical? At this point in time nobody actually knows why. This theory answers that question (they're all indistinguishable because *they're all the same electron*) at the expense of raising tons of other, much more challenging, questions.

INDEX

soybean, 73
soy sauce, 144
speakers, 280, 280f
specialization, 36, 120, 127
spectroscope, 170–71
spectrum
 electromagnetic, 276–77, 276f
 sound, 343, 345
 visible light, 276, 276n
speed of light, 194–95, 194f, 399t
speed of sound, 399t
spermicides, 229, 229n
sphere, glass, 170n
spinning wheel, 224–27, 227f
splint, 338
spoken language, 11t, 12–15
sprouted grains, 143
square sail, 284, 284n, 285f
stabilizers (in aircraft), 297, 297f
Staphylococcus bacteria, 152–53
steam engines, 185–92, 188f
steam-powered aircraft, 298
steam-powered boats, 287n
steel, 240–45
steering oars, 283
stellar fusion, 311
stencils, 252, 252n
stethoscope, 155–56
stitches (surgical), 339
stoneware, 164
straw (drinking), 145
stringed instruments, 341–42
subtraction, 361–62, 362n
sugar
 for preserving foods, 136–37
 sources of, 74
sugarcane, 74
sulfur, 73
sulfuric acid
 in batteries, 195–96, 196n–97n
 in hydrogen production, 292–93, 293f
 in phosphate production, 52
 production and uses of, 389–90
sull-coating, 244–45
sum, 369–70
sun, 206, 269
sundial, 205
Sun Simiao, 108
supernovae, 311
sweet orange, 74–75
sweet potato, 79
syllogism, 300–305, 302t, 303n

symbiotic logical reasoning, 393–94
symbols, 301
Symphony No. 9, "Ode to Joy," 351–52
syphilis, 232

tacking, in sailing, 285–86, 286f
tally marks, 21, 21f
tambourine, 341n
tangent, 396, 397t–98t
tannic acid, 223
tanning, 221–24
tannins, 69, 222–23
tap dancing, 341n
tar, 120, 283
tea, 75
technologies
 five fundamental, 10–34
 language as, 7
 for natural resources, 157–74
 problems solved by, 113–305
 tree of, 378–79
telescope, 170, 171f
temperature
 measurement of, 40, 210n
 normal body, 330
templates
 for measurement, 43f
 for quadrant, 270, 270f
temporal psychosis, 331
ten
 base, 23, 23t
 powers of, 41, 41t, 42t
terminal, 196
Tesla, Nikola, 200
Tetris®, catchy tune from, 357
theory, 33
thermodynamics, laws of, 186, 186n
thermometers, 210–14, 211n–12n
thermoscopes, 211–12
thread, 225
three-field crop rotation, 49–50, 49t
three-point perspective, 319–20, 320f
throat-and-girth harness, 128–29, 130n
thrust, 191, 296–97
tides, 273n
time, measurement of, 42, 44
time machine, xi–xii, xiif, xvi
time period flowchart, 1–5
time signature, 350
time travel, xi–xvi, xiiif, xivf
time zones, 210
tin cans, 137–38, 174

Discard
EH